Norbert Peters Bernd Rogg (Eds.)

Reduced Kinetic Mechanisms for Applications in Combustion Systems

Springer-Verlag Berlin Heidelberg GmbH

Editors

Norbert Peters
Institut für Technische Mechanik, RWTH Aachen
Templergraben 64, W-5100 Aachen, Fed. Rep. of Germany

Bernd Rogg
University of Cambridge, Department of Engineering
Trumpington Street, Cambridge CB2 1PZ, U.K.

ISBN 978-3-662-13952-3 ISBN 978-3-540-47543-9 (eBook)
DOI 10.1007/978-3-540-47543-9

Originally published by Springer-Verlag Berlin Heidelberg New York in 1993
Softcover reprint of the hardcover 1st edition 1993

Typesetting: Camera ready by author/editor using the TEX/LATEX macro package from Springer-Verlag Berlin Heidelberg GmbH
58/3140-543210 - Printed on acid-free paper

Lecture Notes in Physics

New Series m: Monographs

The Editorial Policy for Monographs

The series Lecture Notes in Physics reports new developments in physical research and teaching - quickly, informally, and at a high level. The type of material considered for publication in the New Series m includes monographs presenting original research or new angles in a classical field. The timeliness of a manuscript is more important than its form, which may be preliminary or tentative. Manuscripts should be reasonably self-contained. They will often present not only results of the author(s) but also related work by other people and will provide sufficient motivation, examples, and applications.
The manuscripts or a detailed description thereof should be submitted either to one of the series editors or to the managing editor. The proposal is then carefully refereed. A final decision concerning publication can often only be made on the basis of the complete manuscript, but otherwise the editors will try to make a preliminary decision as definite as they can on the basis of the available information.
Manuscripts should be no less than 100 and preferably no more than 400 pages in length. Final manuscripts should preferably be in English, or possibly in French or German. They should include a table of contents and an informative introduction accessible also to readers not particularly familiar with the topic treated. Authors are free to use the material in other publications. However, if extensive use is made elsewhere, the publisher should be informed.
Authors receive jointly 50 complimentary copies of their book. They are entitled to purchase further copies of their book at a reduced rate. As a rule no reprints of individual contributions can be supplied. No royalty is paid on Lecture Notes in Physics volumes. Commitment to publish is made by letter of interest rather than by signing a formal contract. Springer-Verlag secures the copyright for each volume.

The Production Process

The books are hardbound, and quality paper appropriate to the needs of the author(s) is used. Publication time is about ten weeks. More than twenty years of experience guarantee authors the best possible service. To reach the goal of rapid publication at a low price the technique of photographic reproduction from a camera-ready manuscript was chosen. This process shifts the main responsibility for the technical quality considerably from the publisher to the author. We therefore urge all authors to observe very carefully our guidelines for the preparation of camera-ready manuscripts, which we will supply on request. This applies especially to the quality of figures and halftones submitted for publication. Figures should be submitted as originals or glossy prints, as very often Xerox copies are not suitable for reproduction. In addition, it might be useful to look at some of the volumes already published or, especially if some atypical text is planned, to write to the Physics Editorial Department of Springer-Verlag direct. This avoids mistakes and time-consuming correspondence during the production period.
As a special service, we offer free of charge LaTeX and TeX macro packages to format the text according to Springer-Verlag's quality requirements. We strongly recommend authors to make use of this offer, as the result will be a book of considerably improved technical quality. The typescript will be reduced in size (75% of the original). Therefore, for example, any writing within figures should not be smaller than 2.5 mm.
Manuscripts not meeting the technical standard of the series will have to be returned for improvement.
For further information please contact Springer-Verlag, Physics Editorial Department II, Tiergartenstrasse 17, W-6900 Heidelberg, FRG.

Preface

In general, combustion is a spatially three-dimensional, highly complex physico-chemical process of transient nature. Models are therefore needed that simplify a given combustion problem to such a degree that it becomes amenable to theoretical or numerical analysis but that are not so restrictive as to distort the underlying physics or chemistry. In particular, in view of worldwide efforts to conserve energy and to control pollutant formation, models of combustion chemistry are needed that are sufficiently accurate to allow confident predictions of flame structures. Reduced kinetic mechanisms, which are the topic of the present book, represent such combustion-chemistry models.

Historically combustion chemistry was first described as a global one-step reaction in which fuel and oxidizer react to form a single product. Even when detailed mechanisms of elementary reactions became available, empirical one-step kinetic approximations were needed in order to make problems amenable to theoretical analysis. This situation began to change in the early 1970s when computing facilities became more powerful and more widely available, thereby facilitating numerical analysis of relatively simple combustion problems, typically steady one-dimensional flames, with moderately detailed mechanisms of elementary reactions. However, even on the fastest and most powerful computers available today, numerical simulations of, say, laminar, steady, three-dimensional reacting flows with reasonably detailed and hence realistic kinetic mechanisms of elementary reactions are not possible. Even worse than for laminar reacting flows are the prospects for the numerical simulation of turbulent combustion problems with reasonably detailed kinetic mechanisms of elementary reactions. This is where systematically reduced kinetic mechanisms come in. However, not only are these mechanisms useful for realizing numerical simulations of laminar and turbulent flows under the variety of practically relevant operating conditions of small-scale and large-scale combustion systems, they also open avenues for new theoretical approaches to analysis of combustion problems. The so-called "rate-ratio asymptotics", which have now supplemented activation-energy asymptotics, are an example of one such new theoretical approach.

Whilst the concept of steady-state and partial-equilibrium approximations has been known for a long time, the history of deriving reduced kinetic mechanisms with these approximations in a systematic manner is relatively short.

The first mechanisms were derived and published in the mid-1980s for premixed and non-premixed methane flames, and, shortly thereafter, they were used in asymptotic and numerical analysis. In these "early years" from 1985 to 1988 a number of research groups focused attention on methane flames because these flames provided a good testing ground for the development of strategies that are useful in the systematic reduction of detailed kinetic mechanisms to only a few kinetic steps. In particular, workshops on the derivation of reduced kinetic mechanisms for methane flames were held in Sydney in 1987, in Yale in 1988, and in La Jolla at UCSD in 1989. During the 1990 workshop in Cambridge (UK) it emerged that the first "learning period" was over, and that the acquired methodologies and skills of various research groups had become sufficiently uniform for the reduction procedure to be cast into an unambiguous form and to apply it to a variety of fuels and flames. It is the main intention of the present book to provide the reader with an overview of the state-of-the-art of these now generally accepted reduction techniques and of the reduced mechanisms that they can produce for a variety of fuels. In this sense the book may be viewed as an "album of reduced mechanisms" for flame calculations.

In summary, the result of the research reported in the present book is that in many of the chapters more than one reduced mechanism is proposed for the same fuel. Two-step mechanisms are appropriate for hydrogen–air premixed (Chap. 3) and diluted and non-diluted diffusion flames (Chaps. 10 and 11), whilst two-step and three-step mechanisms are used for wet carbon-monoxide premixed and diffusion flames (Chaps. 4 and 12). For methane–air premixed flames (Chap. 5) the four-step mechanism previously used turns out to be the best choice. This is also demonstrated by a comparison between a five-step and a four-step mechanism for methane–air diffusion flames (Chap. 13). Four-step mechanisms also perform well for ethylene and ethane premixed flames (Chap. 6). For acetylene premixed flames (Chap. 7), however, this is less evident, and comparisons between a seven-, six-, five- and four-step mechanism are performed showing different levels of accuracy. Then, a five-step mechanism is used for acetylene–air diffusion flames (Chap. 14). For propane premixed flames nine, six or four steps are used (Chap. 8) and nine or seven steps for propane–air diffusion flames (Chap. 15). Methanol requires five steps for premixed (Chap. 9) and five or four steps for diffusion flames (Chap. 16). Finally, for a mixture of methane and ethane burning in a diffusion flame, a minimum of five steps is needed but nine steps are required for a sufficiently accurate calculation of NOx production, which requires two additional steps. In many of the chapters burning velocities of premixed flame calculations are compared to experimental data presented in Chap. 2.

A further intention of the book is to present existing software for reducing mechanisms and flame calculations and to define a general, uniform format for the collection of flamelet libraries. These descriptions are found in the Appendices A to C. In particular, for many of the flames studied herein, flamelet libraries are available as is the Cambridge Laminar Flamelet code.

This book would not have been possible without the help of various people. Specifically, the manuscripts of all chapters were collected and carefully read and prepared for printing by Josef Göttgens and Peter Terhoeven. Ute Bennerscheidt, Heike Jackisch, Andrea Dressen and Robert Korff typed and retyped many of the chapters and inserted many of the figures. The editors are highly indebted to them for their assistance.

Aachen
May 1992

N. Peters
B. Rogg

Contents

Part I

Fundamentals

1. Flame Calculations with Reduced Mechanisms — An Outline

N. Peters
Institut für Technische Mechanik, RWTH Aachen, W-5100 Aachen, Germany

1.1 Introduction

The exciting phenomenon called a flame results from the interaction of convection and molecular diffusion with many chemical reactions on small length scales. This interaction can be described on the basis of balance equations for continuity, momentum, energy and mass fractions of the chemical species involved. The balance equations are highly non-linear differential equations and require specific numerical solution techniques. Such techniques have been developed for one-dimensional flame problems by different groups since the early 1970's [1.1]–[1.19] and are quite efficient and reliable today [1.20]–[1.22]. During the same time and stimulated by this working tool, the detailed kinetic mechanisms for hydrocarbon flames have continuously been refined and improved, resulting, however, in a continuously growing number of reactions and species.

For the oxidation of hydrocarbons up to propane of the order of hundred reactions and around 30 species are believed to be necessary for a sufficiently accurate calculation of the chemistry. While such a detailed description is indeed necessary as a starting point, it is hard to believe that all these reactions are rate-determining and that all the species need accurately to be calculated in order to obtain global properties such as the burning velocity for premixed flames and the extinction strain rate for diffusion flames. Furthermore, if one wants to extend the procedure to multi-dimensional or even turbulent flames, the power of even the largest super-computer would be insufficient to resolve of the order of 30 non-linear partial differential equations. Therefore, the idea of reducing the detailed chemical mechanism to a few steps involving the reactants, products and only a few reaction intermediates, has always been attractive. A definite procedure to do so, based essentially on introducing steady state assumptions for intermediates, has been developed recently [1.23]–[1.24]. A subsequent asymptotic analysis for premixed methane flames was performed in [1.25]. Partial reviews on the subject are given in [1.25] and [1.27].

This book intends to provide an up-to data summary of the work by different research groups using the same reduction strategy on hydrogen, carbon-monoxide and various hydrocarbon flames.

The same detailed kinetic mechanism should be the common basis for hydrogen and hydrocarbons up to propane. This mechanism of 82 reactions is summarized in Table 1 and will be explained below.

1.2 Governing Equations

A common feature of all flame calculations summarized in this book is the use of essentially the same governing equations. They are the one-dimensional balance equations for continuity, momentum, energy and the chemical species.

Unstretched Premixed Flames

Here we consider a planar steady state flame configuration normal to the x-direction with the unburnt mixture at $x \to -\infty$, and the burnt gas at $x \to +\infty$.

The equation governing conservation of overall mass, species mass and energy are:

Continuity

$$\frac{d(\rho u)}{dx} = 0, \tag{1.1}$$

Species

$$\rho u \frac{dY_i}{dx} = -\frac{dj_i}{dx} + \dot{m}_i, \tag{1.2}$$

Energy

$$\rho u c_p \frac{dT}{dx} = \frac{d}{dx}\left(\lambda \frac{dT}{dx}\right) - \sum_{i=1}^{n} h_i \dot{m}_i - \sum_{i=1}^{n} c_p j_i \frac{dT}{dx}. \tag{1.3}$$

The continuity equation may be integrated once to yield

$$\rho u = \rho_u s_L, \tag{1.4}$$

where the subscript u denotes conditions in the fresh, unburnt mixture, and where s_L denotes the burning velocity. The latter is an eigenvalue, which must be determined as part of the solution. For the low Mach-number flows considered therein, the pressure is taken as constant, an assumption that is also made for the counterflow diffusion flames below.

Counterflow Diffusion Flames

Here we consider a planar or axi-symmetric counterflow configuration with the fuel coming from $y = -\infty$ and the oxidizer from $y = +\infty$. There exists a similarity solution for this flow field resulting in a set of one-dimensional equations in the y direction. Introducing $u = Gx$ one obtains the following governing equations:

Continuity

$$\frac{\partial \rho v}{\partial y} + (j+1)\rho G = 0, \tag{1.5}$$

Momentum

$$\rho v \frac{dG}{dy} = -\rho G^2 + P' + \frac{d}{dy}\left(\mu \frac{dG}{dy}\right) = 0\,, \tag{1.6}$$

Species

$$\rho v \frac{dY_i}{dy} = \frac{dj_i}{dy} + \dot{m}_i\,, \tag{1.7}$$

Energy

$$\rho v c_p \frac{dT}{dy} = \frac{d}{dy}\left(\lambda \frac{dT}{dy}\right) - \sum_{i=1}^{n} h_i m_i - \sum_{i=1}^{n} c_{pi} j_i \frac{dT}{dy}\,, \tag{1.8}$$

Mixture fraction

$$\rho v \frac{dZ}{dy} = \frac{d}{dy}\left(\frac{\lambda}{c_p}\frac{dZ}{dy}\right)\,. \tag{1.9}$$

Here $j = 0$ applies for the planar and $j = 1$ for axi-symmetric configuration. The parameter P' is defined as

$$P' = \rho_\infty a^2\,,$$

where $a = (\partial u/\partial x)_\infty$ is the velocity gradient in the oxidizer stream. This parameter represents the stretch rate and is in general prescribed. If the oxidizer flow issues from a burner sufficiently close to the flame the boundary condition $u = 0$ must be calculated with the solution [1.28]. In the equation for the mixture fraction, λ/c_p is calculated locally from the flame solution. The mixture fraction Z thereby defined may therefore be viewed to represent a normalized enthalpy based on the assumption of unity Lewis numbers. It is introduced here to be able to plot the scalar profiles unambiguously over the mixture fraction as the independent variable.

In the above equations u and v are the velocity components in x and y directions, respectively. Furthermore, ρ is the density, Y_i the mass fraction of species i, j_i the diffusion flux, $\dot{m}_i$ the chemical production rate of species i per unit mass, T the temperature, c_p the heat capacity at constant pressure, λ the thermal conductivity and h_i the specific enthalpy of species i. The diffusion flux j_i may be related to the diffusion velocity V_i by

$$j_i = \rho_i V_i = \rho Y_i V_i, \tag{1.10}$$

where $\rho_i = \rho Y_i$ is the partial density. The chemical production rate $\dot{m}_i$ contains contributions from all reactions, i.e.,

$$\dot{m}_i = M_i \sum_{k=1}^{r} \nu_{ik}\, w_k, \tag{1.11}$$

where M_i is the molecular weight of species i and the reaction rates are given by

$$w_k = k_{fk}(T) \prod_{j=1}^{n} \left(\frac{\rho Y_j}{M_j}\right)^{\nu'_{kj}} - k_{bk}(T) \prod_{j=1}^{n} \left(\frac{\rho Y_j}{M_i}\right)^{\nu''_{kj}} . \tag{1.12}$$

The stoichiometric coefficients of the forward and backward step for species j in reaction k are denoted by ν'_{kj} and ν''_{kj}. The rate coefficients $k_{fk}(T)$ and $k_{bk}(T)$ are expressed in the form

$$k_k = A_k\, T^{n_k} \exp\left(-\frac{E_k}{RT}\right) \tag{1.13}$$

where A_k is the frequency factor, n_k is the pre-exponential temperature exponent, and E_k is the activation energy of reaction k.

1.3 The Detailed Kinetic Mechanism for Fuels up to Propane

The kinetic mechanism to be used as a common basis for all flame calculations is listed in Table 1. These data are compilations from the recent literature [1.29]. The units of A_k are

$$\left[\frac{1}{\mathrm{sec\,K}^{n_k}}\right], \quad \left[\frac{\mathrm{cm}^3}{\mathrm{mole\,sec\,K}^{n_k}}\right], \quad \text{or} \quad \left[\frac{\mathrm{cm}^6}{\mathrm{mole}^2\,\mathrm{sec\,K}^{n_k}}\right]$$

for a uni-molecular, bi-molecular or tri-molecular reaction, respectively. The activation energy is given in [kJ/mole], and the universal gas constant is $R = 8.3147$ J/mole K. The data of reactions 34, 36, 51 and 58 are expressed as in [1.30], in which the rate coefficient is calculated as a function of temperature and pressure in the form

$$k = F \cdot k_\infty \cdot k_L . \tag{1.14}$$

Here

$$k_L = \frac{k_0\,[\mathrm{M}]/k_\infty}{1 + k_0\,[\mathrm{M}]/k_\infty} \tag{1.15}$$

is the Lindemann form; F is approximated by

$$\log_{10} F = \log_{10} F_c \times \frac{1}{1 + (\log_{10}[k_0\,[\mathrm{M}]/k_\infty]\,1/N)^2}, \tag{1.16}$$

where

$$N = 0.75 - 1.27 \log_{10} F_c \tag{1.17}$$

represents the corrections. Note that k_L and F are equal for forward and backward reactions; $[\mathrm{M}] = \sum_{i=1}^{n} \rho Y_i/\mathrm{M}_i = p/RT$ is the molar density of the mixture. For reactions 34, 36, 51 and 58 the following expressions are used:

reaction 34:

$$F_c = 0.577 \exp\left[-\frac{T}{2370.0}\right], \tag{1.18}$$

reaction 36:

$$F_c = 0.38 \exp\left[-\frac{T}{73.0}\right] + 0.62 \exp\left[-\frac{T}{1180.0}\right], \tag{1.19}$$

reaction 51:

$$F_c = 0.35, \tag{1.20}$$

reaction 58:

$$F_c = 0.411 \exp\left[\frac{-73.4}{T}\right] + \exp\left[-\frac{T}{422.8}\right], \tag{1.21}$$

where T is in degrees Kelvin.

Equations for transport properties are given, for instance, in [1.31]. Their choice, however, is left to the disgression of the groups contributing the individual chapters. It should be noted that only transport properties for non-steady state species are needed in calculations based on reduced mechanisms. Therefore the sometimes unknown properties of the steady-state species are of minor importance only, even in the calculations based on detailed mechanisms. Thermochemical properties are based on NASA polynomials [1.32], for instance, but their choice too is left to the disgression of the individual groups.

Table 1.1.

Nr.	Reaction	A mole, cm^3, sec	n	E kJ/mole
1.1 H_2/O_2 Chain Reactions				
1f	$O_2+H \rightarrow OH+O$	2.000E+14	0.00	70.30
1b	$OH+O \rightarrow O_2+H$	1.568E+13	0.00	3.52
2f	$H_2+O \rightarrow OH+H$	5.060E+04	2.67	26.30
2b	$OH+H \rightarrow H_2+O$	2.222E+04	2.67	18.29
3f	$H_2+OH \rightarrow H_2O+H$	1.000E+08	1.60	13.80
3b	$H_2O+H \rightarrow H_2+OH$	4.312E+08	1.60	76.46
4f	$OH+OH \rightarrow H_2O+O$	1.500E+09	1.14	0.42
4b	$H_2O+O \rightarrow OH+OH$	1.473E+10	1.14	71.09
1.2 HO_2 Formation and Consumption				
5f	$O_2+H+M' \rightarrow HO_2+M'$	2.300E+18	-0.80	0.00
5b	$HO_2+M' \rightarrow O_2+H+M'$	3.190E+18	-0.80	195.39
6	$HO_2+H \rightarrow OH+OH$	1.500E+14	0.00	4.20
7	$HO_2+H \rightarrow H_2+O_2$	2.500E+13	0.00	2.90
8	$HO_2+OH \rightarrow H_2O+O_2$	6.000E+13	0.00	0.00
9	$HO_2+H \rightarrow H_2O+O$	3.000E+13	0.00	7.20
10	$HO_2+O \rightarrow OH+O_2$	1.800E+13	0.00	-1.70
1.3 H_2O_2 Formation and Consumption				
11	$HO_2+HO_2 \rightarrow H_2O_2+O_2$	2.500E+11	0.00	-5.20
12f	$OH+OH+M' \rightarrow H_2O_2+M'$	3.250E+22	-2.00	0.00
12b	$H_2O_2+M' \rightarrow OH+OH+M'$	1.692E+24	-2.00	202.29
13	$H_2O_2+H \rightarrow H_2O+OH$	1.000E+13	0.00	15.00
14f	$H_2O_2+OH \rightarrow H_2O+HO_2$	5.400E+12	0.00	4.20
14b	$H_2O+HO_2 \rightarrow H_2O_2+OH$	1.802E+13	0.00	134.75
1.4 Recombination Reactions				
15	$H+H+M' \rightarrow H_2+M'$	1.800E+18	-1.00	0.00
16	$OH+H+M' \rightarrow H_2O+M'$	2.200E+22	-2.00	0.00
17	$O+O+M' \rightarrow O_2+M'$	2.900E+17	-1.00	0.00

Table 1.1. continued

Nr.	Reaction	A mole, cm^3, sec	n	E kJ/mole
		2. CO/CO_2 Mechanism		
18f	$CO+OH \rightarrow CO_2+H$	4.400E+06	1.50	-3.10
18b	$CO_2+H \rightarrow CO+OH$	4.956E+08	1.50	89.76
		3.1 CH Consumption		
19	$CH+O_2 \rightarrow CHO+O$	3.000E+13	0.00	0.00
20	$CO_2+CH \rightarrow CHO+CO$	3.400E+12	0.00	2.90
		3.2 CHO Consumption		
21	$CHO+H \rightarrow CO+H_2$	2.000E+14	0.00	0.00
22	$CHO+OH \rightarrow CO+H_2O$	1.000E+14	0.00	0.00
23	$CHO+O_2 \rightarrow CO+HO_2$	3.000E+12	0.00	0.00
24f	$CHO+M' \rightarrow CO+H+M'$	7.100E+14	0.00	70.30
24b	$CO+H+M' \rightarrow CHO+M'$	1.136E+15	0.00	9.97
		3.3 CH_2 Consumption		
25f	$CH_2+H \rightarrow CH+H_2$	8.400E+09	1.50	1.40
25b	$CH+H_2 \rightarrow CH_2+H$	5.830E+09	1.50	13.08
26	$CH_2+O \rightarrow CO+H+H$	8.000E+13	0.00	0.00
27	$CH_2+O_2 \rightarrow CO+OH+H$	6.500E+12	0.00	6.30
28	$CH_2+O_2 \rightarrow CO_2+H+H$	6.500E+12	0.00	6.30
		3.4 CH_2O Consumption		
29	$CH_2O+H \rightarrow CHO+H_2$	2.500E+13	0.00	16.70
30	$CH_2O+O \rightarrow CHO+OH$	3.500E+13	0.00	14.60
31	$CH_2O+OH \rightarrow CHO+H_2O$	3.000E+13	0.00	5.00
32	$CH_2O+M' \rightarrow CHO+H+M'$	1.400E+17	0.00	320.00
		3.5 CH_3 Consumption		
33f	$CH_3+H \rightarrow CH_2+H_2$	1.800E+14	0.00	63.00
33b	$CH_2+H_2 \rightarrow CH_3+H$	3.680E+13	0.00	44.30
34	$CH_3+H \rightarrow CH_4$ k_∞	2.108E+14	0.00	0.00
	k_0	6.257E+23	-1.80	0.00

Table 1.1. continued

Nr.	Reaction	A mole, cm^3, sec	n	E kJ/mole
35	$CH_3+O \rightarrow CH_2O+H$	7.000E+13	0.00	0.00
36	$CH_3+CH_3 \rightarrow C_2H_6$ k_∞ k_0	3.613E+13 1.270E+41	0.00 -7.00	0.00 11.56
37	$CH_3+O_2 \rightarrow CH_2O+OH$	3.400E+11	0.00	37.40
38f	$CH_4+H \rightarrow CH_3+H_2$	2.200E+04	3.00	36.60
38b	$CH_3+H_2 \rightarrow CH_4+H$	8.391E+02	3.00	34.56
39	$CH_4+O \rightarrow CH_3+OH$	1.200E+07	2.10	31.90
40f	$CH_4+OH \rightarrow CH_3+H_2O$	1.600E+06	2.10	10.30
40b	$CH_3+H_2O \rightarrow CH_4+OH$	2.631E+05	2.10	70.92
	4.1 C_2H Consumption			
41f	$C_2H+H_2 \rightarrow C_2H_2+H$	1.100E+13	0.00	12.00
41b	$C_2H_2+H \rightarrow C_2H+H_2$	5.270E+13	0.00	119.95
42	$C_2H+O_2 \rightarrow CHCO+O$	5.000E+13	0.00	6.30
	4.2 CHCO Consumption			
43f	$CHCO+H \rightarrow CH_2+CO$	3.000E+13	0.00	0.00
43b	$CH_2+CO \rightarrow CHCO+H$	2.361E+12	0.00	-29.39
44	$CHCO+O \rightarrow CO+CO+H$	1.000E+14	0.00	0.00
	4.3 C_2H_2 Consumption			
45	$C_2H_2+O \rightarrow CH_2+CO$	4.100E+08	1.50	7.10
46	$C_2H_2+O \rightarrow CHCO+H$	4.300E+14	0.00	50.70
47f	$C_2H_2+OH \rightarrow C_2H+H_2O$	1.000E+13	0.00	29.30
47b	$C_2H+H_2O \rightarrow C_2H_2+OH$	9.000E+12	0.00	-15.98
48	$C_2H_2+CH \rightarrow C_3H_3$	2.100E+14	0.00	-0.50
	4.4 C_2H_3 Consumption			
49	$C_2H_3+H \rightarrow C_2H_2+H_2$	3.000E+13	0.00	0.00
50	$C_2H_3+O_2 \rightarrow C_2H_2+HO_2$	5.400E+11	0.00	0.00
51f	$C_2H_3 \rightarrow C_2H_2+H$ k_∞ k_0	2.000E+14 1.187E+42	0.00 -7.50	166.29 190.40
51b	$C_2H_2+H \rightarrow C_2H_3$ k_∞	1.053E+14	0.00	3.39

Table 1.1. continued

Nr.	Reaction		A mole, cm^3, sec	n	E kJ/mole
4.5 C_2H_4 Consumption					
52f	$C_2H_4+H \rightarrow C_2H_3+H_2$		1.500E+14	0.00	42.70
52b	$C_2H_3+H_2 \rightarrow C_2H_4+H$		9.605E+12	0.00	32.64
53	$C_2H_4+O \rightarrow CH_3+CO+H$		1.600E+09	1.20	3.10
54f	$C_2H_4+OH \rightarrow C_2H_3+H_2O$		3.000E+13	0.00	12.60
54b	$C_2H_3+H_2O \rightarrow C_2H_4+OH$		8.283E+12	0.00	65.20
55	$C_2H_4+M' \rightarrow C_2H_2+H_2+M'$		2.500E+17	0.00	319.80
4.6 C_2H_5 Consumption					
56f	$C_2H_5+H \rightarrow CH_3+CH_3$		3.000E+13	0.00	0.00
56b	$CH_3+CH_3 \rightarrow C_2H_5+H$		3.547E+12	0.00	49.68
57	$C_2H_5+O_2 \rightarrow C_2H_4+HO_2$		2.000E+12	0.00	20.90
58f	$C_2H_5 \rightarrow C_2H_4+H$	k_∞ k_0	2.000E+13 1.000E+17	0.00 0.00	166.00 130.00
58b	$C_2H_4+H \rightarrow C_2H_5$	k_∞	3.189E+13	0.00	12.61
4.7 C_2H_6 Consumption					
59	$C_2H_6+H \rightarrow C_2H_5+H_2$		5.400E+02	3.50	21.80
60	$C_2H_6+O \rightarrow C_2H_5+OH$		3.000E+07	2.00	21.40
61	$C_2H_6+OH \rightarrow C_2H_5+H_2O$		6.300E+06	2.00	2.70
5.1 C_3H_3 Consumption					
62	$C_3H_3+O_2 \rightarrow CHCO+CH_2O$		6.000E+12	0.00	0.00
63	$C_3H_3+O \rightarrow C_2H_3+CO$		3.800E+13	0.00	0.00
64f	$C_3H_4 \rightarrow C_3H_3+H$		5.000E+14	0.00	370.00
64b	$C_3H_3+H \rightarrow C_3H_4$		1.700E+13	0.00	19.88
5.2 C_3H_4 Consumption					
65	$C_3H_4+O \rightarrow C_2H_2+CH_2O$		1.000E+12	0.00	0.00
66	$C_3H_4+O \rightarrow C_2H_3+CHO$		1.000E+12	0.00	0.00
67	$C_3H_4+OH \rightarrow C_2H_3+CH_2O$		1.000E+12	0.00	0.00
68	$C_3H_4+OH \rightarrow C_2H_4+CHO$		1.000E+12	0.00	0.00

Table 1.1. continued

Nr.	Reaction	A mole, cm^3, sec	n	E kJ/mole
	5.3 C_3H_5 Consumption			
69f	$C_3H_5 \rightarrow C_3H_4+H$	3.980E+13	0.00	293.10
69b	$C_3H_4+H \rightarrow C_3H_5$	1.267E+13	0.00	32.48
70	$C_3H_5+H \rightarrow C_3H_4+H_2$	1.000E+13	0.00	0.00
	5.4 C_3H_6 Consumption			
71f	$C_3H_6 \rightarrow C_2H_3+CH_3$	3.150E+15	0.00	359.30
71b	$C_2H_3+CH_3 \rightarrow C_3H_6$	2.511E+12	0.00	-34.69
72	$C_3H_6+H \rightarrow C_3H_5+H_2$	5.000E+12	0.00	6.30
	5.5 C_3H_7 Consumption			
73	n-$C_3H_7 \rightarrow C_2H_4+CH_3$	9.600E+13	0.00	129.80
74f	n-$C_3H_7 \rightarrow C_3H_6+H$	1.250E+14	0.00	154.90
74b	$C_3H_6+H \rightarrow$ n-C_3H_7	4.609E+14	0.00	21.49
75	i-$C_3H_7 \rightarrow C_2H_4+CH_3$	6.300E+13	0.00	154.50
76	i-$C_3H_7+O_2 \rightarrow C_3H_6+HO_2$	1.000E+12	0.00	20.90
	5.6 C_3H_8 Consumption			
77	$C_3H_8+H \rightarrow$ n-$C_3H_7+H_2$	1.300E+14	0.00	40.60
78	$C_3H_8+H \rightarrow$ i-$C_3H_7+H_2$	1.000E+14	0.00	34.90
79	$C_3H_8+O \rightarrow$ n-C_3H_7+OH	3.000E+13	0.00	24.10
80	$C_3H_8+O \rightarrow$ i-C_3H_7+OH	2.600E+13	0.00	18.70
81	$C_3H_8+OH \rightarrow$ n-$C_3H_7+H_2O$	3.700E+12	0.00	6.90
82	$C_3H_8+OH \rightarrow$ i-$C_3H_7+H_2O$	2.800E+12	0.00	3.60
	6.1 CH_2OH Consumption			
83	$CH_2OH+H \rightarrow CH_2O+H_2$	3.000E+13	0.00	0.00
84	$CH_2OH+O_2 \rightarrow CH_2O+HO_2$	1.000E+13	0.00	30.10
85	$CH_2OH+M' \rightarrow CH_2O+H+M'$	1.000E+14	0.00	105.10
	6.2 CH_3OH Consumption			
86	$CH_3OH+H \rightarrow CH_2OH+H_2$	4.000E+13	0.00	25.50
87	$CH_3OH+OH \rightarrow CH_2OH+H_2O$	1.000E+13	0.00	7.10

$[M'] = 6.5[CH_4]+6.5[H_2O]+1.5[CO_2]+0.75[CO]+0.4[O_2]+0.4[N_2]+1.0[Other]$

References

[1.1] Spalding, D. B., Stephenson, D. L., Taylor, R. G.: A Calculation Procedure for the Prediction of Laminar Flame Speeds, Comb. and Flame, **17**, p. 55, 1971.

[1.2] Wilde, K. A.: Boundary-Value Solutions of the One-Dimensional Laminar Flame Propagation Equations, Comb. and Flame, **18**, p. 43, 1972.

[1.3] Bledjian, L.: Computation of Time-Dependent Laminar Flame Structure, Comb. and Flame, **20**, p. 5, 1973.

[1.4] Peters, N.: The Dissociation of Water Vapor on a Hot Porous Surface, Deuxième Symposium Europeen sur la Combustion, pp. 142–47, Orléans 1975.

[1.5] Peters, N.: Berechnung einer Methan-Luft-Diffusionsflamme im örtlichen Gleichgewicht und im Nichtgleichgewicht, VDI-Bericht, **246**, pp. 5–12, 1975.

[1.6] Peters, N.: Analysis of a Laminar Flat Plate Boundary Layer Diffusion Flame, Int. J. Heat Mass Transf., **19**, pp. 385–93, 1976.

[1.7] Margolis, S. B.: Time-Dependent Solution of a Premixed Laminar Flame, J. Comp. Phys., **27**, p. 410, 1978.

[1.8] Warnatz, J.: Calculation of the Structure of Laminar Flat Flames I; Flame Velocity of Freely Propagating Ozone Decomposition Flames, Ber. Bunsenges. Phys. Chem., **82**, p. 193, 1978.

[1.9] Warnatz, J.: Calculation of the Structure of Laminar Flat Flames II; Flame Velocity and Structure of Freely Propagating HydrogenOxygen and Hydrogen-Air-Flames, Ber. Bunsenges. Phys. Chem., **82**, p. 643, 1978.

[1.10] Dixon-Lewis, G.: Kinetic Mechanism, Structure and Properties of Premixed Flames in Hydrogen-Oxygen-Nitrogen Mixtures, Phil. Trans. of the Royal Soc. London, **292**, p. 45, 1979.

[1.11] Westbrook, C. K., Dryer, F. L.: A Comprehensive Mechanism for Methanol Oxidation, Comb. Sci. and Tech., **20**, p. 125, 1979.

[1.12] Westbrook, C. K., Dryer, F. L.: Prediction of Laminar Flame Properties of Methanol Air Mixtures, Comb. and Flame, **37**, p. 171, 1980.

[1.13] Coffee, T. P., Heimerl, J. M.: The Detailed Modeling of Premixed, Laminar Steady-State-Flames. I. Ozone, Comb. and Flame, **39**, p. 301, 1980.

[1.14] Miller, J. A., Mitchell, R. E., Smooke, M. D., Kee, R. J.: Toward a Comprehensive Chemical Kinetic Mechanism for the Oxidation of Acetylene: Comparison of Model Predictions with Results from Flame and Shock Tube Experiments, Nineteenth Symposium (International) on Combustion, p. 181, Reinhold, New York 1982.

[1.15] Smooke, M. D.: Solution of Burner Stabilized Premixed Laminar Flames by Boundary Value Methods, J. Comp. Phys., **48**, p. 72, 1982.

[1.16] Smooke, M. D., Miller, J. A., Kee, R. J.: On the Use of Adaptive Grids in Numerically Calculating Adiabatic Flame Speeds, Numerical Methods in Laminar Flame Propagation, N. Peters and J. Warnatz (Eds.), Friedr. Vieweg und Sohn, Wiesbaden, 1982.

[1.17] Smooke, M. D., Miller, J. A., Kee, R. J.: Determination of Adiabatic Flame Speeds by Boundary Value Methods, Comb. Sci. and Tech., **34**,p. 39, 1983.

[1.18] Smooke, M. D.: On the Use of Adaptive Grids in Premixed Combustion, AIChE J., **32**, p. 1233, 1986.

[1.19] Dixon-Lewis, G., David, T., Haskell, P. H., Fukutani, S., Jinno, H., Miller, J. A., Kee, R. J., Smooke, M. D., Peters, N., Effelsberg, E., Warnatz, J., Behrendt, F.: Calculation of the Structure and Extinction Limit of a Methane-Air Counterflow Diffusion Flame in the Forward Stagnation Region of a Porous Cylinder, Twentieth Symposium (Int.) on Combustion, p. 1893, Reinhold, New York 1985.

[1.20] Smooke, M. D., Premixed and Non-Premixed Test Problem Results, in: Reduced Kinetic Mechanisms and Asymptotic Approximations for Methane-Air Flames, (M. D. Smooke, Ed.).

[1.21] Rogg, B.: Sensitivity Analysis of Laminar Premixed CH_4-Air Flames Using Full and Reduced Kinetic Mechanisms, in M. D. Smooke, editor, Reduced Kinetic Mechanisms and Asymptotic Approximations for Methane-Air Flames, pp. 159–192, Springer, Berlin 1991.
[1.22] Rogg, B., RUN-IDL: A Computer Program for the Simulation of One-Dimensional Chemically Reacting Flows, Report CUED/A-THERMO/TR39, Cambridge University Engineering Department, April 1991.
[1.23] Peters, N.: Numerical and Asymptotic Analysis of Systematically Reduced Reaction Schemes for Hydrocarbon Flames, in Numerical Simulation of Combustion Phenomena, R. Glowinski et al., Eds., Lecture Notes in Physics, Springer-Verlag, p. 90, 1985.
[1.24] Peters, N., Kee, R. J.: The Computation of Stretched Laminar Methane-Air Diffusion Flames Using a Reduced Four-Step Mechanism, Comb. and Flame, **68**, p. 17, 1987.
[1.25] Peters, N., Williams, F. A.: The Asymptotic Structure of Stoichiometric Methane Air Flames, Comb. and Flame, **68**, p. 185, 1987.
[1.26] Peters, N.: Systematic Reduction of Flame Kinetics, in Principles and Details, Dynamics of Reactive Systems, Progress in Astronautics and Aeronautics, **113**, pp. 67–86, 1988.
[1.27] Peters, N.: Reducing Mechanisms, in Reduced Kinetic Mechanisms and Asymptotic Approximations for Methane-Air Flames, (M. D. Smooke, Ed.), Lecture Notes in Physics, **384**, pp. 48–67, Springer, 1981.
[1.28] Behrendt, F., Warnatz, J.: Numerical Study of NO Formation in Pure and Partially Premixed CH_4/Air and C_3H_8/Air, Joint Meeting of the British and the French Section of the Combustion Institute, pp. 25–28, Rouen, April 1989.
[1.29] Warnatz, J.: Private Communication, 1991.
[1.30] Gardiner, W. C., Troe, J.: Rate Coefficients of Thermal Dissociation, Isomerization and Recombination Reactions, Combustion Chemistry (W. C. Gardiner, Ed.), pp. 173–195, Springer, 1984.
[1.31] Oran, E., Boris, J.: Detailed Modeling of Combustion Systems, Prog. Energy Combustion Sci., **7**, pp. 1–72.
[1.32] Burcat, A.: Thermochemical Data for Combustion Calculations, Combustion Chemistry (W. C. Gardiner, Ed.), pp. 456–504, Springer, 1984.

2. A Compilation of Experimental Data on Laminar Burning Velocities

C. K. Law
Department of Mechanical and Aerospace Engineering,
Princeton University, Princeton, NJ 08544, U.S.A.

2.1 Introduction

The pioneering works of Dixon-Lewis [2.1], Warnatz [2.2], and Westbrook and Dryer [2.3] have demonstrated the usefulness of studying flame kinetics through numerical simulation of the flame structure and the subsequent comparison of the calculated results with experimental data. The comparison can be conducted either at the global level of the bulk flame responses such as the laminar burnig velocities, s_L, and the extinction stretched rates, or at the detailed level which includes not only the bulk flame responses but also the temperature and species profiles.

While the profile comparison obviously provides a significantly more stringent test for the kinetic scheme adopted in the simulation, very few experimental data have been determined for the profiles. On the contrary, a large body of experimental laminar burning velocities of a great variety of mixtures, under diverse conditions, have been accumulated over the years [2.4]. Thus much of the experimental comparisons have been conducted by using s_L. The rationale being that while an agreement does not validate a kinetic mechanism, a disagreement either in trend or of sufficient magnitude would imply the existence of certain deficiency in the mechanism. Thus an agreement in s_L is the minimum requirement for the validation of a kinetic mechanism.

Earlier comparisons were complicated by the scatter and uncertain accuracy of the experimental data [2.4]. The most serious source of error was probably caused by the coupled effects of stretch and preferential diffusion, which could lead to systematic shifts in s_L depending on the specific experimental methodology and even mixture stoichiometry [2.5]. Unaware of such experimental and phenomenological subtleties, some kinetic mechanisms have been either incorrectly "validated" or unjustifiably "calibrated".

Recognizing the importance of stretch, a counterflow flame technique has been developed [2.6], [2.7] from which nearly stretch-free s_L data of mixtures of hydrogen, methane, the C_2-hydrocarbons, propane, and methanol have been determined with extensive variations in stoichiometry and pressure [2.6]–[2.13]. These data have been consolidated and re-plotted in a uniform format, and are presented herein as a centralized source of reference for comparisons.

2.2 Experimental Methodology and Accuracy of Data

The experiment involves establishing two nearly planar and adiabatic flames in a nozzle-generated counterflow, and the determination of the axial velocity profile along the centerline of the flow by using laser Doppler velocimetry (Fig. 2.1). By identifying the minimum point of the velocity profile as a reference upstream burning rate, $s_{L,\mathrm{ref}}$, and the velocity gradient ahead of it as the imposed local stretch rate a, $s_{L,\mathrm{ref}}$ can be determined as a function of a for the given mixture. By linearly extrapolating $s_{L,\mathrm{ref}}(a)$ to $a = 0$ yields a value which should closely approximate the stretch-free burning rate s_L.

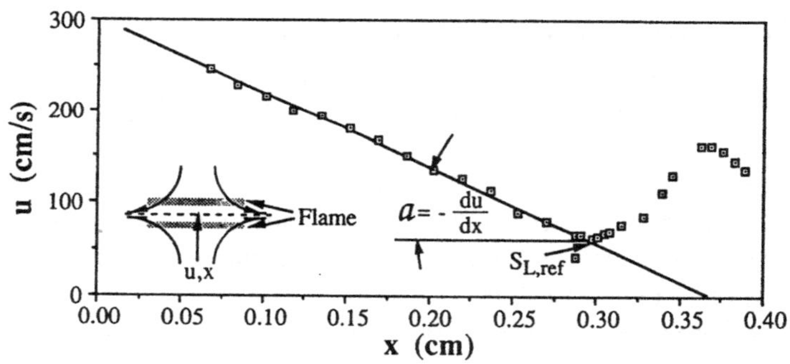

(a)

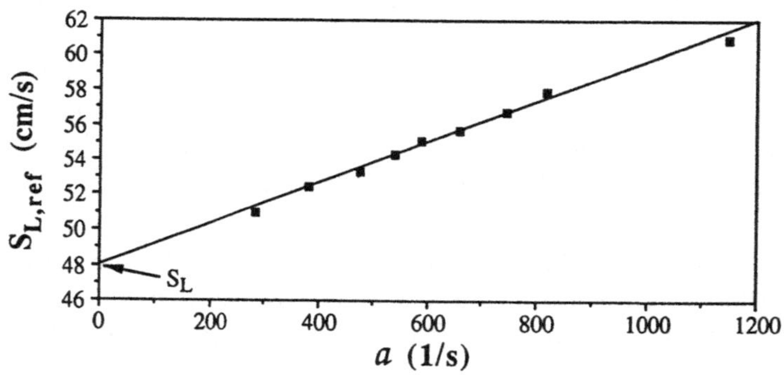

(b)

Fig. 2.1. (a) Typical axial velocity profile across a stagnation flame, showing the definitions of a and $s_{L,\mathrm{ref}}$. (b) Dependence of the burning velocity $s_{L,\mathrm{ref}}$ on the strain rate a.

There are several additional advantages with this technique. First, the upstream and downstream heat losses are minimized because the flow is nozzle-generated and because of symmetry. Furthermore, buoyancy effects are automatically included in the flow-field influences because the LDV-determined velocities are the local values experienced by the specific flame segment under measurement. Finally, because of the positively-stretched nature of the flame, flamefront diffusional-thermal instabilities are not favored [2.14]. Indeed, this technique produced smooth flame surfaces for the diffusionally-unstable lean hydrogen-air flames such that the burning velocities of ultra-lean mixtures were successfully determined.

The major sources of inaccuracy are the following. The LDV-determination of velocity is believed to be accurate to about 5% for the range of velocities encountered. The equivalence ratio, ϕ, is reproducible to about 0.01, being limited by the accuracies in metering the various mixture components.

Perhaps the most severe source of inaccuracy is the linear extrapolation of $s_{L,\mathrm{ref}}$ to zero a to yield s_L. Recent computational [2.15] and analytical [2.16] results have shown that the plot of $s_{L,\mathrm{ref}}$ versus a is not linear as a approaches zero; instead it bends down slightly. These are higher order effects and could result in an over-determination of s_L by at most 10% for the present experimental configurations. This point should be taken into consideration when attempting to achieve closer agreements between calculated and experimental results. It is also important to recognize that, because of the global nature of the laminar burning velocity, an agreement between calculation and experiment over a limited range of equivalence ratio is not adequate for the validation of a proposed kinetic mechanism. Thus our experimental-determined s_L cover as extensively as possible the mixture composition, from very lean to very rich. Furthermore, s_L has also been determined as a function of pressure, p, because of its relatively strong influence on the three-body termination reactions as compared to the two-body branching reactions.

Since a change in the equivalence ratio involves the simultaneous variations in the mixture stoichiometry as well as the flame temperature, T_{ad}, in some experiments T_{ad} was held fixed while ϕ was varied through substitution of a certain amount of the nitrogen in the air by either argon or carbon dioxide. Figure 2.17 shows the contents of the inerts needed to achieve the fixed flame temperatures. Such a mapping, involving the systematic and independent variations of ϕ, p, and T_{ad}, is expected to reflect well the influences of the three thermodynamic factors that affect elementary reactions, namely the local values of the concentration of at least one of the reactants, the density and thereby the pressure to indicate the frequency of collision, and the temperature to indicate the energetics of collision.

The last point to note is that since the conserved quantity in crossing the flame is the mass burning rate $\dot{m} = \rho_u s_L$ instead of s_L, where ρ_u is the density of the unburned mixture, strictly speaking $\dot{m}$ is a more suitable parameter to be compared. However, bowing to tradition we shall present most of our data as s_L. Care should be exercised in interpreting the influence of pressure on s_L because it includes effects of both the density change and the kinetics. For ultra-lean hydrogen-air and nitrogen-diluted methane-air mixtures, we have also presented the data on $\dot{m}$ because of its non-monotonic variation with pressure caused by the competition between two-body branching and three-body termination reactions.

The experimental data are presented as graphs, from Fig. 2.2 to 2.16. Additional specifications and the references from which the data were originally reported are given in the figure captions.

Acknowledgement

This work was supported in part by the Division of Basic Energy Sciences of the U.S.Department of Energy, and by the U.S. Air Force Office of Scientific Research. Professor F.N. Egolfopoulos generously assisted the preparation of this manuskript by consolidating and replotting all the data compiled herein.

Figures

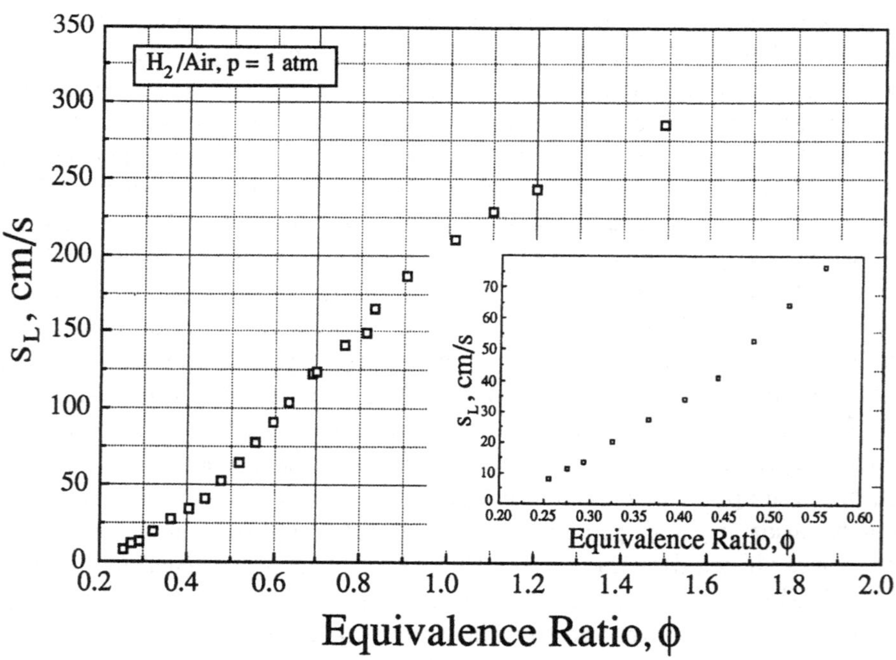

Fig. 2.2. Laminar burnig velocities of mixtures of hydrogen and air at one atmosphere pressure as function of equivalence ratio [2.12].

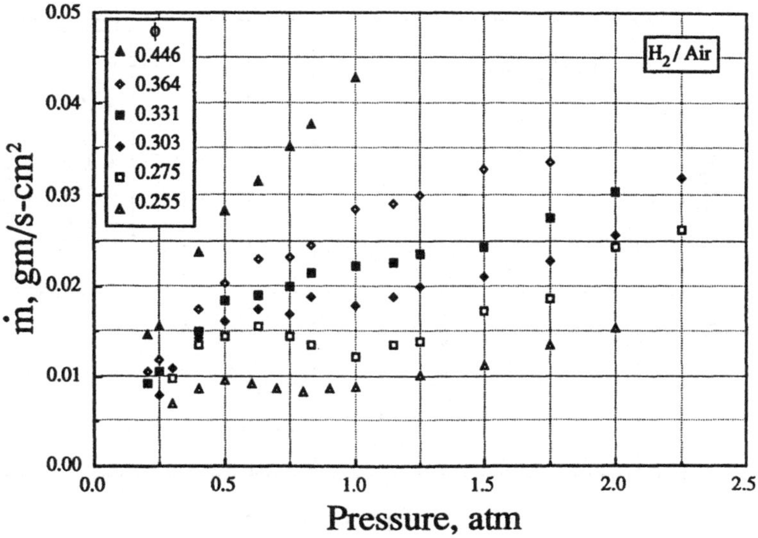

Fig. 2.3. Laminar burning rates of mixtures of hydrogen and air as functions of equivalence ratio and pressure [2.12].

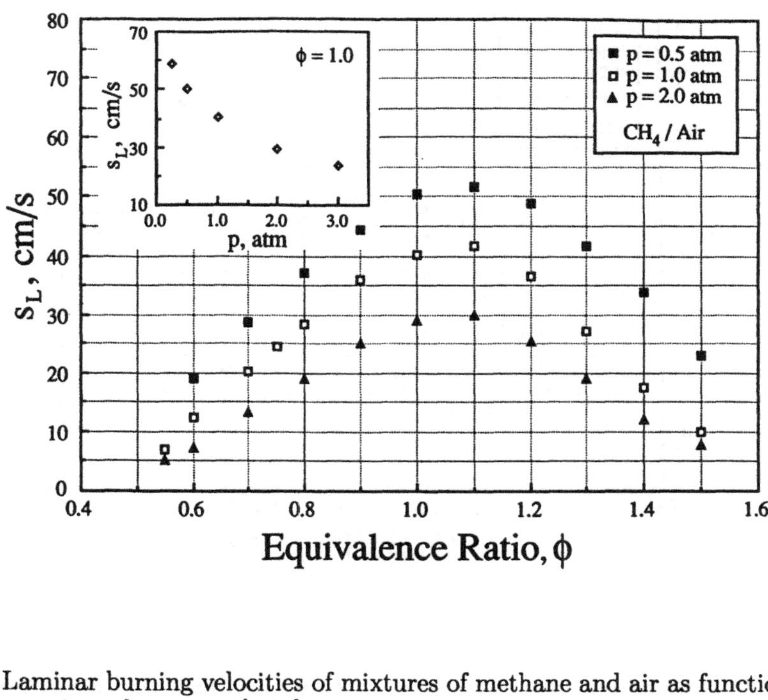

Fig. 2.4. Laminar burning velocities of mixtures of methane and air as functions of equivalence ratio and pressure [2.11].

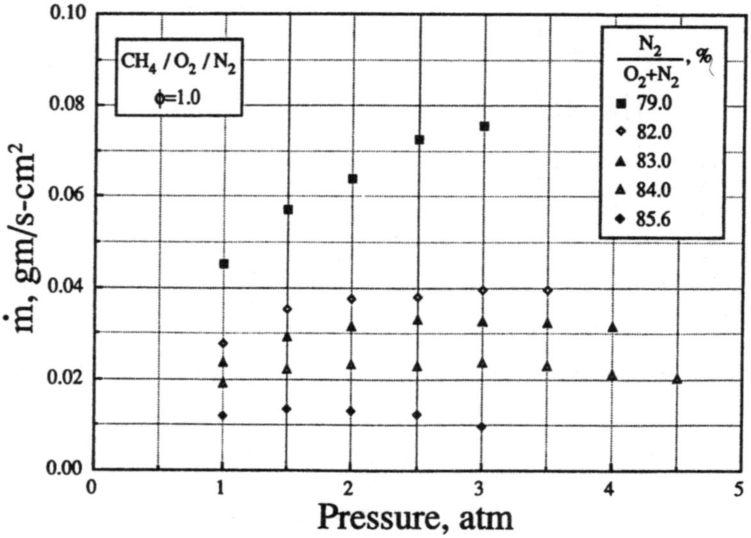

Fig. 2.5. Laminar burning rates of mixtures of methane, oxygen and nitrogen as functions of pressure and nitrogen dilution [2.10].

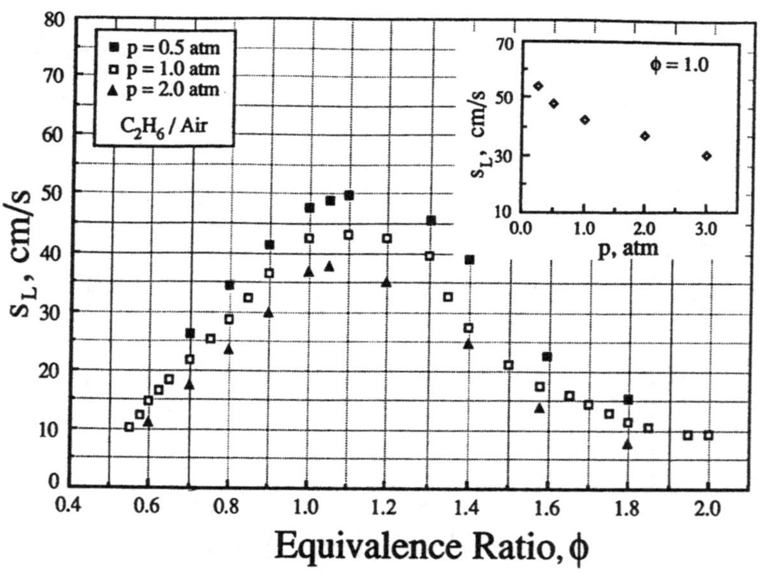

Fig. 2.6. Laminar burning velocities of mixtures of ethane and air as functions of equivalence ratio and pressure [2.11].

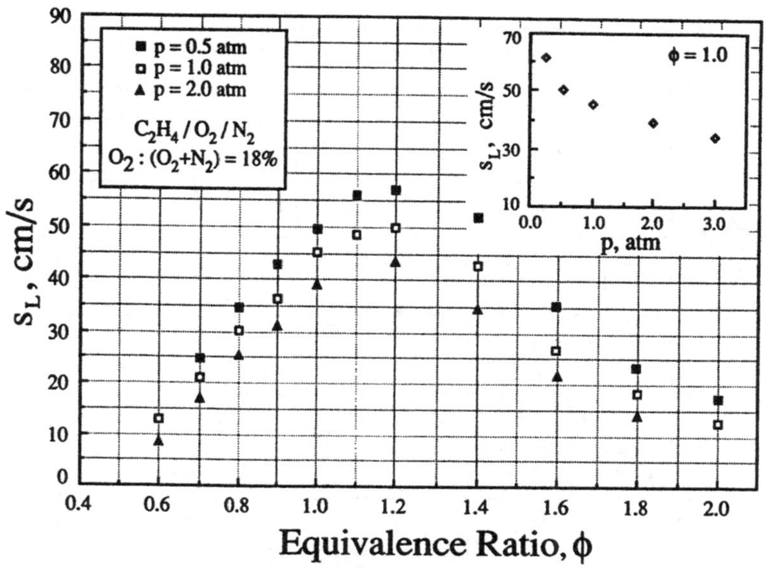

Fig. 2.7. Laminar burning velocities of mixtures of ethylene, oxygen and nitrogen as functions of equivalence ratio and pressure [2.11].

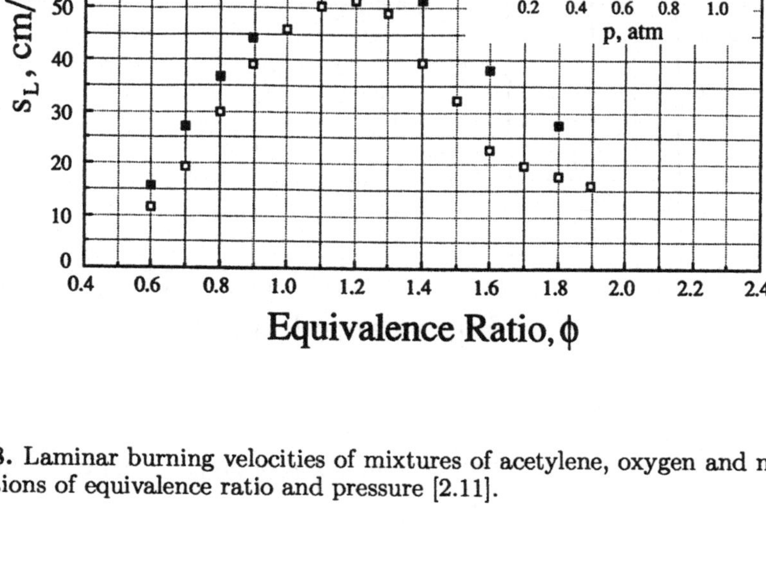

Fig. 2.8. Laminar burning velocities of mixtures of acetylene, oxygen and nitrogen as functions of equivalence ratio and pressure [2.11].

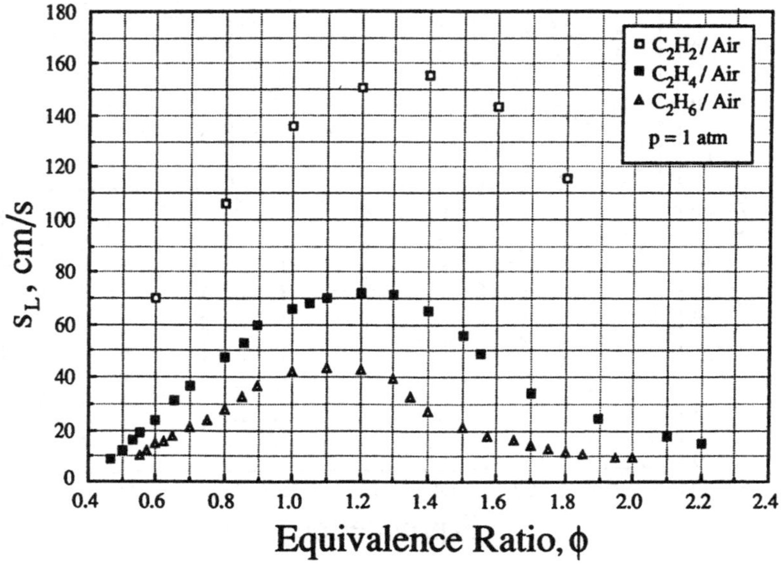

Fig. 2.9. Laminar burning velocities of mixtures of ethane, ethylene, and acetylene with air, at one atmosphere pressure, as functions of equivalence ratio [2.11].

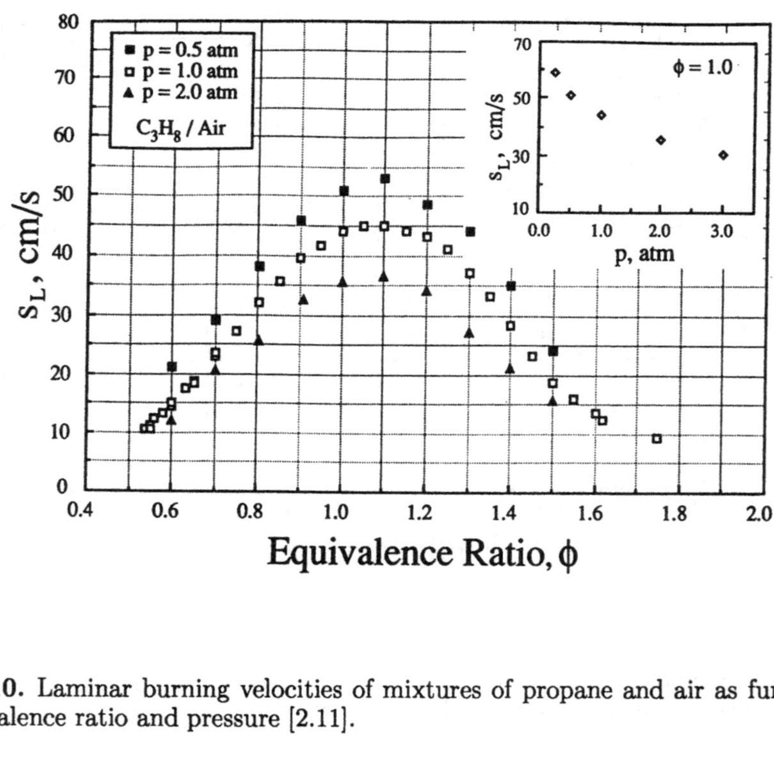

Fig. 2.10. Laminar burning velocities of mixtures of propane and air as functions of equivalence ratio and pressure [2.11].

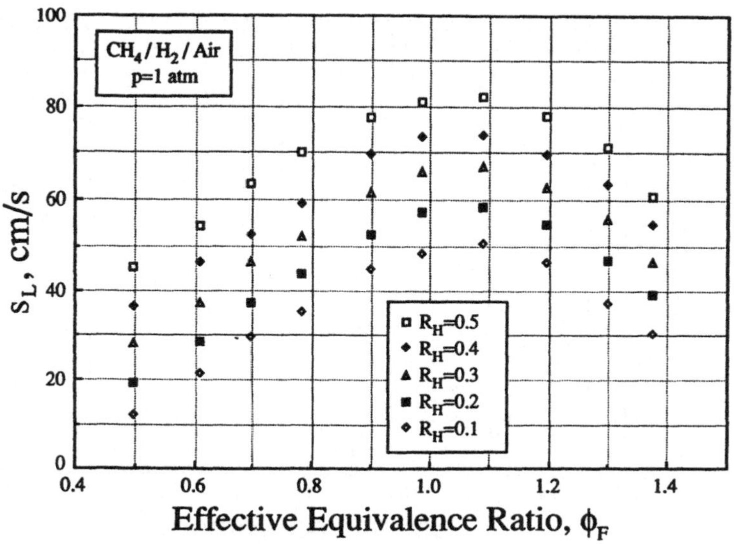

Fig. 2.11. Laminar burning velocities of mixtures of methane, hydrogen, and air, at one atmosphere pressure, as functions of R_H and an effective equivalence ratio ϕ_F. Here $R_H = \{X_H + X_H/(X_H/X_A)_{st}\}/\{X_F + [X_A - X_H/(X_H/X_A)_{st}]\}$, $\phi_F = \{X_F/[X_A - X_H/(X_H/X_A)_{st}]\}/\{(X_F/X_A)_{st}\}$, where X_i is the mole fraction and the subscripts F, H, A, st respectively designate the hydrocarbon fuel, hydrogen, air and the stoichiometric state [2.8].

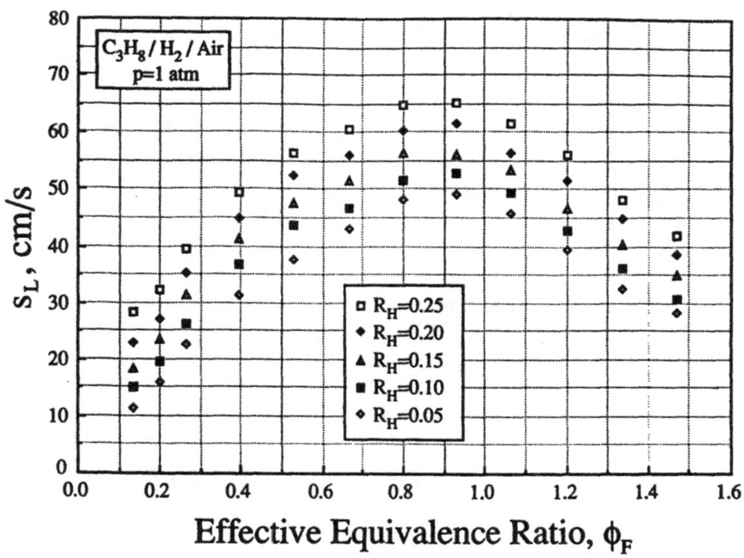

Fig. 2.12. Laminar burning velocities of propane, hydrogen, and air, at one atmospheric pressure, as functions of R_H and an effective equivalence ratio ϕ_F. See Fig. 2.11. for the definitions of R_H and ϕ_F [2.8].

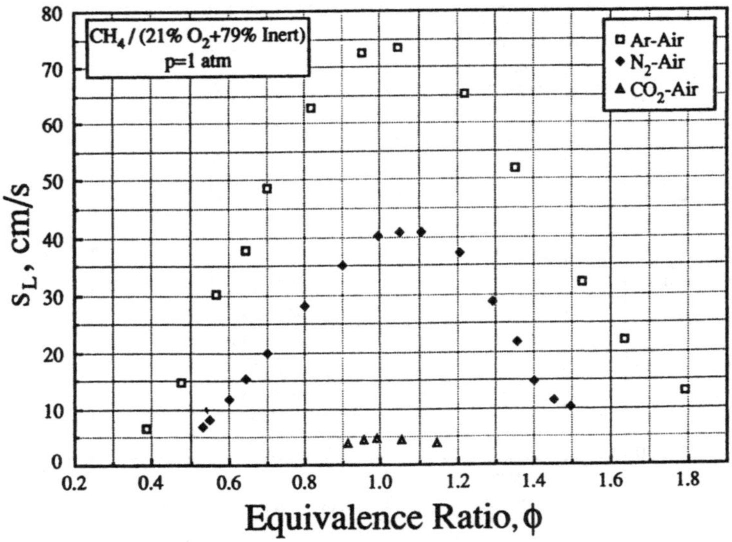

Fig. 2.13. Laminar burning velocities of mixtures of methane and various inert-substituted air, at one atmosphere pressure, as functions of equivalence ratio [2.7].

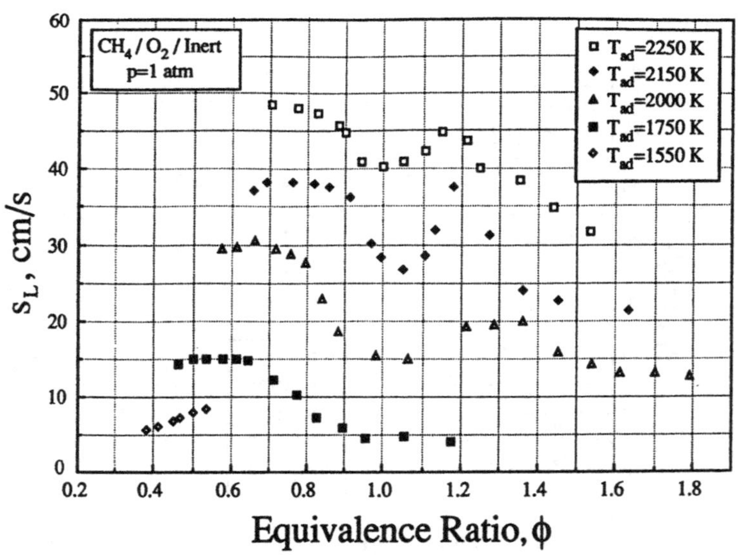

Fig. 2.14. Laminar burning velocities of mixtures of methane, oxygen, and (nitrogen, argon, and carbon dioxide), at one atmosphere pressure, as functions of equivalence ratio and flame temperature. See Fig. 2.17. for inert content [2.7].

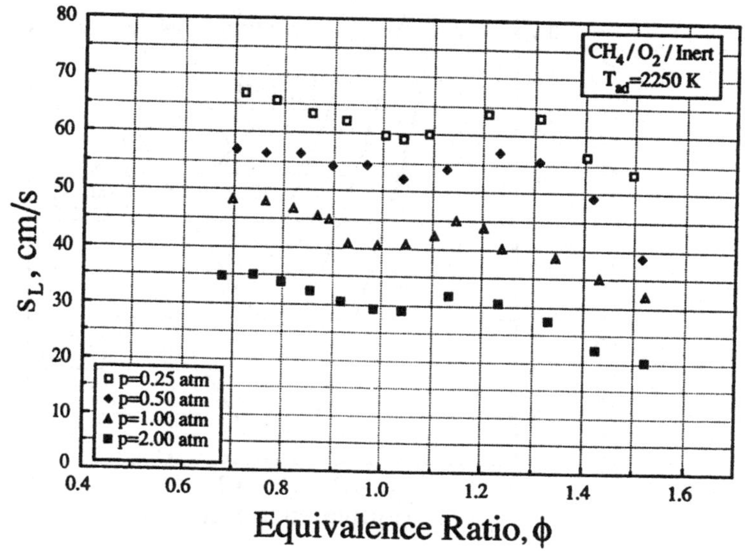

Fig. 2.15. Laminar burning velocities of mixtures of methane, oxygen, and (nitrogen, argon, carbon dioxide), at 2,250 K flame temperature, as functions of equivalence ratio and pressure. See Fig. 2.17. for inert content [2.7].

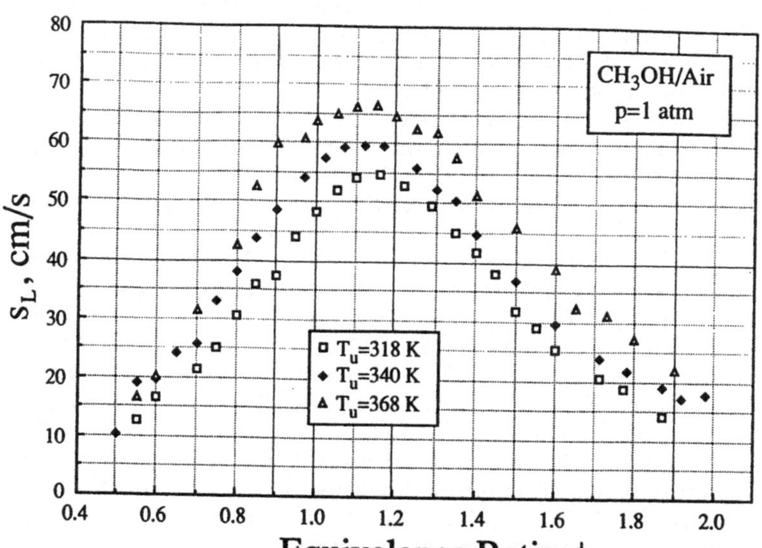

Fig. 2.16. Laminar burning velocities of mixtures of methane and air, at one atmosphere pressure, as functions of equivalence ratio and freestream temperature T_u [2.13].

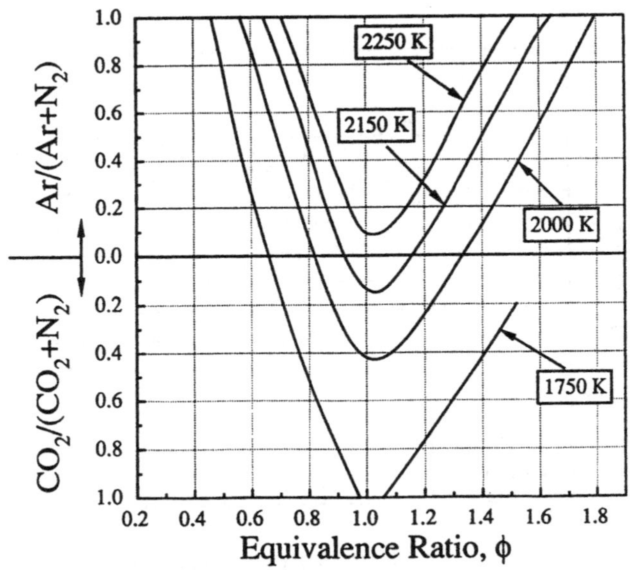

Fig. 2.17. The inert content used in obtaining the flames with constant flame temperatures shown in Figs. 2.14. and 2.15.

References

[2.1] Dixon-Lewis, G., Phil. Trans. Roy. Soc. Lond., **A292**, p. 45, 1979.

[2.2] Westbrook, C. K., Dryer, F. L., Eighteenth Symposium (International) on Combustion, pp. 749–767, The Combustion Institute, 1981.

[2.3] Warnatz, J., Eighteenth Symposium (International) on Combustion, pp. 369–384, The Combustion Institute, 1981.

[2.4] Andrews, G. E., Bradley, D., Combust. Flame, **18**, 133, 1972.

[2.5] Law, C. K., Twenty-Second Symposium (International) on Combustion, pp. 1381–1402, The Combustion Institute, 1988.

[2.6] Wu, C. K., Law, C. K., Twentieth Symposium (International) on Combustion, pp. 1941–1949, The Combustion Institute, 1984.

[2.7] Zhu, D. L., Egolfopoulos, F. N., Law, C. K., Twenty-Second Symposium (International) on Combustion, pp. 1537–1545, The Combustion Institute, 1988.

[2.8] Yu, G., Law, C. K., Wu, C. K., Combust. Flame, **63**, p. 339, 1986.

[2.9] Egolfopoulos, F. N., Cho, P., Law, C. K., Combust. Flame, **76**, p. 375, 1989.

[2.10] Egolfopoulos, F. N., Law, C. K., Combust. Flame, **80**, p. 7, 1990.

[2.11] Egolfopoulos, F. N., Zhu, D. L., Law, C. K., Twenty-Third Symposium (International) on Combustion, The Combustion Institute, pp. 471–478, 1990.

[2.12] Egolfopoulos, F. N., Law, C. K., Twenty-Third Symposium (International) on Combustion, The Combustion Institute, pp. 333–340, 1990.

[2.13] Egolfopoulos, F. N., Du, D. X., Law, C. K., A Comprehensive Study of Methanol Kinetics in Freely-Propagating and Burner-Stabilized Flames, Flow and Statics Reactors and Shock Tubes, Combust. Sci. Technol. 83, p.33, 1992.

[2.14] Sivashinsky, G. I., Law, C. K., Joulin, G., Combust. Sci. Technol., **28**, p. 155, 1982.

[2.15] Dixon-Lewis, G., Twenty-Third Symposium (International) on Combustion, The Combustion Institute, pp. 305–324, 1990.

[2.16] Tien, J. H., Matalon, M., Combust. Flame, **84**, p. 238, 1991.

Part II

Unstretched Premixed Flames

3. Reduced Kinetic Mechanisms for Premixed Hydrogen Flames

F. Mauss and N. Peters
Institut für Technische Mechanik, RWTH Aachen, W-5100 Aachen, West Germany

B. Rogg
Department of Engineering, University of Cambridge, Trumpington Street, Cambridge CB2 1PZ, England

F. A. Williams
Department of Applied Mechanics and Engineering Sciences, University of California, San Diego, La Jolla, CA 92093, U.S.A.

3.1 Introduction

There have been a number of studies of structures and propagation velocities of premixed hydrogen-oxygen flames through numerical integrations of one-dimensional, adiabatic conservation equations. Especially notable among these are the very extensive and pioneering works of Dixon-Lewis [3.1], [3.2] and the more recent thorough studies of Warnatz [3.3]–[3.5]. These references should be consulted for additional literature on the subject. These cited works accomplished more than merely the computation of flame structures. They helped to improve knowledge of rates of important elementary steps and also identified accuracies of various steady-state and partial-equilibrium approximations throughout the flames. Moreover, they reviewed and evaluated available experimental results sufficiently thoroughly, so that in the present study it will be adequate to focus attention on theoretical and numerical comparisons with the more recent experimental data presented in Chap. 2 [3.6].

Although the earlier work produced a fairly good understanding of the chemistry in H_2-O_2-N_2 mixtures, it is the development of simplified descriptions and reduction of the chemistry to the smallest number of global steps from which burning velocities can reasonably be calculated that can further improve understanding of these flames. Derivation of such reduced mechanisms has been achieved recently for methane [3.7] and methane, methanol and propane flames [3.8] for example, through systematic development of reduced kinetic mechanisms, using steady state assumptions. Since the full kinetics of methane flames are considerably more complicated than those of hydrogen flames, the hydrogen-flame problem might be thought to be the simpler and therefore more readily solvable. However, the larger numbers of steps in hydrocarbon flames can contribute to accuracy of simplified approximations for radical pools by keeping radical concentrations sufficiently low. This is not the case for the hydrogen flame and, therefore, in many respects the hydrogen flame has been found to be the more difficult one for which to obtain asymptotic descriptions [3.9]. The full chemical-kinetic mechanism itself for this flame is not excessively complicated, involving only 8 reacting species (H_2, O_2, H, O, OH, HO_2, H_2O_2, and H_2O). Attempts to further simplify this mechanism have shown that each step in reduction of the mechanism entails a clearly identificable degradation in accuracy in at least some part of the flame. In view of the fact that the 8-species detailed mechanism, involving the first 17 elementary reactions in Table 1 of Chap. 1, is still too complicated for analytical descriptions, there is motivation for deriving reduced mechanisms and for applying asymptotics to obtain simplified descriptions of the flame structures. Some progress in this direction has been made for hydrogen-air diffusion flames [3.10] but no results have yet been reported for the premixed flame. The intent of the present investigation is to present a sufficiently accurate 2-step reduced mechanism suitable for numerical calculations of premixed hydrogen flames. This mechanism will form the basis of an asymptotic analysis of the flame structure to be developed in the future.

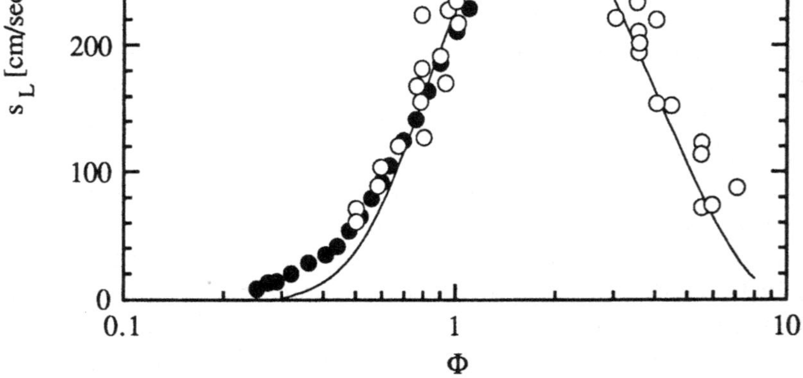

Fig. 3.1. The numerical solution with the starting mechanism, several data compiled by Warnatz [3.3] and recent data by Law [3.6] for atmospheric H_2-air flames.

3.2 Reduced kinetic mechanisms

Shown in Table 1 of Chap. 1 is the full mechanism selected for hydrogen, hydrocarbon and methanol flames. The subset to be used here are the first 10 reactions and reactions 15–17 which involve only species of the H_2-O_2 system, with H_2O_2 excluded. Only those elementary steps have been included since sensitivity analysis has shown that reactions 11–14 are unimportant for the problem addressed here. This 13-step mechanism is called the starting mechanism. Burning velocities for atmospheric flames calculated with this mechanism are compared in Fig. 3.1 with experimental data compiled by Warnatz [3.3]. Recent data from Chap. 2 of this book are also included in this presentation. Besides the fuel H_2, oxygen O_2, and nitrogen N_2 (assumed inert), this mechanism involves the main product H_2O, the members of the hydrogen-oxygen radical pool H, O, and OH, and the hydroperoxyl radical HO_2. For the seven reacting species, the following system of balance equations is derived

$$\begin{aligned}
L([\mathrm{H}]) &= -w_1 + w_2 + w_3 - w_5 - w_6 - w_7 - w_9 - 2\times w_{15} - w_{16} \ , \\
0 = L([\mathrm{OH}]) &= w_1 + w_2 - w_3 - 2\times w_4 + 2\times w_6 - w_8 + w_{10} - w_{16} \ , \\
0 = L([\mathrm{O}]) &= w_1 - w_2 + w_4 + w_9 - w_{10} - 2\times w_{17} \ , \\
L([\mathrm{H_2}]) &= -w_2 - w_3 + w_7 + w_{15} \ , \\
L([\mathrm{O_2}]) &= -w_1 - w_5 + w_7 + w_8 + w_{10} + w_{17} \ , \\
L([\mathrm{H_2O}]) &= w_3 + w_4 + w_8 + w_9 + w_{16} \ , \\
0 = L([\mathrm{HO_2}]) &= w_5 - w_6 - w_7 - w_8 - w_9 - w_{10} \quad .
\end{aligned} \tag{3.1}$$

Here $L([X_i])$ represents the convective-diffusive operator in the balance equations. For the 4 intermediate species H, O, OH and HO_2 the terms accounting for convection, diffusion, chemical production and consumption are shown in Figs. 3.2–3.5 as a function of the non-dimensional coordinate

$$x^* = (\rho_u \, s_L) \int_0^x \frac{c_p}{\lambda} \, dx \, . \tag{3.2}$$

Except for hydrogen, production and consumption are dominating throughout the flame as compared to convection and diffusion. Therefore, the steady-state approximation is introduced for the last 3 species. With these three steady states and 2 atom-conservation conditions (for H and O) among the 7 reacting species, a total number of two global steps should describe the chemistry. The following system of balance equations is derived by eliminating the reaction rates w_2 for O, w_3 for OH and w_7 for HO_2 as those which are the fastest to consume these species, respectively. By the procedure described by Peters [3.11] one obtains the following system of balance equations for the remaining non-steady-state species H, H_2, O_2 and H_2O:

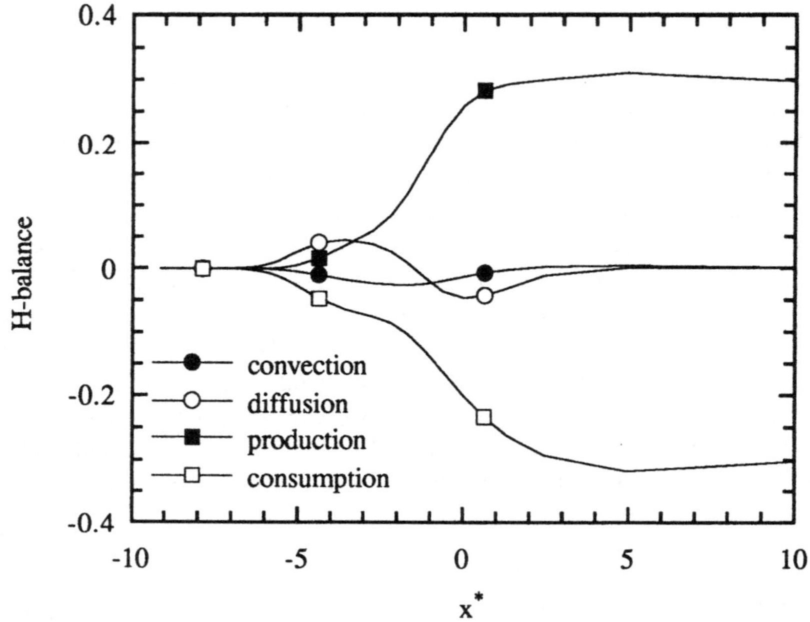

Fig. 3.2. Balance of convection, diffusion, production and consumption of H radical for atmospheric, stoichiometric H_2-air flames.

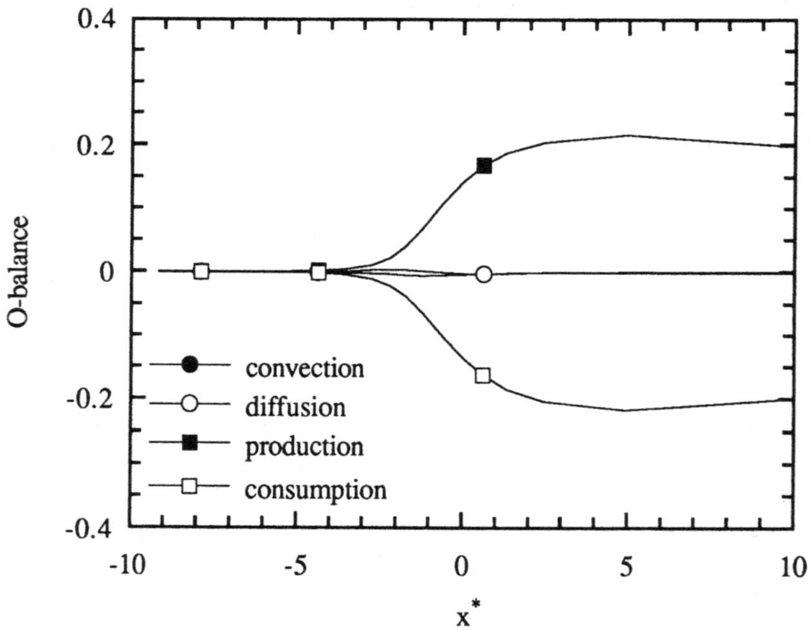

Fig. 3.3. Balance of convection, diffusion, production and consumption of O radical for atmospheric, stoichiometric H_2-air flames.

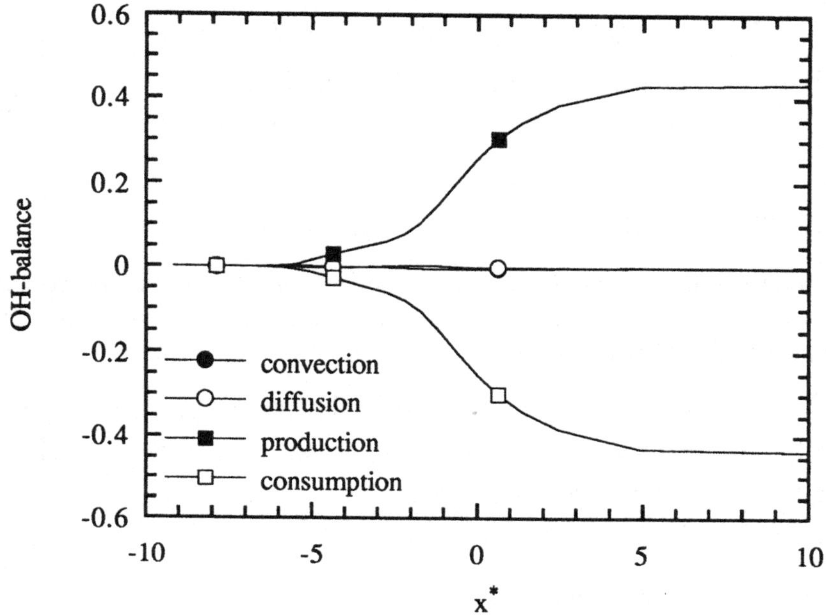

Fig. 3.4. Balance of convection, diffusion, production and consumption of OH radical for atmospheric, stoichiometric H_2-air flames.

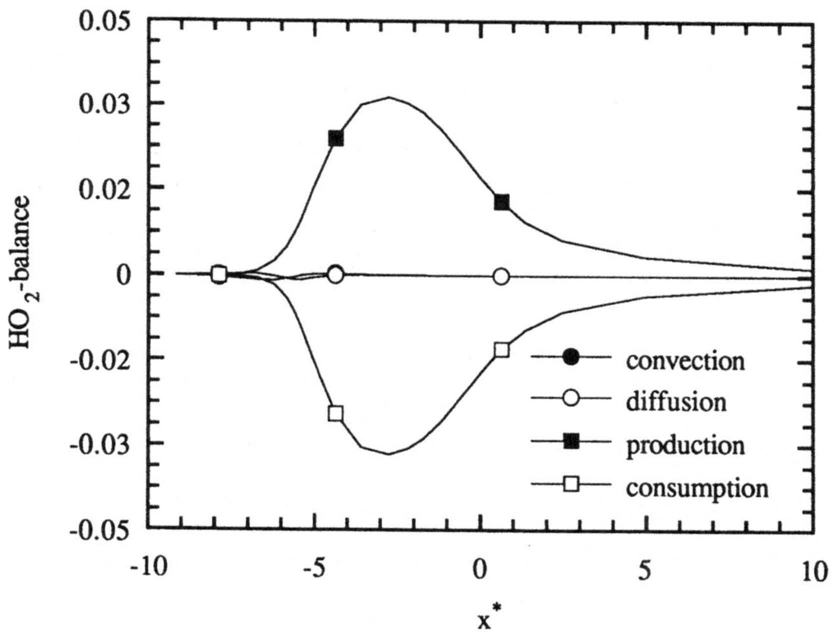

Fig. 3.5. Balance of convection, diffusion, production and consumption of HO_2 radical for atmospheric, stoichiometric H_2-air flames.

$$
\begin{aligned}
&L([H]) + \{L([OH]) + 2 \times L([O]) - L([HO_2])\} \\
&\qquad = 2 \times (w_1 - w_5 + w_6 + w_9 - w_{15} - w_{16} - 2 \times w_{17}) \ , \\
&L([H_2]) + \{-L([OH]) - 2 \times L([O]) + L([HO_2])\} \\
&\qquad = -3 \times w_1 + w_5 - 3 \times w_6 - 3 \times w_9 + w_{15} + w_{16} + 4 \times w_{17} \ , \\
&L([O_2]) + \{L([HO_2])\} = -w_1 - w_6 - w_9 + w_{17} \ , \\
&L([H_2O]) + \{L([OH]) + L([O])\} = 2 \times (w_1 + w_6 + w_9 - w_{17}) \ .
\end{aligned}
\tag{3.3}
$$

The convective-diffusive operators in the curly brackets correspond to those of steady state species and can be neglected. Grouping rates with equal stoichiometric coefficients one obtains

$$
\begin{aligned}
L([H]) \quad &= 2 \times (w_1 + w_6 + w_9 - w_{17}) - 2 \times (w_5 + w_{15} + w_{16} + w_{17}) \ , \\
L([H_2]) \quad &= -3 \times (w_1 + w_6 + w_9 - w_{17}) + (w_5 + w_{15} + w_{16} + w_{17}) \ , \\
L([O_2]) \quad &= -(w_1 + w_6 + w_9 - w_{17}) \ , \\
L([H_2O]) \quad &= 2 \times (w_1 + w_6 + w_9 - w_{17}) \quad .
\end{aligned}
\tag{3.4}
$$

This leads to the 2-step global mechanism

(I) $3H_2 + O_2 = 2H + 2H_2O$,

(II) $2H + M = H_2 + M$,

with the global rates

$$
\begin{aligned}
w_{I} &= w_1 + w_6 + w_9 - w_{17} \ , \\
w_{II} &= w_5 + w_{15} + w_{16} + w_{17} \ .
\end{aligned}
\tag{3.5}
$$

The steady state assumptions for O, OH and HO_2, i.e. $(3.1)_2$, $(3.1)_3$ and $(3.1)_7$ can be used to derive algebraic relations for these species, namely

$$
[O] = \frac{k_{1f}\,[H][O_2] + k_{2b}\,[OH][H] + k_{4f}\,[OH^2] + k_9\,[HO_2][H]}{k_{1b}\,[OH] + k_{2f}\,[H_2] + k_{4b}\,[H_2O] + k_{10}\,[HO_2]}
\tag{3.6}
$$

$$
[OH] = \frac{\sqrt{B^2 + 8 \times k_{4f} \times C} - B}{4 \times k_{4f}}
\tag{3.7}
$$

where here and below reaction 17 is neglected, and where

$$
B = k_{1b}\,[O] + k_{2b}\,[H] + k_{3f}\,[H_2] + k_8\,[HO_2] + k_{16}\,[H][M]
\tag{3.8}
$$

$$
\begin{aligned}
C = {} & k_{1f}\,[H][O_2] + k_{2f}\,[O][H_2] + k_{3b}\,[H][H_2O] + 2k_{4b}\,[O][H_2O] \\
& + 2 \times k_6\,[H][HO_2] + k_{10}\,[O][HO_2]
\end{aligned}
\tag{3.9}
$$

and

$$
[HO_2] = \frac{k_{5f}\,[H][O_2]\,[M]}{(k_6 + k_7 + k_9) \times [H] + k_8\,[OH] + k_{10}\,[O] + k_{5b}[M]} \ .
\tag{3.10}
$$

In Eqn. (3.7) we neglected reaction 17, to avoid a quadratic formulation for [O]. Although these equations are explicit, they are coupled and must be solved iteratively using the principle of inner iteration. This principle is discussed in detail in Chapter 6 of the present book. Numerical iterations lead to convergence in a few steps. It is the aim of the present paper, however, to present an explicit system of algebraic equations that can be used in subsequent asymptotic analysis. This will be done by truncating the steady state equations. Before entering into the truncation procedure the consequences of the steady state assumption shall be analysed. Note that in all calculations thermal diffusion was not taken into account.

Figs. 3.6–3.8 show the burning velocity s_L as a function of the equivalence ratio for atmospheric H_2-air flames at a preheat temperature of 298 K, as a function of pressure for stoichiometric H_2-air flames at the same preheat temperature and for atmospheric, stoichiometric flames at preheat temperatures up to 500 K, respectively. Two versions of the reduced 2-step mechanism are shown: the first is the standard one where the steady state species are neglected in the elemental balance such that the mass fractions of the non-steady state species add up to unity. This one shows a much larger burning velocity for stoichiometric flames over the entire pressure and preheat temperature range. The second one accounts for the fact that a significant part of the elemental mass is stored in the steady state species. The correct element mass fractions should be the weighted sum of the mass fractions of all species according to

$$Z_{\mathrm{H}} = \mathrm{W_H}\left(\frac{Y_{\mathrm{H}}}{\mathrm{W_H}} + \frac{2Y_{\mathrm{H_2}}}{\mathrm{W_{H_2}}} + \frac{Y_{\mathrm{OH}}}{\mathrm{W_{OH}}} + \frac{Y_{\mathrm{HO_2}}}{\mathrm{W_{HO_2}}} + \frac{2Y_{\mathrm{H_2O}}}{\mathrm{W_{H_2O}}}\right) , \tag{3.11}$$

$$Z_{\mathrm{O}} = \mathrm{W_O}\left(\frac{Y_{\mathrm{O}}}{\mathrm{W_O}} + \frac{2Y_{\mathrm{O_2}}}{\mathrm{W_{O_2}}} + \frac{Y_{\mathrm{OH}}}{\mathrm{W_{OH}}} + \frac{2Y_{\mathrm{HO_2}}}{\mathrm{W_{HO_2}}} + \frac{Y_{\mathrm{H_2O}}}{\mathrm{W_{H_2O}}}\right) . \tag{3.12}$$

Comparing these with the element mass fraction obtained when terms for O, OH and HO_2 are neglected in (3.11) and (3.12) an element mass fraction deficit equal to

$$\begin{aligned} \Delta_{\mathrm{H}} &= \mathrm{M_H}\left(\frac{Y_{\mathrm{OH}}}{\mathrm{W_{OH}}} + \frac{Y_{\mathrm{HO_2}}}{\mathrm{W_{HO_2}}}\right) , \\ \Delta_{\mathrm{O}} &= \mathrm{M_O}\left(\frac{Y_{\mathrm{O}}}{\mathrm{W_O}} + \frac{Y_{\mathrm{OH}}}{\mathrm{W_{OH}}} + 2\frac{Y_{\mathrm{HO_2}}}{\mathrm{W_{HO_2}}}\right) \end{aligned} \tag{3.13}$$

is calculated. The large burning velocities for the standard case are attributed to the fact that the elemental deficit is not taken into account. This is seen in Fig. 3.9 showing much larger concentrations of H and O_2 as compared to the starting mechanism for this case, while the concentration of the other non-steady state species are nearly the same. Therefore, the ad-hoc corrections

$$\begin{aligned} Y_{\mathrm{H,new}} &= Y_{\mathrm{H,old}} - a \times \Delta_{\mathrm{H}} , \\ Y_{\mathrm{O_2,new}} &= Y_{\mathrm{O_2,old}} - a \times \frac{1}{2}\Delta_{\mathrm{O}} \end{aligned} \tag{3.14}$$

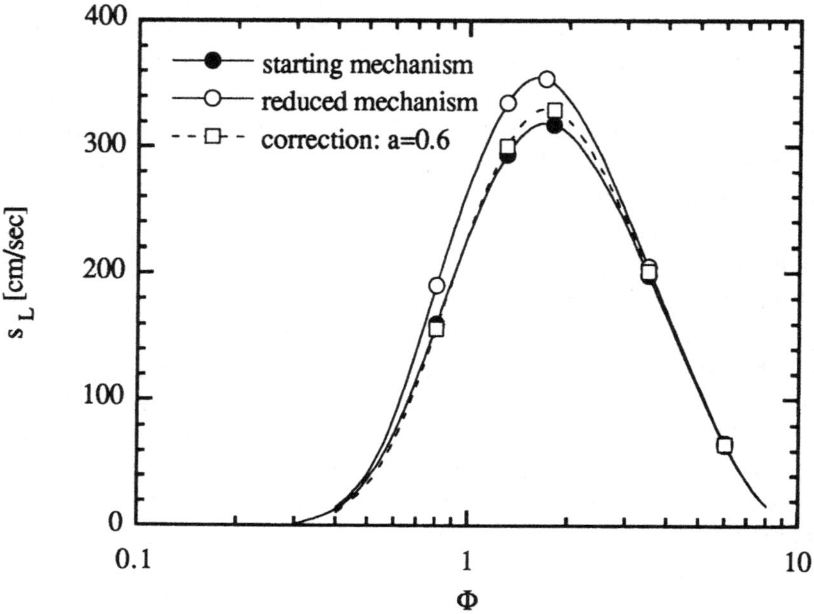

Fig. 3.6. Numerical solution with the starting mechanism, the reduced mechanism and the corrected, reduced mechanism for atmospheric H_2-air flames.

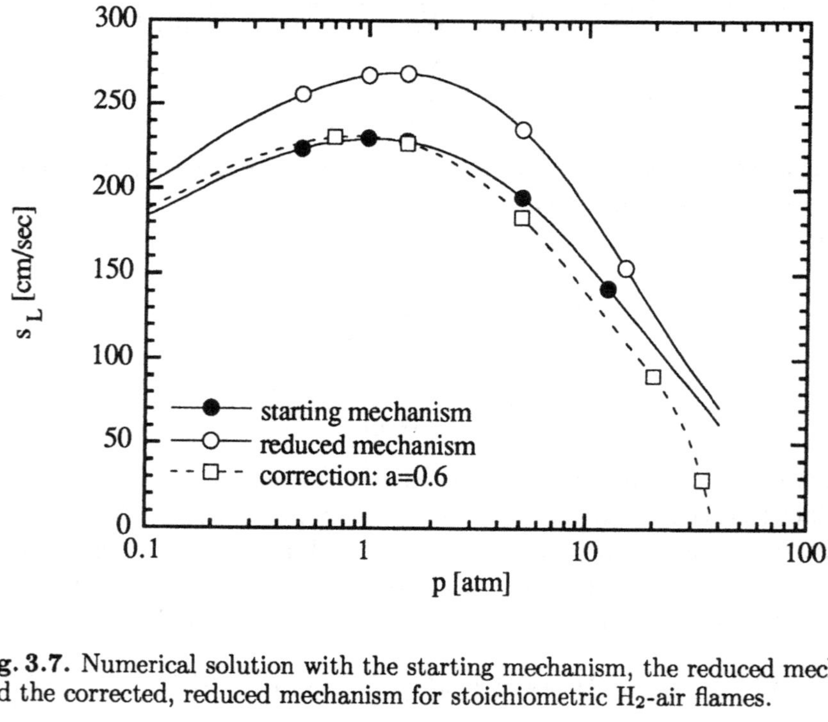

Fig. 3.7. Numerical solution with the starting mechanism, the reduced mechanism and the corrected, reduced mechanism for stoichiometric H_2-air flames.

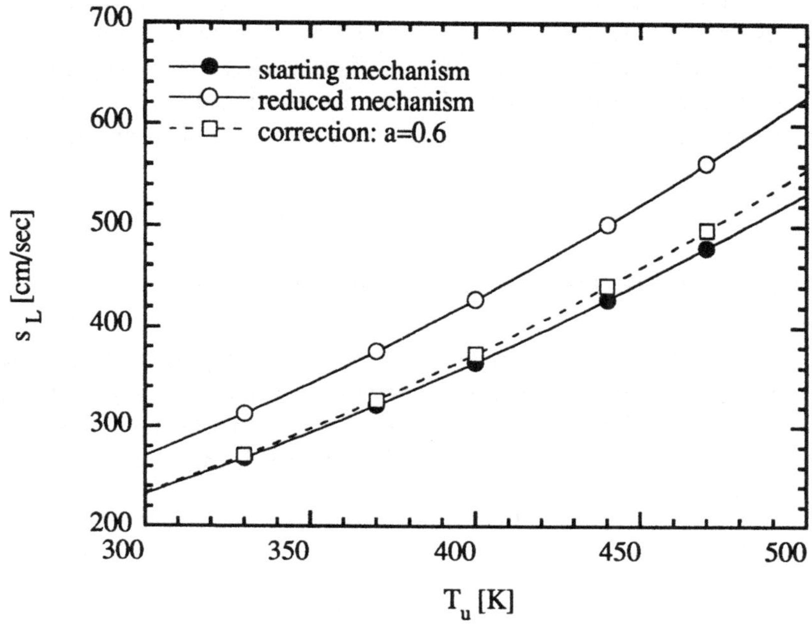

Fig. 3.8. Numerical solution with the starting mechanism, the reduced mechanism and the corrected, reduced mechanism for atmospheric, stoichiometric H_2-air flames.

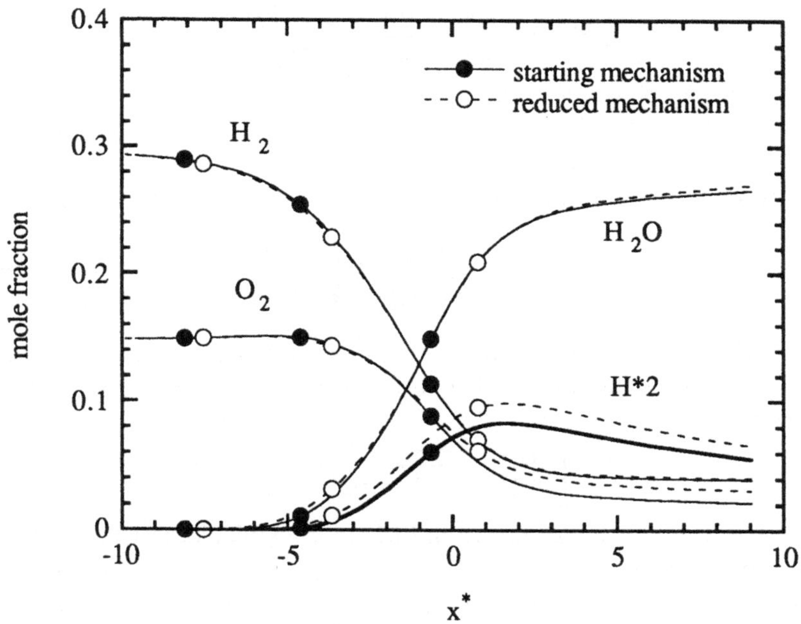

Fig. 3.9. Numerically calculated mole fractions of H_2, O_2, H and H_2O with the starting mechanism and the reduced mechanism for atmospheric, stoichiometric H_2-air flames.

are used in calculating the global reaction rates in (3.6). An optimum value of the fitting factor a was found to be $a = 0.6$ for a wide range of ϕ, p and T_u. This version of the 2-step mechanism shows a more satisfactory agreement with the results using the starting mechanism, as shown in Figs. 3.6–3.8. In the following, however, we will use the reduced 2-step mechanism without these ad-hoc corrections.

3.3 Truncation of steady state assumptions

In order to obtain explicit expressions for steady-state concentrations, it is necessary to analyse the relative importance of all terms in $(3.1)_2$, $(3.1)_3$ and $(3.1)_7$. It is found that the rates w_{4f}, w_{10} and w_{17} can be neglected in all steady state relations and w_9 in those for O and OH, leading to

$$L([\mathrm{O}]) = w_1 - w_2 = 0 \tag{3.15}$$

and

$$L([\mathrm{OH}]) = 2 \times w_1 - w_3 + 2 \times w_6 = 0 \tag{3.16}$$

where again (3.15) has been used and, furthermore, w_8 and w_{16} have been neglected. To replace w_6 in the latter equation, it is sufficient to use the steady state relation for HO_2 with w_8 neglected leading to

$$[\mathrm{OH}] = \frac{2\,w_{1f} + w_{3b} + 2\,w_5\,Z_6}{2\,k_{1b}\,[\mathrm{O}] + k_{3f}\,[\mathrm{H_2}]}\,, \tag{3.17}$$

where

$$Z_6 = k_6/(k_6 + k_7 + k_9)\,. \tag{3.18}$$

The steady state relation for O resulting from (3.15) may be written as

$$[\mathrm{O}] = [\mathrm{H}]\,\frac{k_{1f}\,[\mathrm{O_2}] + k_{2b}\,[\mathrm{OH}]^*}{k_{1b}\,[\mathrm{OH}]^* + k_{2f}\,[\mathrm{H_2}]}\,. \tag{3.19}$$

It was found to be sufficiently accurate to express the concentration of OH here by a balance of the dominating terms in (3.16), which are w_3 and $2\,w_6$. This leads to

$$[\mathrm{OH}]^* = \frac{w_{3b} + 2\,w_5\,Z_6}{k_{3f}\,[\mathrm{H_2}]}\,. \tag{3.20}$$

This form is compared to that from (3.17) for a stoichiometric atmospheric H_2-air flame in Fig. 3.10. It is seen that the agreement is satisfactory in the entire flame structure. When compared with the OH concentration calculated with the starting mechanism, it is seen that the reduced mechanism overpredicts the OH concentration in the downstream part of the flame.
Finally, in the steady state relation for HO_2,

$$L([\mathrm{HO_2}]) = w_5 - w_6 - w_7 - w_8 - w_9 = 0\,, \tag{3.21}$$

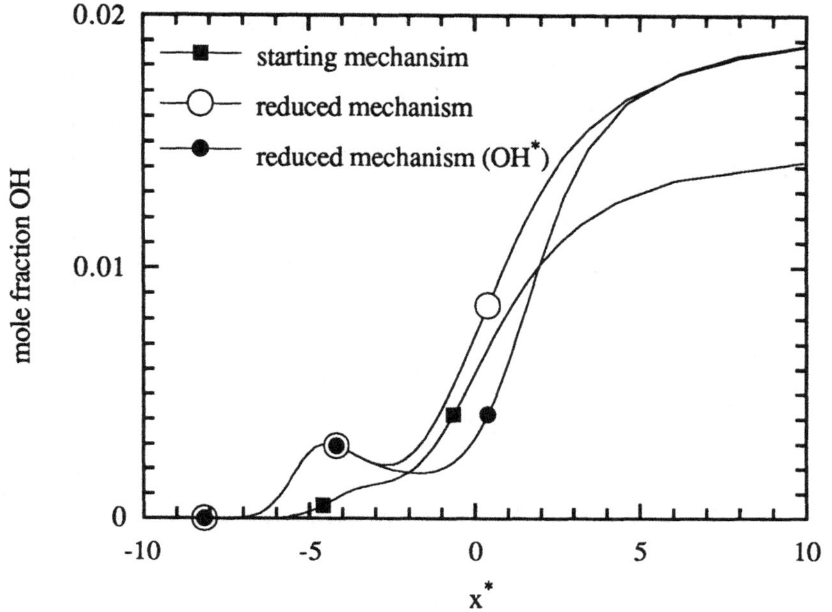

Fig. 3.10. Numerically calculated mole fraction of OH and OH* with the starting mechanism and the reduced mechanism for atmospheric, stoichiometric H_2-air flames.

only the second term in the numerator of (3.20) needs to be used to express the OH concentration needed in w_8. This is due to the fact that w_8 is important only in the upstream region of the flame where $2\,w_5\,Z_6$ dominates over w_{3b}. This leads to

$$[HO_2] = \frac{k_5[H][O_2]\,[M]}{(k_6 + k_7 + k_9)\,[H] + 2\,k_8\,w_5\,Z_6\,/\,(k_{3f}\,[H_2])}\,. \qquad (3.22)$$

3.4 Comparison with experimental data for hydrogen flames in diluted air

Egolfopoulos and Law [3.12] have recently measured the burning velocity of ultra-lean to moderately rich hydrogen flames in diluted air. These data are reproduced in Chap. 2 of this book. Figures 3.11–3.14 show some of these data which were used for comparison with the starting mechanism and the reduced mechanism. Figure 3.11 shows this comparison for a fixed equivalence ratio of $\phi = 1.058$ for different dilutions expressed in percent mole fraction of the oxidizer (the rest being nitrogen). The agreement is satisfactory, both for the starting mechanism and the reduced mechanism. In Figures 3.12–3.14 burning velocity data for fixed dilutions are plotted over the equivalence ratio. In all

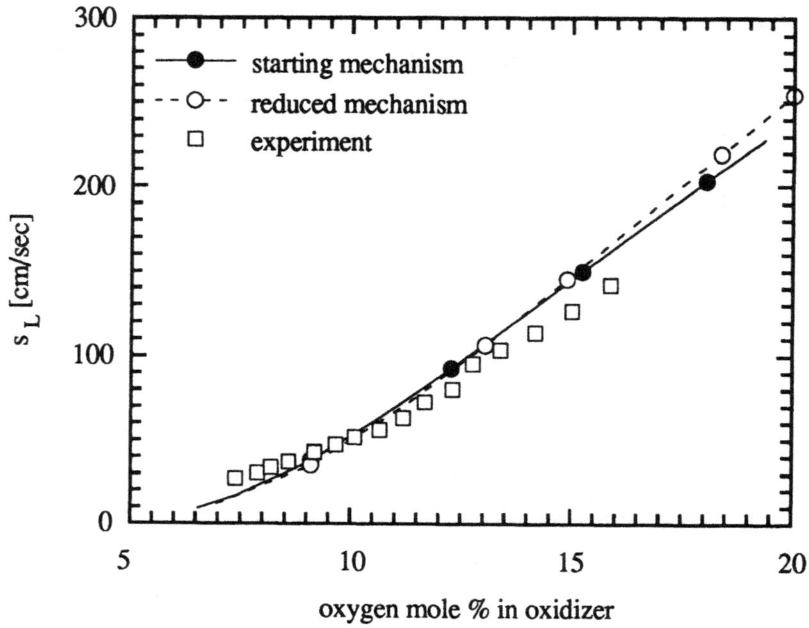

Fig. 3.11. Numerical solution with the starting mechanism and the reduced mechanism and recent data by Law [3.6] for atmospheric, near stoichiometric ($\phi = 1.058$) H_2-air flames.

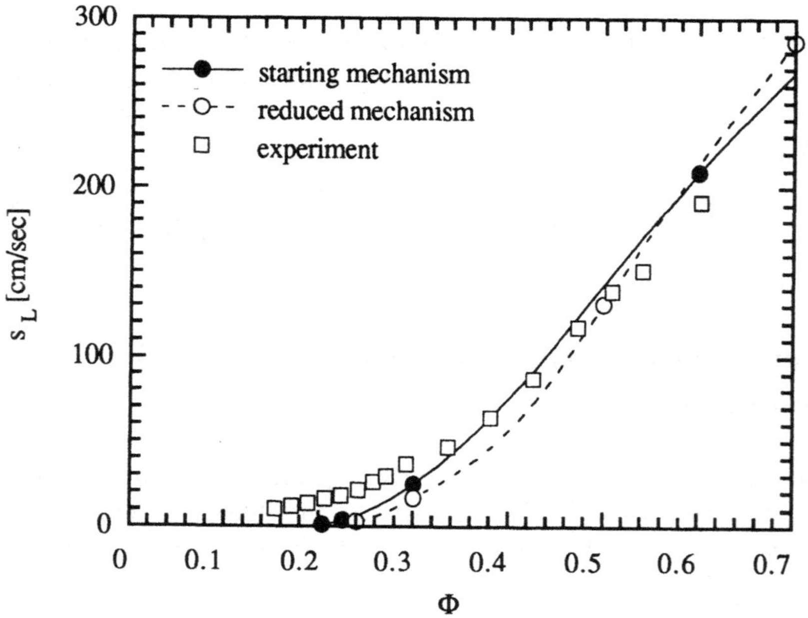

Fig. 3.12. Numerical solution with the starting mechanism and the reduced mechanism and recent data by Law [3.6] for diluted ($O_2/(O_2+N_2) = 30\%$) atmospheric and stoichiometric H_2-air flames.

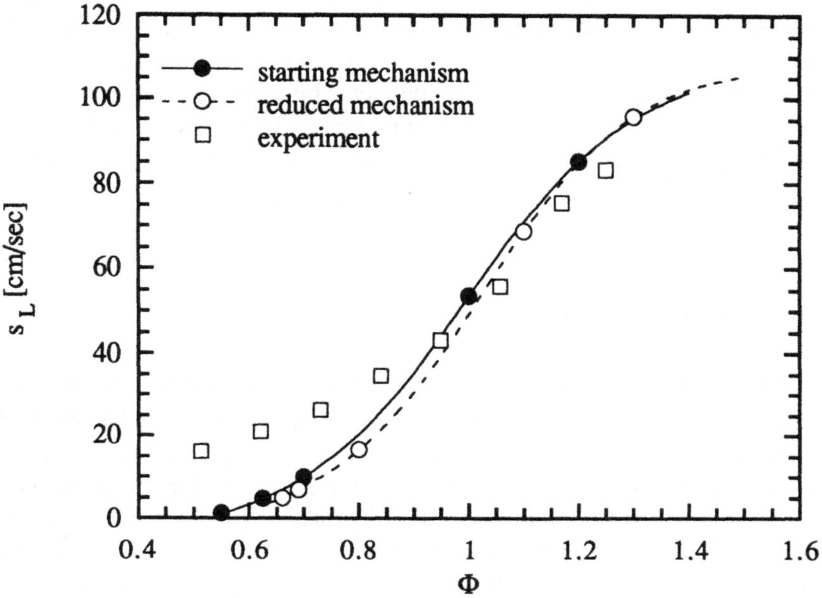

Fig. 3.13. Numerical solution with the starting mechanism and the reduced mechanism and recent data by Law [3.6] for diluted ($O_2/(O_2+N_2) = 10.7\%$), atmospheric and stoichiometric H_2-air flames.

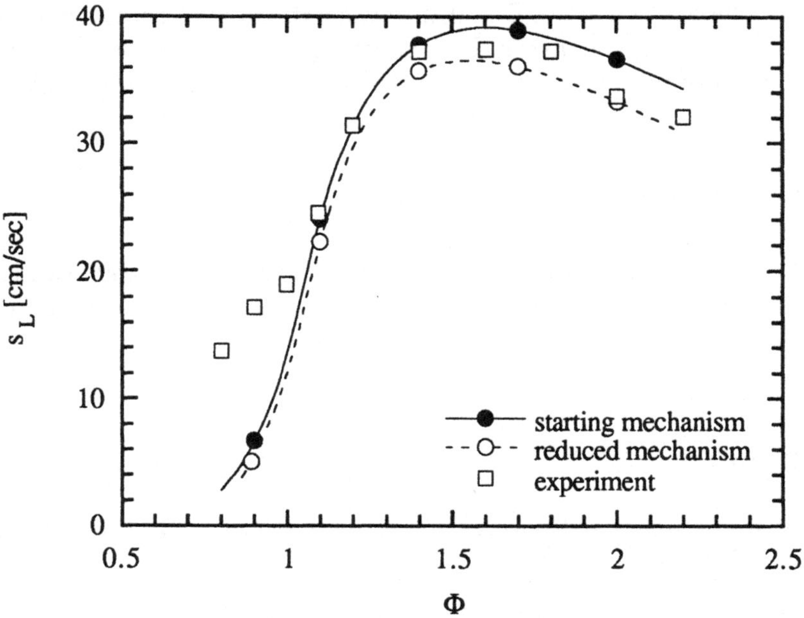

Fig. 3.14. Numerical solution with the starting mechanism and the reduced mechanism and recent data by Law [3.6] for diluted ($O_2/O_2+N_2 = 7.7\%$), atmospheric and stoichiometric H_2-air flames.

these plots there is the phenomenon that the burning velocity for very lean flames is significantly larger than the prediction, both by the starting mechanism and by the reduced mechanism. This does not necessarily invalidate the kinetic mechanisms, since it is known that lean hydrogen flames tend to show a cellular structure which is not taken into account in these one-dimensional calculations. Again, the overall comparison appears to be satisfactory for these cases.

3.5 Conclusions

A 2-step mechanism has been derived for premixed hydrogen flames, for use at pressures between 1 and 40 atm, preheat temperatures up to 500 K and various degrees of dilutions. This mechanism involves as reacting species only the reactants H_2 and O_2, the product H_2O and the radical H as intermediate species. The steady state assumption for the other species considered here, namely O, OH and HO_2, was well justified. Numerical calculations of flame structures, employing this mechanism as well as a 13-step elementary mechanism, were performed to investigate the accuracies of the reduced mechanisms. The additional effect of an element mass fraction defect, which is implicitly associated with the steady state assumptions, causes the concentration of some of the non-steady state species to be too large. This leads to a non-negligible increase of the burning velocity calculated from the 2-step mechanism for close-to-stochiometric flames. Detailed considerations were necessary in order to derive an explicit formulation by truncation of the steady state relations. The agreement of the 2-step mechanism with experimental data is satisfactory.

Acknowledgements

This work was funded by a grant from the Deutsche Forschungsgemeinschaft. The contribution to this chapter of B. Rogg and F.A. Williams was made possible through NATO Grant No 0101/89.

References

[3.1] Dixon-Lewis, G., Goldsworthy, F. A. and Greenberg, J. B., Proc. Roy. Soc. Lond. **A346**, p. 261, 1975.
[3.2] Dixon-Lewis, G., Phil. Trans. Roy. Soc. Lond. **A292**, p. 45, 1979.
[3.3] Warnatz, J., Ber. Bunsenges. Phys. Chem. **83**, p. 950, 1979.
[3.4] Warnatz, J., Eighteenth Symposium (International) on Combustion, p. 369–384, The Combustion Institute, 1981.
[3.5] Warnatz, J., Combust. Sci. Tech. **26**, p. 203, 1981.
[3.6] Law, C. K., A Compilation of Recent Experimental Data of Premixed Laminar Flames, Chap. 2 of this book.

[3.7] Peters, N. in: Numerical Simulation of Combustion Phenomena. Lecture Notes in Physics, **241**, pp. 90–109, 1985.

[3.8] Paczko, G., Lefdal, P. M., Peters, N., Twentyfirst Symposium (International) on Combustion, The Combustion Institute, pp. 739–748, 1988.

[3.9] Williams, F. A., Combustion Theory, 2nd Ed., Addison-Wesley, Menlo Park, CA, p. 178, 1985.

[3.10] Gutheil, E., Williams, F. A., Twenty-Third Symposium (International) on Combustion, The Combustion Institute, Pittsburgh, to appear 1991.

[3.11] Peters, N. in: Reduced Kinetic Mechanisms and Asymptotic Approximations for Methane-Air Flames, (M.D. Smooke, Ed.), Springer 1991.

[3.12] Egolfopoulos, F. N., Law, C. K., Twenty-Third Symposium (International) on Combustion, to appear 1991.

4. Reduced Kinetic Mechanisms for Wet CO Flames

W. Wang and B. Rogg
Department of Engineering, University of Cambridge, Trumpington Street, Cambridge CB2 1PZ, England

F.A. Williams
Department of Applied Mechanics and Engineering Sciences, University of California at San Diego, La Jolla, CA 92093, U.S.A.

4.1 Introduction

Although deflagrations of dry mixtures of carbon monoxide and oxygen can occur at sufficiently high pressures, under practical conditions there are large enough amounts of hydrogen-containing species present for the step OH + CO $\rightarrow CO_2$ + H to dominate hydrogen-free steps. Thus, there is practical interest in investigating deflagrations in wet CO mixtures. Moreover, at low hydrogen content the hydrogen-containing species may be expected to achieve steady states and, thereby, produce a one-step reduced mechanism that describes the flame propagation with high accuracy. This great degree of simplification is not anticipated for the other reactant mixtures considered in the present volume. Therefore, the wet CO flames addressed here offer a unique testing ground for reduced mechanisms.

In a previous paper [4.1] we addressed wet CO flames and reduced the chemistry to a three-step mechanism that gave reasonable agreement with predictions of a full mechanism comprising 67 elementary steps among 12 reacting species; also references to earlier works are given in [4.1]. The conditions selected for the numerical simulations were those corresponding to experiments reported by Lewis and von Elbe [4.2]. The same conditions are selected for the present study, but the starting reaction mechanism and rate parameters are taken from Chap. 1 of the present volume. We shall not go all the way to a one-step mechanism; instead we shall stop at a two-step description. Our objective here is only to bring the problem to the attention of the reader by moving somewhat beyond our earlier [4.1] work.

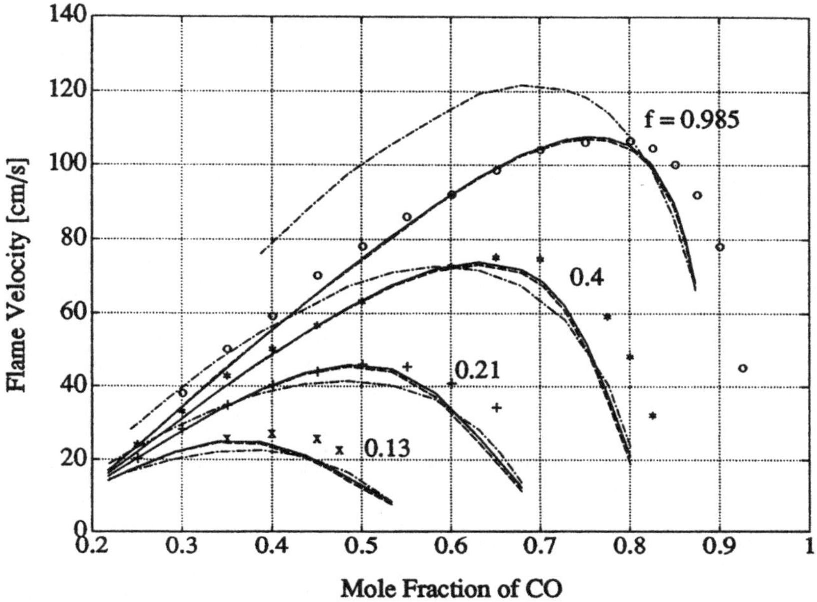

Fig. 4.1. Burning velocities of atmospheric-pressure wet CO flames as a function of initial mole-fraction of fuel (CO: 97.15%, H_2: 1.5% and H_2O: 1.35%). Symbols: experimental data from ref. [2]; solid lines: detailed mechanism; dashed lines: short mechanism; dashed-dotted lines: two-step mechanism.

4.2 Detailed Kinetic Mechanisms

In the present paper two detailed kinetic mechanisms are employed. The first, termed the "full mechanism", comprises reactions 1f to 24b of the detailed mechanism presented in Chap. 1 with the exception of steps 19 and 20. The latter steps are disregarded because they consume the CH radical which in none of steps 1f to 24b is produced. The second detailed kinetic mechanism is termed the "short mechanism"; it is obtained from the full mechanism by disregarding steps 11 to 14b, in which H_2O_2 is either formed or consumed, as well as the recombination step 17, and steps 5b, 22 and 23. The short mechanism comprises ten reacting species and three elements.

In our previous study [4.1] we investigated laminar premixed flames of $CO/O_2/N_2$ fuels with small amounts of hydrogen (1.5% on a molar basis) and water vapour (1.35%) added. Numerical simulations of these flames using the full and, alternatively, the short kinetic mechanism identified above with the rate data specified in Chap. 1 of this book show that with both mechanisms results are obtained for flame structures and burning velocities that are in excellent agreement. Results for the burning velocity are shown in Fig. 4.1. In this figure the solid lines refer to the full mechanism, and the dashed lines to

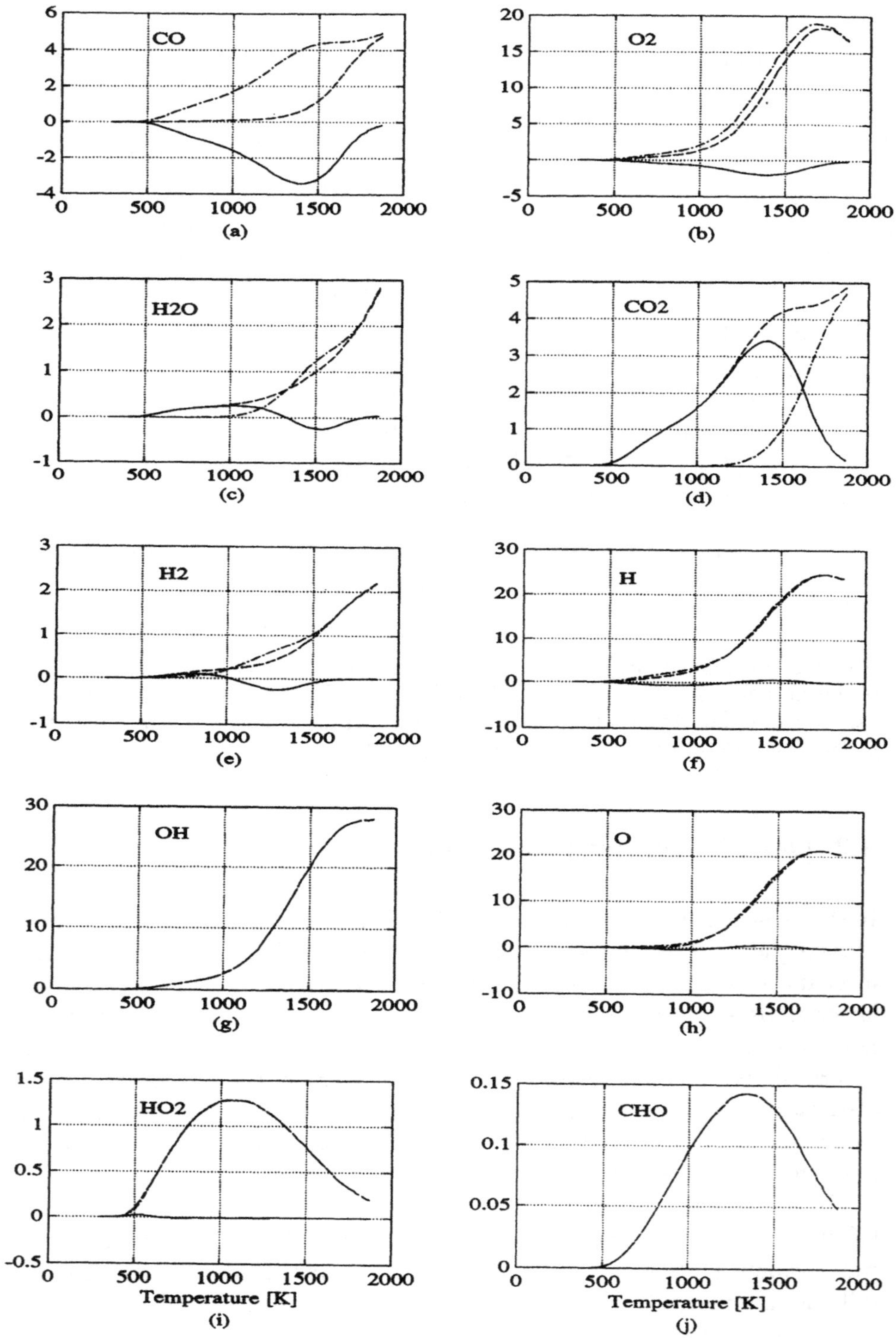

Fig. 4.2. Production, consumption and net rates (in kmol/m^{3}s) of species in an atmospheric-pressure wet CO flame calculated with short mechanism. Initial temperature 300 K; initial mole fraction of fuel 0.3; $f = 0.21$. Solid lines: net rate; dashed lines: production rate; dashed-dotted lines: consumption rate.

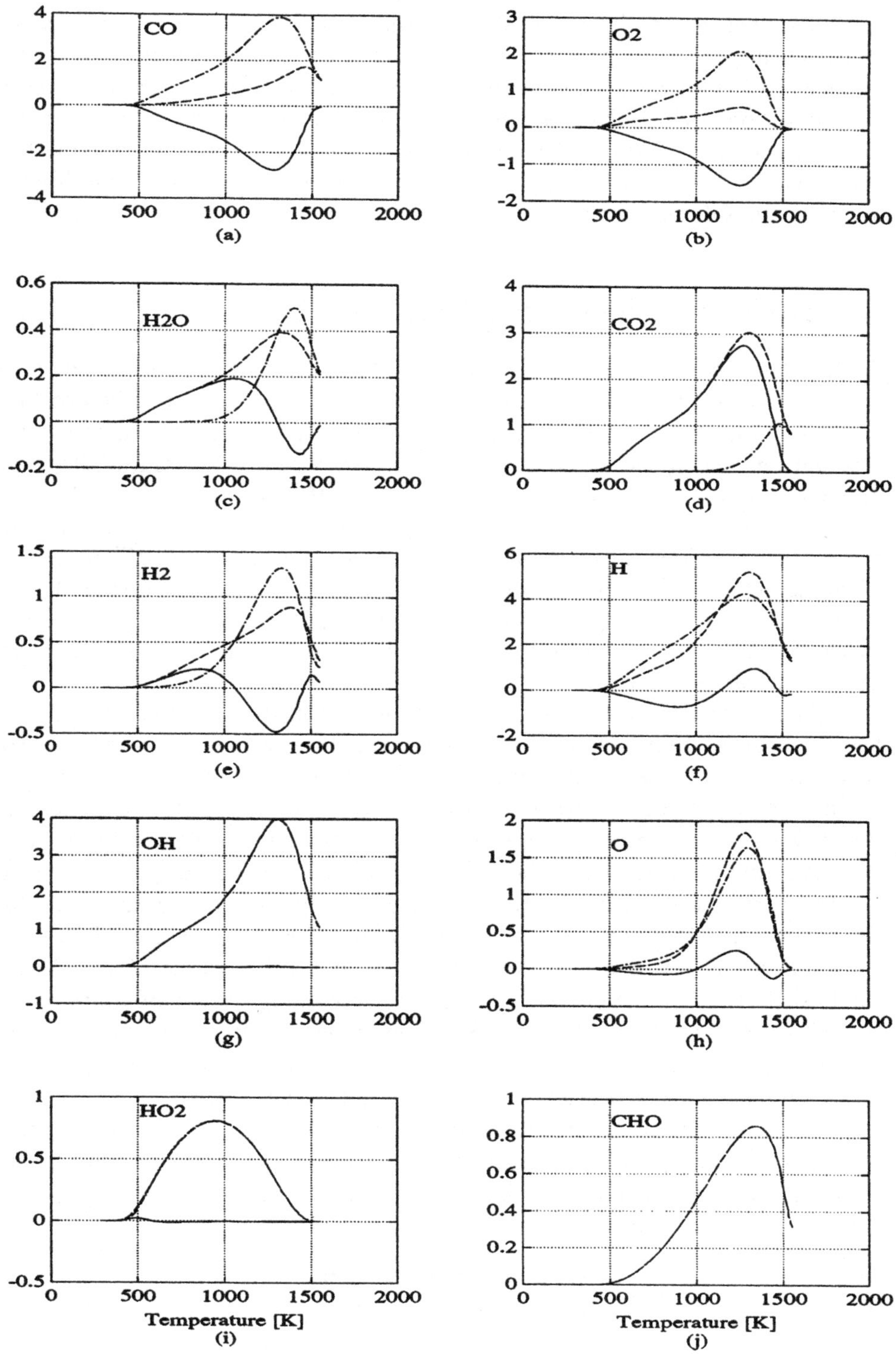

Fig. 4.3. Production, consumption and net rates (in kmol/m^{3}s) of species in an atmospheric-pressure wet CO flame calculated with short mechanism. Initial temperature 300 K; initial mole fraction of fuel 0.65; f = 0.21. Solid lines: net rate; dashed lines: production rate; dashed-dotted lines: consumption rate.

the short mechanism. The dashed-dotted lines can be ignored for the moment because they represent results obtained with a two-step mechanism that will be derived below; the symbols represent experimental data taken from Lewis and von Elbe [4.2]. In the following the short mechanism is taken as a starting point for the derivation of a three-step and a two-step reduced kinetic mechanism because it is seen in Fig. 4.1 to produce results that agree very well with those of the full mechanism.

4.3 Reduced Kinetic Mechanisms

Reduced mechanisms arise when steady-state approximations are good. Figures 4.2 and 4.3 test these steady states for all species at two different fuel mole fractions in air, using the short mechanism. It is seen from these figures that all intermediates maintain good steady states for lean flames, although the approximations become poor for H, H_2 and H_2O at very rich conditions. Hence there is some question of the accuracy of a one-step approximation at very rich conditions, although a two-step approximation may still be good. This further motivates the present emphasis on a two-step description. These conclusions are not foolproof because the dominant terms may be the same in more than one of the steady states, but they do provide an indication that such reduced mechanism are worth trying.

With increasing degrees of reduction of a mechanism the complexity of the algebraic system of equations arising from the steady states increases. To solve these equations conveniently a process termed "truncation" therefore often is employed. In truncation, certain terms in the steady-state expressions are deleted to aid in solving the algebraic system. In an earlier work [4.1] partial equilibrium of step 3, $H_2 + OH \rightarrow H_2O + H$, was employed as part of the truncation. Since partial equilibria often are applied in truncation methods, they are tested with the short mechanism for lean and rich flames in Figs. 4.4 and 4.5. It is seen from these figures that although step 1 maintains the best partial equilibrium for lean flames, over the full range of stoichiometry step 3, on balance, is the best. However, even this approximation is seen in these figures to be somewhat inaccurate. Therefore, in the present study the approximations of partial equilibria will not be introduced.

Since there are $10 - 3 = 7$ independent chemical steps in the short mechanism, with four species in steady state, a three-step kinetic mechanism is obtained. If HCO, O, OH and HO_2 are selected as the steady-state species, the three-step mechanism

(I) $$CO + H_2O \rightleftharpoons CO_2 + H_2\,,$$

(II) $$2H + M \rightarrow H_2 + M\,,$$

(III) $$O_2 + 3H_2 \rightleftharpoons 2H_2O + 2H$$

can be derived. In this mechanism, step I is the overall CO consumption step which neither creates nor destroys reaction intermediaries. Step II represents an overall recombination step, and step III an overall radical-production,

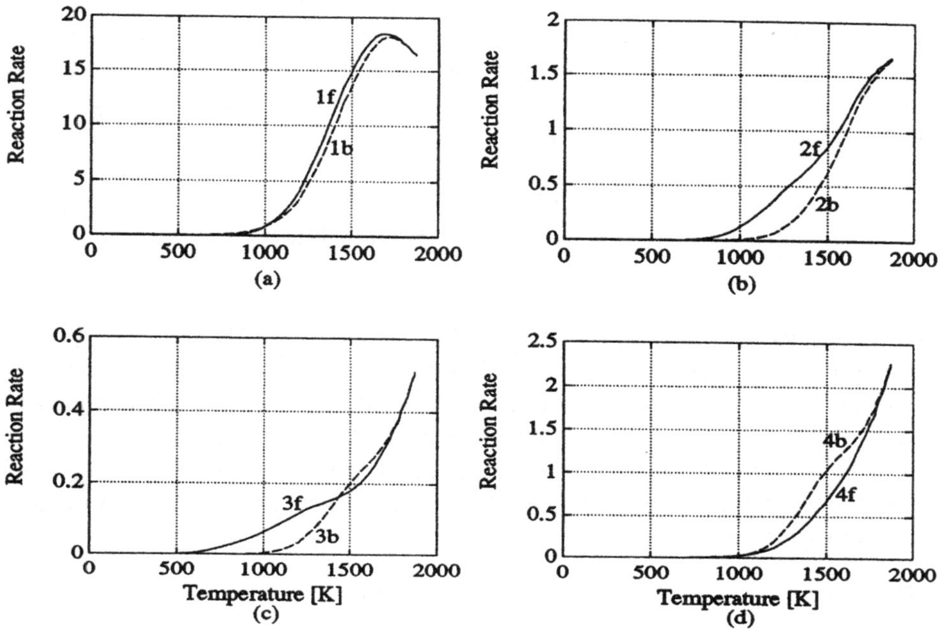

Fig. 4.4. Forward and backward rates (in kmol/m^{3}s) of reactions 1, 2, 3 and 4 in an atmospheric-pressure wet CO flame calculated with the short mechanism. Initial temperature 300 K; initial mole fraction of fuel 0.3; $f = 0.21$.

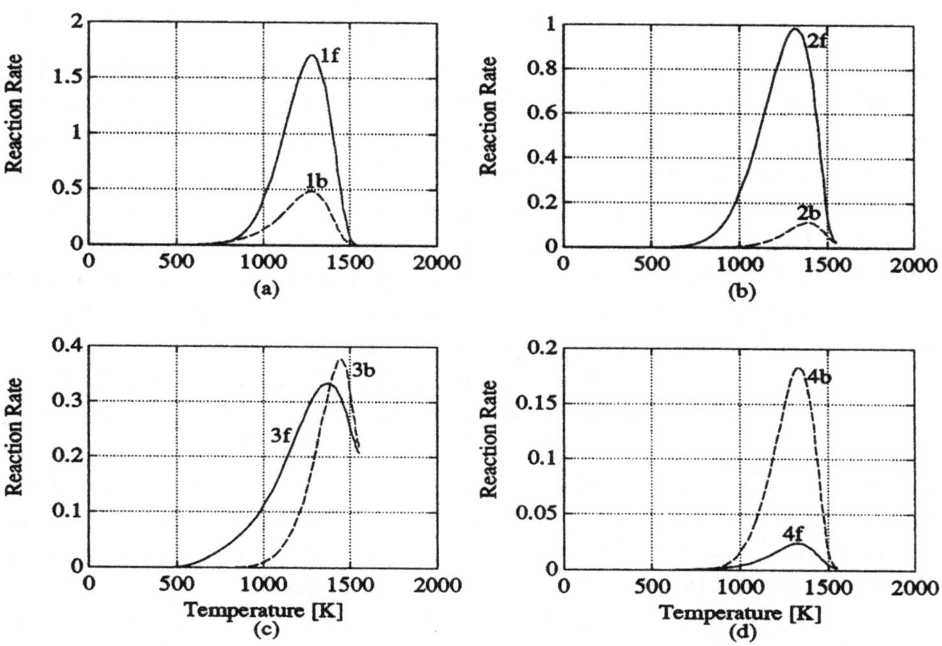

Fig. 4.5. Forward and backward rates (in kmol/m^{3}s) of reactions 1, 2, 3 and 4 in an atmospheric-pressure wet CO flame calculated with the short mechanism. Initial temperature 300 K; initial mole fraction of fuel 0.65; $f = 0.21$.

oxygen-consumption step. Note that this three-step mechanism is formally identical to that derived by Rogg and Williams [4.1] for wet CO flames and by Chen et al. [4.3] for CO/air diffusion flames. In terms of the rates of the elementary steps contained in the short mechanism, the global rates of the three-step mechanism can be written as

$$w_{\mathrm{I}} = w_{18}\,, \tag{4.1}$$

$$w_{\mathrm{II}} = w_5 + w_{15} + w_{16} - w_{24}\,, \tag{4.2}$$

$$w_{\mathrm{III}} = w_1 + w_6 + w_9\,. \tag{4.3}$$

If in (4.3) ω_9 is neglected, the global rates used by Rogg and Williams [4.1] are recovered. Since the three-step mechanism for wet CO flames was discussed in detail in [4.1], herein it is not considered further. Instead, subsequently the three-step mechanism is used as the basis for further reduction.

If H atoms are assumed to be in steady state, the above three-step mechanism becomes a two-step mechanism which, after standard manipulations, can be written as

(I′) $$\mathrm{CO} + \frac{1}{2}\mathrm{O_2} \rightarrow \mathrm{CO_2}\,,$$

(II′) $$\mathrm{H_2} + \frac{1}{2}\mathrm{O_2} \rightarrow \mathrm{H_2O}\,.$$

The global rates of this mechanism are given by

$$w_{\mathrm{I}} = w_{18f} - w_{18b}\,, \tag{4.4}$$

$$w_{\mathrm{II}} = w_{3f} - w_{3b} + [w_{4f}] - w_{4b} + [w_8] + [w_9] + w_{16}\,. \tag{4.5}$$

In deriving (4.4) and (4.5) use has been made of the steady-state relationships for the steady-state species identified above. On the r.h.s. of (4.5) rates of elementary steps in brackets, such as $[w_{4f}]$, have been neglected *in the computations* based on the reduced mechanism. It is important to note that the truncation decisions to neglect certain rates of elementary step in the rate of this global step have not been made in an arbitrary manner but are the result of a carefully conducted numerical sensitivity analysis which aimed to finely tune the reduced mechanism so as to give best results for flame structure and burning velocity for a wide range of stoichiometries.

In the elementary rates appearing on the right-hand-sides of (4.4) and (4.5), the concentrations of the steady-state species must be expressed in terms of the concentrations of the species appearing explicitly in the two-step mechanism. Therefore, the global rates are algebraically more or less complex expressions containing rate data of many of the elementary steps of the short mechanism; the degree of complexity of the global rates depends, of course, on the specific assumptions introduced in their derivation.

In the following we assume that the concentrations of the species appearing explicitly in the two-step mechanism, i.e., $[\mathrm{CO}]$, $[\mathrm{O_2}]$, $[\mathrm{H_2}]$, $[\mathrm{CO_2}]$ and $[\mathrm{H_2O}]$, are known by solving numerically the conservation equations for premixed, laminar flames as summarized in Chap. 1 of this book. The concentrations of the steady-state species H, OH, O, HO_2 and HCO, which do not appear explicitly in the reduced mechanism but are required to evaluate the global rates according to (4.4) and (4.5), are expressed in terms of the known species concentrations as follows. Based on numerical explorations, herein full steady states are adopted for O, HO_2 and HCO, and truncated steady states for H and O atoms. Specifically, in the steady-state relationship for H atoms the elementary steps 3b and 7 are omitted, in that for O atoms steps 2b and 10. The resulting steady-state relationships are

$$\begin{aligned} 0 &= -w_{1f} + w_{1b} + w_{2f} - w_{2b} + w_{3f} - w_{5f} - w_6 - w_9 \\ &\quad - 2w_{15} - w_{16} + w_{18f} - w_{18b} - w_{21} + w_{24f} - w_{24b}\,, \end{aligned} \tag{4.6}$$

$$\begin{aligned} 0 &= w_{1f} - w_{1b} + w_{2f} - w_{2b} - w_{3f} + w_{3b} - 2w_{4f} + 2w_{4b} \\ &\quad + 2w_6 - w_8 + w_{10} - w_{16} - w_{18f} + w_{18b}\,, \end{aligned} \tag{4.7}$$

$$0 = w_{1f} - w_{1b} - w_{2f} + w_{4f} - w_{4b} + w_9\,, \tag{4.8}$$

$$0 = w_{5f} - w_6 - w_7 - w_8 - w_9 - w_{10}\,, \tag{4.9}$$

$$0 = w_{21} + w_{24f} - w_{24b}\,. \tag{4.10}$$

Equations (4.6)–(4.10) can be solved for the concentrations of the steady-state species, viz.,

$$\begin{aligned} [\mathrm{H}] &= \frac{N}{D}\,, \\ N &= k_{1b}[\mathrm{O}][\mathrm{OH}] + \mathrm{k_{2f}}[\mathrm{H_2}][\mathrm{O}] + \mathrm{k_{3f}}[\mathrm{H_2}][\mathrm{OH}] \\ &\quad + k_{18f}[\mathrm{CO}][\mathrm{OH}] + \mathrm{k_{24f}}[\mathrm{CHO}][\mathrm{M}]\,, \\ D &= k_{1f}[\mathrm{O_2}] + \mathrm{k_{2b}}[\mathrm{OH}] + \mathrm{k_{5f}}[\mathrm{M}][\mathrm{O_2}] + \mathrm{k_6}[\mathrm{HO_2}] \\ &\quad + k_9[\mathrm{HO_2}] + 2\mathrm{k_{15}}[\mathrm{H}][\mathrm{M}] + \mathrm{k_{16}}[\mathrm{M}][\mathrm{OH}] \\ &\quad + k_{18b}[\mathrm{CO_2}] + \mathrm{k_{21}}[\mathrm{CHO}] + \mathrm{k_{24b}}[\mathrm{CO}][\mathrm{M}]\,, \end{aligned} \tag{4.11}$$

$$\begin{aligned} [\mathrm{OH}] &= \frac{N}{D}\,, \\ N &= k_{1f}[\mathrm{H}][\mathrm{O_2}] + \mathrm{k_{2f}}[\mathrm{H_2}][\mathrm{O}] + \mathrm{k_{3b}}[\mathrm{H}][\mathrm{H_2O}] \\ &\quad + 2k_{4b}[\mathrm{H_2O}][\mathrm{O}] + 2\mathrm{k_6}[\mathrm{H}][\mathrm{HO_2}] \\ &\quad + k_{10}[\mathrm{HO_2}][\mathrm{O}] + \mathrm{k_{18b}}[\mathrm{CO_2}][\mathrm{H}]\,, \\ D &= k_{1b}[\mathrm{O}] + \mathrm{k_{2b}}[\mathrm{H}] + \mathrm{k_{3f}}[\mathrm{H_2}] + 2\mathrm{k_{4f}}\mathrm{OH}] \\ &\quad + k_8[\mathrm{HO_2}] + \mathrm{k_{16}}[\mathrm{H}][\mathrm{M}] + \mathrm{k_{18f}}[\mathrm{CO}]\,, \end{aligned} \tag{4.12}$$

$$[\mathrm{O}] = \frac{k_{1f}[\mathrm{H}][\mathrm{O_2}] + \mathrm{k_{4f}}[\mathrm{OH}][\mathrm{OH}] + \mathrm{k_9}[\mathrm{H}][\mathrm{HO_2}]}{\mathrm{k_{1b}}[\mathrm{OH}] + \mathrm{k_{2f}}[\mathrm{H_2}] + \mathrm{k_{4b}}[\mathrm{H_2O}]}\,, \tag{4.13}$$

$$[\mathrm{CHO}] = \frac{k_{24b}[\mathrm{CO}][\mathrm{H}][\mathrm{M}]}{\mathrm{k_{21}}[\mathrm{H}] + \mathrm{k_{24f}}[\mathrm{M}]}\,, \tag{4.14}$$

$$[\mathrm{HO_2}] = \frac{k_{5f}[\mathrm{H}][\mathrm{O_2}][\mathrm{M}]}{\mathrm{k_6}[\mathrm{H}] + \mathrm{k_7}[\mathrm{H}] + \mathrm{k_8}[\mathrm{OH}] + \mathrm{k_9}[\mathrm{H}] + \mathrm{k_{10}}[\mathrm{O}]}\,. \tag{4.15}$$

Inspection of (4.11)–(4.15) shows that some of these equations are strongly coupled and, therefore, must be solved by "inner iteration". For a detailed discussion of inner iteration Chap. 6 of the present volume should be consulted [4.4]. In order to start the inner iteration for the present problem, the initial concentration of H is prescribed as 10^{-10}; the initial concentrations of OH and O are calculated by assuming partial-equilibrium for reactions 2 and 3, i.e., from

$$[\mathrm{OH}] = \frac{k_{3b}[\mathrm{H}][\mathrm{H_2O}]}{k_{3f}[\mathrm{H_2}]}, \tag{4.16}$$

$$[\mathrm{O}] = \frac{k_{2b}[\mathrm{OH}][\mathrm{H}]}{k_{2f}[\mathrm{H_2}]}. \tag{4.17}$$

It is emphasized here that these partial equilibria are only starting guesses and are not present in the final results.

4.4 Numerical Method

The calculations were performed with the Cambridge Laminar-Flame Code RUN-1DL [4.5] which has been developed for the numerical solution of premixed burner-stabilized and freely propagating flames, strained premixed, non-premixed and partially premixed flames subject to a variety of boundary conditions, and tubular flames. RUN-1DL employs fully self-adaptive gridding; it solves both transient and steady-state problems in physical space and for diffusion flames, alternatively, in mixture-fraction space; models for thermodynamics and molecular transport are implemented that range from trivially simple to very sophisticated; a model of thermal radiation is also implemented. Chemistry models that can be handled range from overall global one-step reactions over reduced kinetic mechanisms to detailed mechanisms comprising an arbitrary number of chemical species and elementary steps. Furthermore, RUN-1DL is able simulate two-phase combustion problems, such as droplet and spray combustion. A copy of RUN-1DL is available from the author upon request [4.5].

4.5 Results and Discussion

For all flames considered herein, atmospheric pressure and an initial temperature of 300 K were adopted, and the fuel was assumed to contain fixed molar percentages of CO (97.15%), H_2 (1.5%) and H_2O (1.35%). Parameters varied systematically in the computations are the initial $O_2/(N_2 + O_2)$ molar ratio f and the initial fuel mole fraction in the mixture X_{CO}.

Shown in Fig. 4.1 is the burning velocity as a function of the initial fuel mole fraction for $f = 0.13$, 0.21, 0.4, and 0.985. Solid lines refer to the full mechanism, dashed lines to the short mechanism and dashed-dotted lines to the two-step mechanism; the symbols represent experimental data taken from Lewis and von Elbe [4.2]. The burning velocities obtained from the two-step

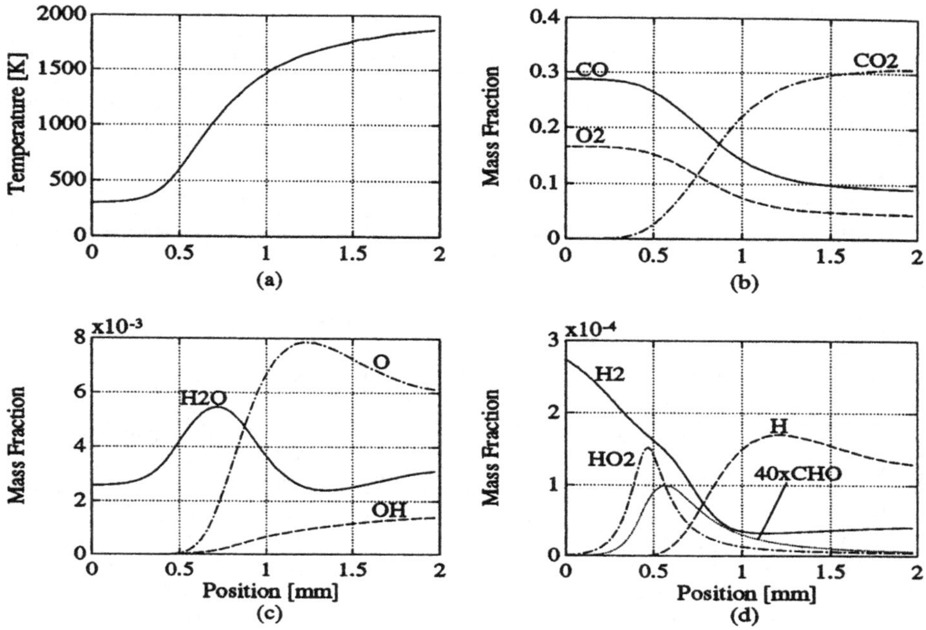

Fig. 4.6. Temperature and species mass-fraction profiles of a lean atmospheric-pressure CO-air flame (initial mole fraction of fuel 0.3, $f = 0.21$) calculated with the short mechanism.

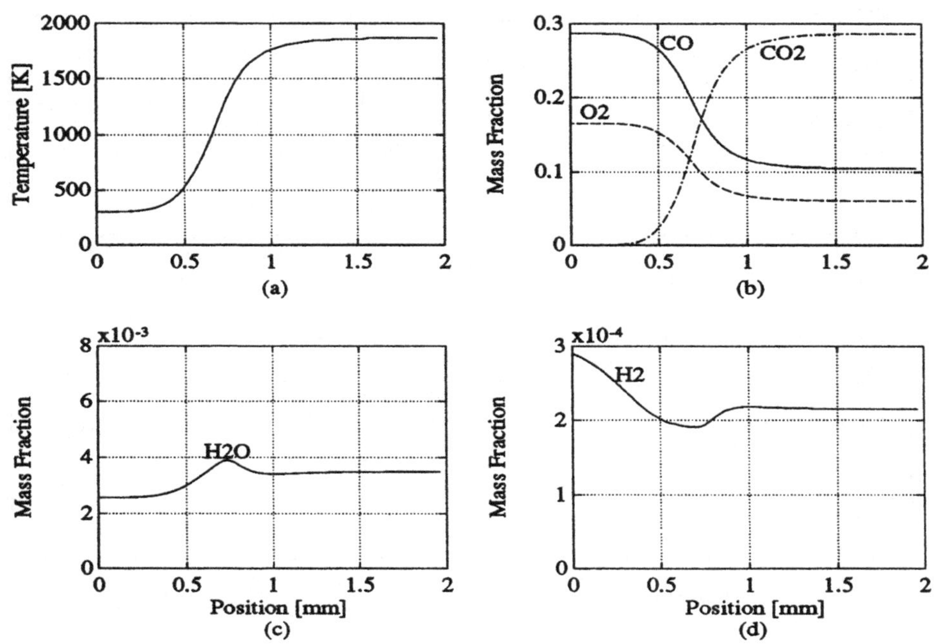

Fig. 4.7. Temperature and species mass-fraction profiles of a lean atmospheric-pressure CO-air flame (initial mole fraction of fuel 0.3, $f = 0.21$) calculated with the two-step reduced mechanism.

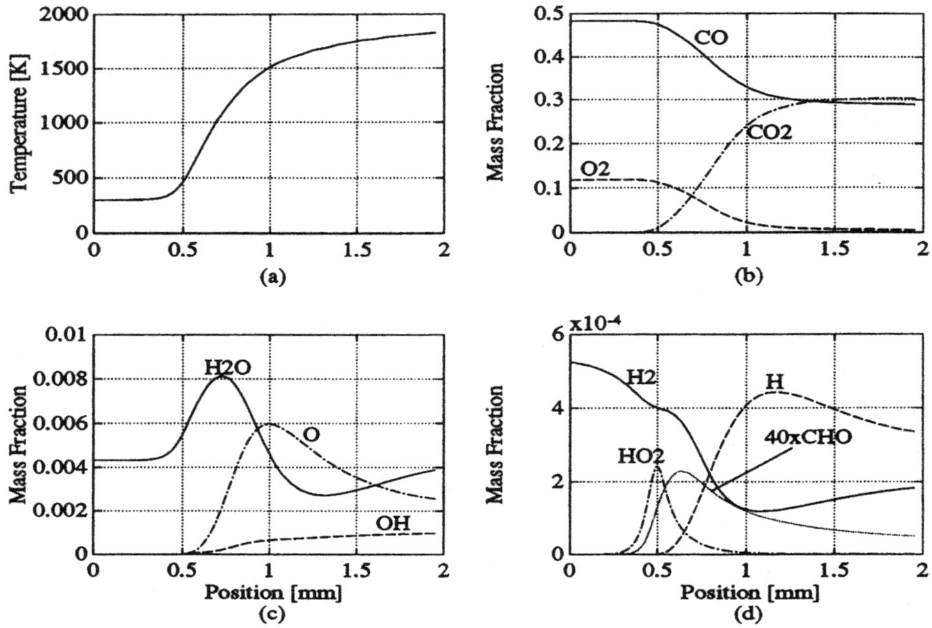

Fig. 4.8. Temperature and species mass-fraction profiles of a lean atmospheric-pressure CO-air flame (initial mole fraction of fuel 0.5, $f = 0.21$) calculated with the short mechanism.

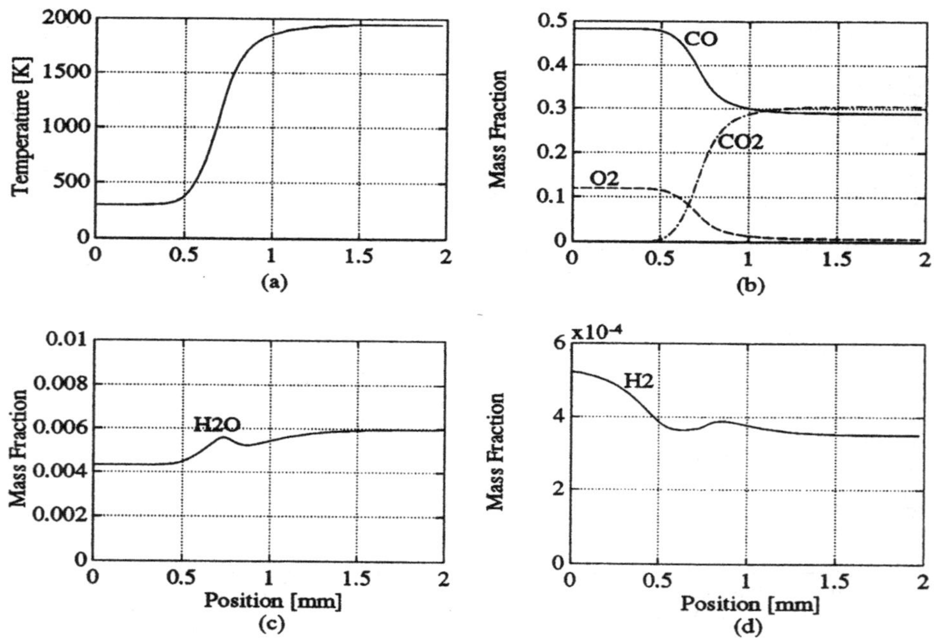

Fig. 4.9. Temperature and species mass-fraction profiles of a lean atmospheric-pressure CO-air flame (initial mole fraction of fuel 0.5, $f = 0.21$) calculated with the two-step reduced mechanism.

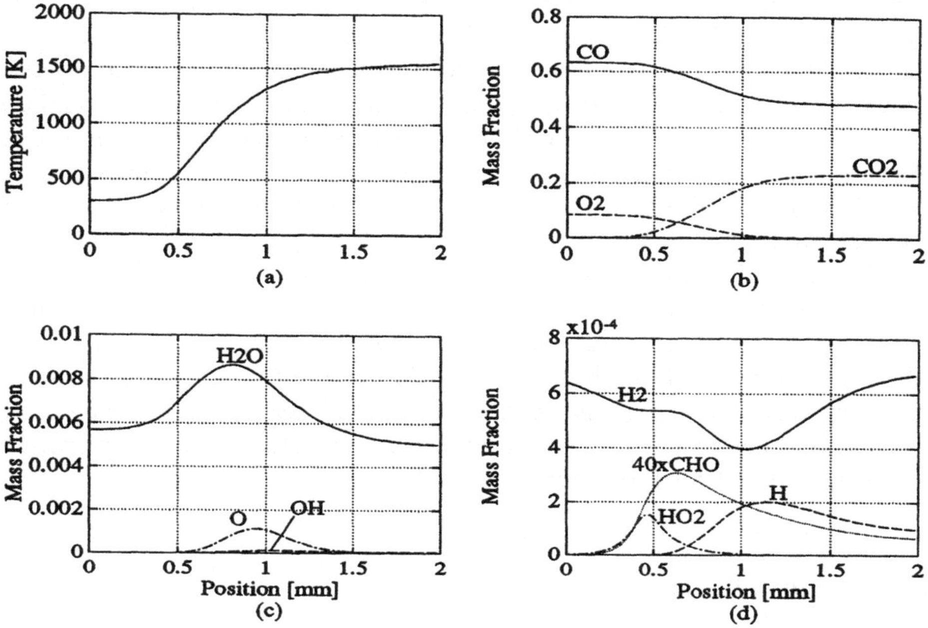

Fig. 4.10. Temperature and species mass-fraction profiles of a lean atmospheric-pressure CO-air flame (initial mole fraction of fuel 0.65, $f = 0.21$) calculated with the short mechanism.

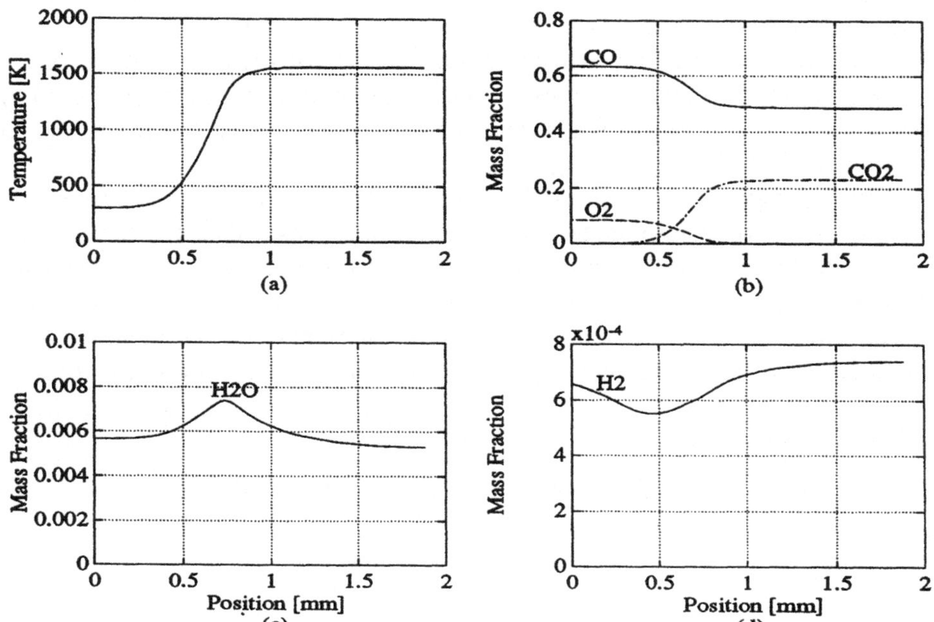

Fig. 4.11. Temperature and species mass-fraction profiles of a lean atmospheric-pressure CO-air flame (initial mole fraction of fuel 0.65, $f = 0.21$) calculated with the two-step reduced mechanism.

mechanism are seen to be rather good except at the highest value of f, which corresponds to nearly no nitrogen and the highest flame temperatures. Since the steady states are good, especially for fuel-lean flames where the discrepancy in burning velocity is greatest, the difference must be attributed to failure of the truncation approximation in the two-step mechanism. Determination of the specific difficulties with the truncation is left for future investigation, since the present intention is merely to exhibit methodology and to expose the problem.

Shown in Figs. 4.6 to 4.11 are numerical results for flame structures obtained with the short and the two-step mechanism. Study of these figures shows that with the two-step mechanism the flames are noticeably thinner. Also, the H_2 and H_2O profiles clearly differ for the two-step and the short mechanism. However, the other profiles are in quite reasonable agreement. As with the burning-velocity discrepancies, the causes of the thickness and profile differences are left for future investigation.

4.6 Conclusions

Wet CO flames are interesting testing grounds for reduced mechanisms because of the initial anticipation of being able to achieve a good one-step approximation. An excellent short mechanism was identified here, and tests of steady states with this mechanism indicated that, except for very fuel-rich flames, a good one-step reduced mechanism should indeed be obtainable so long as dominant steady-state terms are not redundant. Because of its algebraic complications, such a mechanism was not sought here, but instead a two-step mechanism, with various truncation approximations included as well, was developed and tested. The intent was to demonstrate the methods and to call attention to this interesting flame.

Burning velocities predicted by the two-step mechanism differed by as much as 30% from those of the short mechanism. This calls attention to deficiencies in the truncation approximations adopted for the two-step mechanism or underlying redundance in the steady states. These deficiencies are greatest on the lean side of the maximum burning velocity and at the higher oxygen concentrations. Further study of this intriguing flame therefore is warranted.

Acknowledgement

This work was supported by NATO through Grant No. 0102/89.

References

[4.1] Rogg, B. and Williams, F. A., Twentysecond Symposium (International) on Combustion, p. 1441–1451, The Combustion Institute, 1988.

[4.2] Lewis, B. and von Elbe, G., Combustion, Flames and Explosions of Gases, 3rd edition, Academic Press, Orlando, p. 398, 1987.

[4.3] Chen, J.-Y., Dibble, R. W. and Bilger, R. W., Twentythird Symposium (International) on Combustion, pp. 775–780, The Combustion Institute, 1990.

[4.4] Wang, W. and Rogg, B., Premixed Ethylene/Air and Ethane/Air Flames: Reduced Mechanisms Based on Inner Iteration, Chap. 6 of this volume.

[4.5] Rogg, B., RUN-1DL: A Computer Program for the Simulation of One-Dimensional Chemically Reacting Flows, Report CUED/A-THERMO/TR39, Cambridge University Engineering Department, April 1991.

5. Reduced Kinetic Mechanisms for Premixed Methane–Air Flames

F. Mauss and N. Peters
Institut für Technische Mechanik,
RWTH Aachen, W-5100 Aachen, Germany

5.1 Introduction

Methane-air flames have served as the first example for the development of a strategy to systematically reduce kinetic mechanisms. Premixed methane-air flames were first considered in [5.1] and [5.2], methane-air diffusion flames in [5.3]. At last, an entire book, from which further references may be taken, was devoted to the subject of "Reduced Kinetic Mechanisms and Asymptotic Approximations for Methane-Air Flames" [5.4]. The idea there was to use a relatively short kinetic mechanism containing only 25 reactions of the C_1-chain, and to analyse not only different approximations for reduced mechanisms of methane-air flames, but asymptotic formulations as well.

It may at first be surprising that reduced mechanisms were first derived for methane rather than for the kinetically simpler hydrogen flames. The reason is that the radical level in methane flames is lower than in hydrogen flames and that therefore steady state relations are better justified. It is lower because the chain from methane that leads to CH_3, CH_2O, CHO and eventually CO consumes radicals. This part of the mechanism therefore is chain-breaking. A consequence of this chain-breaking effect of the fuel reactions is that the preheat zone in a premixed methane-air flame is chemically inert. Fuel and radicals cannot coexist there, radicals disappear in a thin layer downstream of the preheat zone, the radical consumption zone [5.5]. This effect also simplifies the asymptotic analysis of methane flames.

In this paper, differently form the previous formulations, we will include the C_2-chain in deriving a reduced mechanism for methane-air flames. The first 61 reactions in Table 1, Chap. 1, but without reactions 17 and 48 will be used as the starting mechanism for the reduction procedure. Reaction 17 was excluded since it plays a negligible role in methane-air flames. Because it involves the square of the O-concentration it would have complicated the derivation of an explicit algebraic steady state relation for the O-radical. Reaction 48 was excluded because it leads to the C_3-hydrocarbons. This reaction

is important for the soot formation mechanism which will not be considered here.
The oxidation of methane proceeds through the fast reactions

$$\begin{array}{llll} 38\,\mathrm{f} & CH_4 + H & \rightarrow & CH_3 + H_2 \\ 35 & CH_3 + O & \rightarrow & CH_2O + H \\ 29 & CH_2O + H & \rightarrow & CHO + H_2 \\ 24\,\mathrm{f} & CHO + M' & \rightarrow & CO + H + M' \end{array} \tag{5.1}$$

representing the main chain of the C_1 mechanism. A side chain through the C_2 mechanism proceeds as

$$\begin{array}{llll} 36 & CH_3 + CH_3 + M & \rightarrow & C_2H_6 + M \\ 59 & C_2H_6 + H & \rightarrow & C_2H_5 + H_2 \\ 58\,\mathrm{f} & C_2H_5 + M & \rightarrow & C_2H_4 + H + M \\ 52\,\mathrm{f} & C_2H_4 + H & \rightarrow & C_2H_3 + H_2 \\ 51\,\mathrm{f} & C_2H_3 + M & \rightarrow & C_2H_2 + H + M \\ 45 & C_2H_2 + O & \rightarrow & CH_2 + CO \\ 25\,\mathrm{f} & CH_2 + H & \rightarrow & CH + H_2 \\ 19 & CH + O_2 & \rightarrow & CHO + O \\ 24\,\mathrm{f} & CHO + M' & \rightarrow & CO + H + M'\,. \end{array} \tag{5.2}$$

Assuming the intermediates in steady state and adding up these reactions leads to the global step

$$2\,CH_3 + O_2 \rightarrow 2\,CO + 3\,H_2\,. \tag{5.3}$$

The latter step states that the oxidation of CH_3 via the C_2-chain essentially does not produce nor consume radicals if the reactions above are the fastest side reactions to be considered. This global step has to be compared to the one resulting from the C_1-chain. Adding to the last three reactions in (5.1) the fast shuffling reactions

$$\begin{array}{llll} 2\,\mathrm{b} & OH + H & \rightarrow & H_2 + O \\ 3\,\mathrm{b} & H_2O + H & \rightarrow & H_2 + OH \end{array} \tag{5.4}$$

one obtains the global step (taken twice)

$$2\,CH_3 + 2\,H_2O + 2\,H \rightarrow 2\,CO + 6\,H_2\,. \tag{5.5}$$

Here, two H radicals are consumed. Therefore, when the C_2 chain is added to the methane mechanism, its overall effect is to reduce the chain breaking effect of the C_1 chain. This, of course, is triggered by the relative importance of reactions 35 and 36. For lean methane flames the amount of O-radicals present is always high enough to make reaction 35 dominate over reaction 36. For rich flames, however, this is no longer the case and the C_2 chain becomes important. Because reaction 36 is trimolecular the influence of the C_2-chain shifts with increasing pressure to lean flames.

Figure 5.1 shows the burning velocity over the equivalence ratio calculated with the starting mechanism with and without C_2-species, as compared

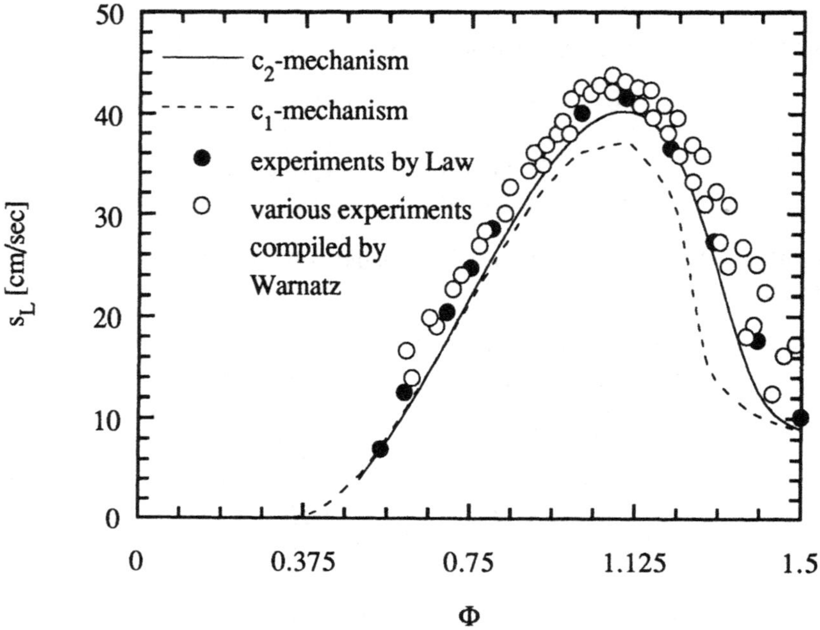

Fig. 5.1. Burning velocities calculated with the starting C_1-mechanism and the starting C_2-mechanism, several data compiled by Warnatz [5.7] and recent data by Law [5.6] for atmospheric methane-air flames

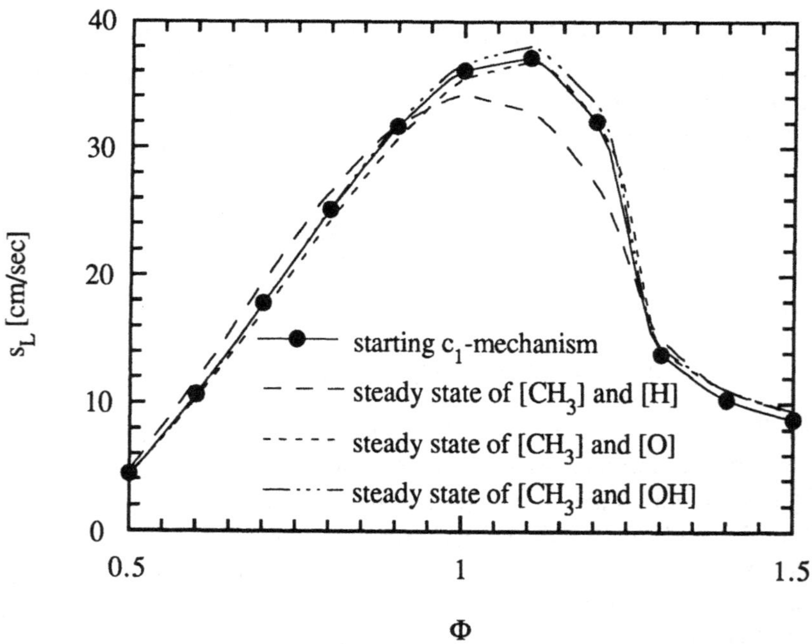

Fig. 5.2. Burning velocities calculated with the starting C_1-mechanism and the three combinations of steady state assumptions for the dominant radicals H, O, OH with CH_3

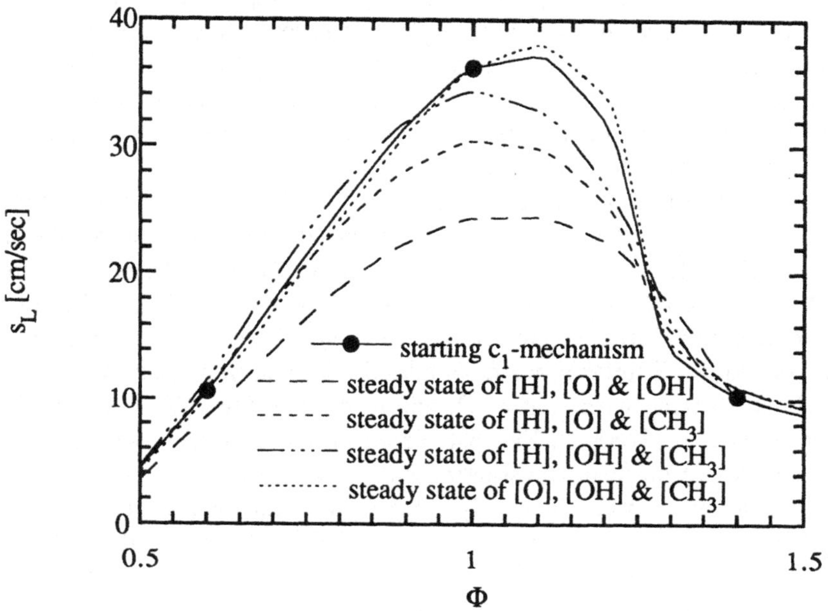

Fig. 5.3. Burning velocities calculated with the starting C_1-mechanism and the permutation of three steady state assumptions for the dominant radicals H, O, OH and CH_3

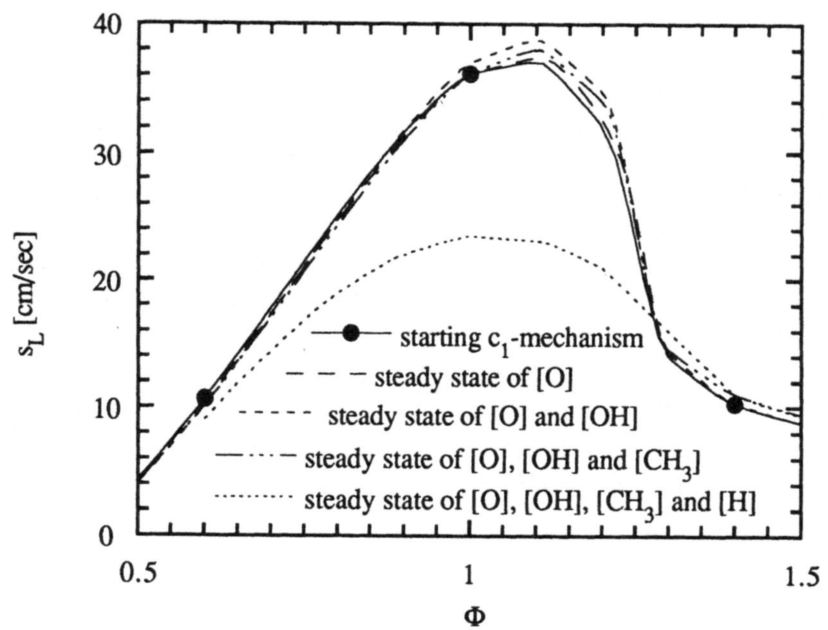

Fig. 5.4. Burning velocities calculated with the starting C_1-mechanism and step by step introduction of steady state assumptions for the dominant radicals O, OH, CH_3 and H

to recent measurements compiled by Law [5.6] (chapter 2 of his book) and to a previous compilation by Warnatz [5.7]. It is seen that the burning velocity calculated with the C_1 mechanism is more than 2 cm/s smaller for rich flames while the starting mechanism including C_2 species shows a good agreement within the range of experimental accuracy, although the calculations are always on the lower side of the data.

To obtain the best choice of steady state assumptions for the dominant radicals H, O, OH and CH_3, we have compared the three combinations of H, O and OH with CH_3 in Fig. 5.2 and permutations of three of them in Fig. 5.3. These steady state approximations were introduced by simply setting the convective-diffusive operator for three species equal to zero in the numerical procedure. Similarly, we have added consecutively more steady state assumptions, starting with that of O, adding that of OH, CH_3 and finally H. This comparison is shown in Fig. 5.4. From all these results one can clearly see that the deviation form the starting mechanism becomes unacceptable, as soon as H is assumed to be in steady state. This confirms the previous choice to always keep H as a non-steady state species in the reduced mechanism. Setting all the four radicals in steady state would lead to a 3-step rather than a 4-step mechanism but the price in terms of accuracy is too large, as seen by the lower curve in Fig. 5.4.

5.2 The four-step mechanism for the C_1- and the C_2-chain

We will now identify the steady state species. As always, steady state assumptions of O, OH, HO_2 and H_2O_2 are used to eliminate reactions 2, 3, 8 and 12, respectively. Following the reasoning above, the steady-state assumptions of $CH, CH_2, CHO, CH_2O, CH_3, C_2H_2, C_2H_3, C_2H_4, C_2H_5, C_2H_6$ are justified by assuming reactions 19, 25f, 24f, 29, 35, 45, 51f, 52f, 58f and 59 to be fast and, therefore, can be eliminated. In addition, reaction 44 is assumed to consume CHCO and reaction 42 to consume C_2H very rapidly. With these assumptions, one obtains the following 4-step mechanism

$$\begin{array}{llcl} \mathrm{I} & CH_4 + 2\,H + H_2O & = & CO + 4\,H_2 \\ \mathrm{II} & CO + H_2O & = & CO_2 + H_2 \\ \mathrm{III} & H + H + M & = & H_2 + M \\ \mathrm{IV} & O_2 + 3\,H_2 & = & 2\,H + 2\,H_2O\,. \end{array} \tag{5.6}$$

For the C_1-mechanism the reaction rates are

$$\begin{aligned} w_{\mathrm{I}}^{c_1} &= w_{38} + w_{39} + w_{40} - w_{34} \\ w_{\mathrm{II}}^{c_1} &= w_{18} - w_{20} + w_{28} \\ w_{\mathrm{III}}^{c_1} &= w_5 - w_{11} + w_{13} + w_{14} + w_{15} + w_{16} \\ &\quad + w_{21} + w_{22} + w_{23} - w_{32} + w_{34} \\ w_{\mathrm{IV}}^{c_1} &= w_1 + w_6 + w_9 + w_{11} - w_{14} - w_{20} - w_{26} + w_{33} + w_{37}\,. \end{aligned} \tag{5.7}$$

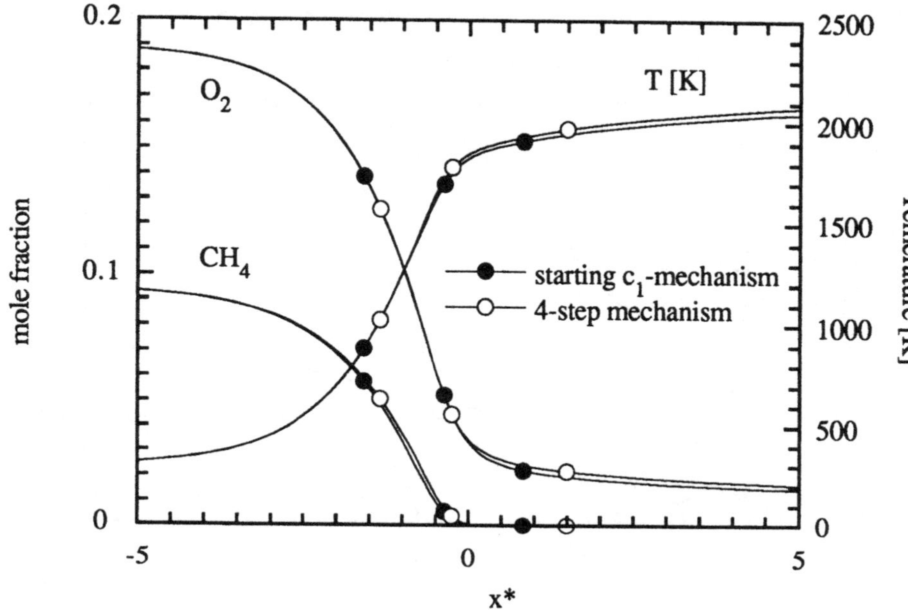

Fig. 5.5a. Numerically calculated mole fractions of O_2 and CH_4 and the temperature profile with the starting C_1-mechanism and the reduced mechanism for atmospheric, stoichiometric methane-air flames

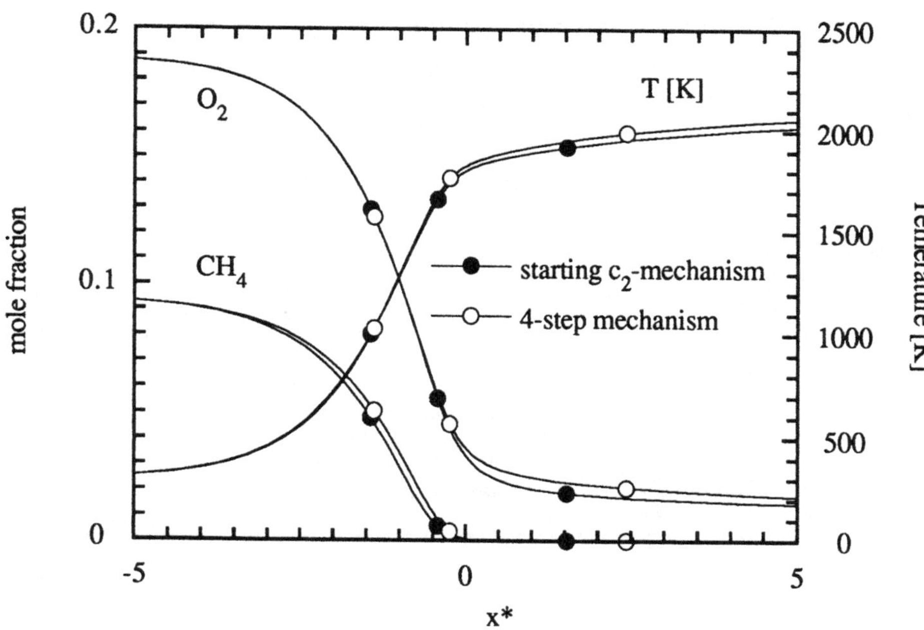

Fig. 5.5b. Numerically calculated mole fractions of O_2 and CH_4 and the temperature profile with the starting C_2-mechanism and the reduced mechanism for atmospheric, stoichiometric methane-air flames

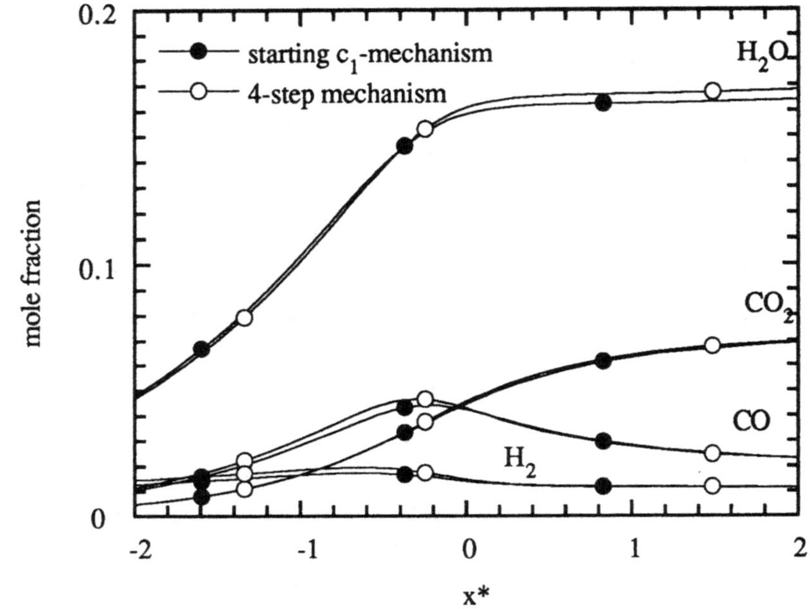

Fig. 5.6a. Numerically calculated mole fractions of H_2, H_2O, CO and CO_2 with the starting C_1-mechanism and the reduced mechanism for atmospheric, stoichiometric methane-air flames

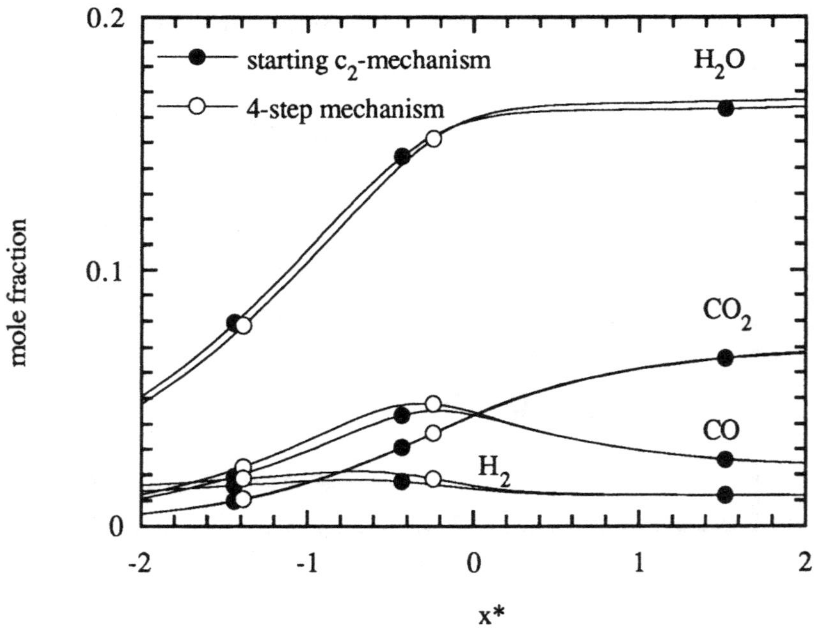

Fig. 5.6b. Numerically calculated mole fractions of H_2, H_2O, CO and CO_2 with the starting C_2-mechanism and the reduced mechanism for atmospheric, stoichiometric methane-air flames

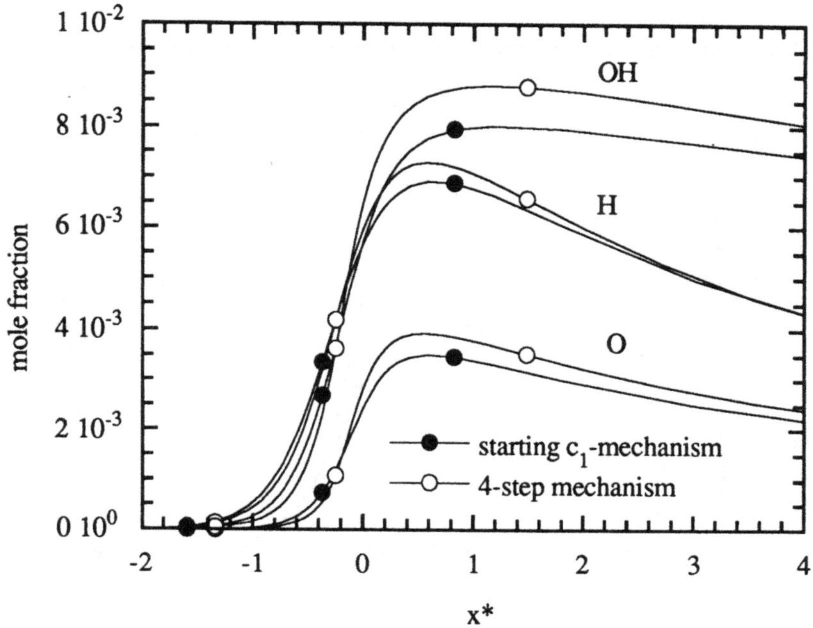

Fig. 5.7a. Numerically calculated mole fractions of H, O and OH with the starting C_1-mechanism and the reduced mechanism for atmospheric, stoichiometric methane-air flames

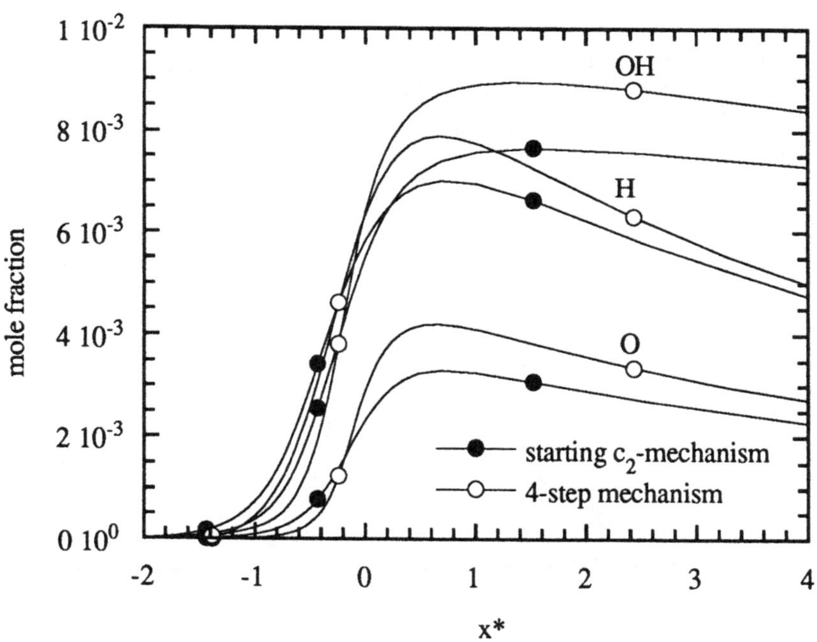

Fig. 5.7b. Numerically calculated mole fractions of H, O and OH with the starting C_2-mechanism and the reduced mechanism for atmospheric, stoichiometric methane-air flames

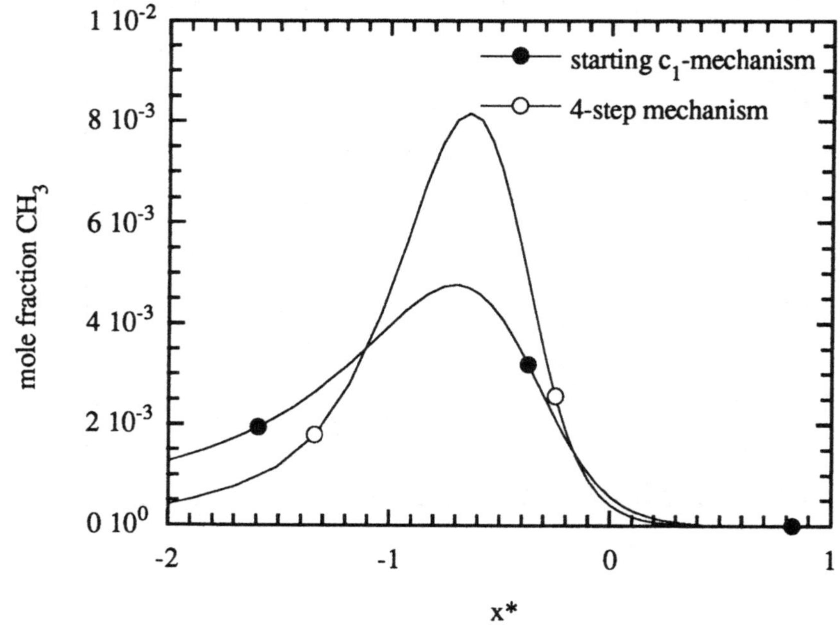

Fig. 5.8a. Numerically calculated mole fraction of CH_3 with the starting C_1-mechanism and the reduced mechanism for atmospheric, stoichiometric methane-air flames

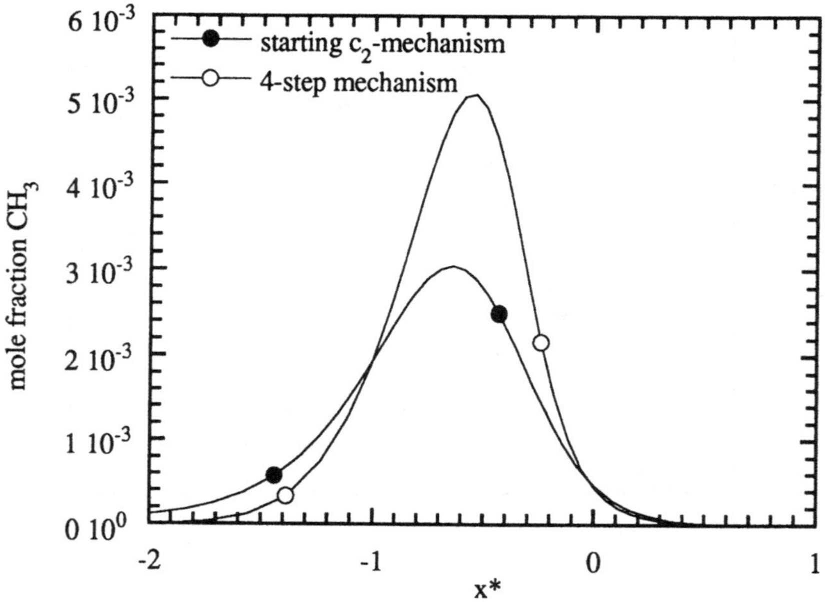

Fig. 5.8b. Numerically calculated mole fraction of CH_3 with the starting C_2-mechanism and the reduced mechanism for atmospheric, stoichiometric methane-air flames

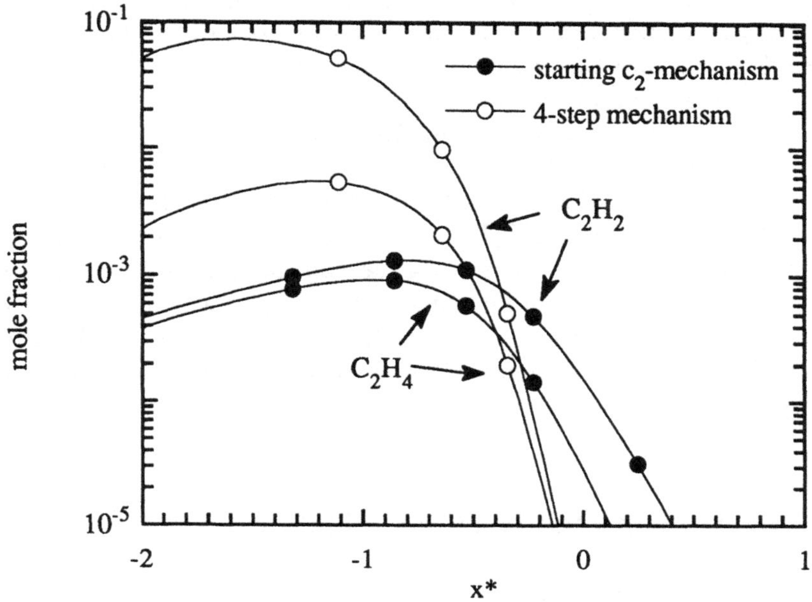

Fig. 5.9. Numerically calculated mole fractions of C_2H_2 and C_2H_4 with the starting C_2-mechanism and the reduced mechanism for atmospheric, stoichiometric methane-air flames

If the C_2-chain is considered, additional side reaction must be included. This leads to

$$\begin{aligned} w_{\mathrm{I}}^{c_2} &= w_{\mathrm{I}}^{c_1}, \quad w_{\mathrm{II}}^{c_2} = w_{\mathrm{II}}^{c_1} \\ w_{\mathrm{III}}^{c_2} &= w_{\mathrm{III}}^{c_1} + w_{49} + w_{50} + w_{56} + w_{57} \\ w_{\mathrm{IV}}^{c_2} &= w_{\mathrm{IV}}^{c_1} + w_{36} + w_{43} - w_{46} - w_{53} - w_{56}\,. \end{aligned} \tag{5.8}$$

It is, in particular, the contribution of w_{36} in the chain breaking reaction IV that increases the burning velocity in Fig. 5.1.

In the following, we will analyse the flame structure by plotting profiles as a function of the non-dimensional coordinate

$$x^* = (\varrho_u s_L) \int_0^x \frac{c_p}{\lambda} dx\,. \tag{5.9}$$

Profiles of the temperature and stable species for a stoichiometric atmospheric-pressure flame at $T_u = 298$ K are shown in Figs. 5.5 and 5.6 for both, the C_1- and the C_2-mechanism as compared with the corresponding starting mechanisms. The agreement is good for both cases, but at first surprisingly, slightly better for the C_1 mechanism. In case of the C_2 mechanism the steady state assumption for the stable species C_2H_2 and C_2H_4 overpredicts the influence of the C_2-chain. Following the arguments from Chapter 5.1 less radicals are consumed by the C_1-chain. This can be seen, when the radical profiles for the two cases are compared in Fig. 5.7. The CH_3 concentration shown in Fig. 5.8 compares slightly better in the upstream part of the flame for the C_2 mechanism, because the recombination of CH_3 via reaction 36 validates the steady state assumption at lower temperatures. But the maximum values, important for the prediction of pollutant formation like soot and prompt NO, are off by a factor of two in both cases. This indicates that the steady state assumption of CH_3 should be reconsidered for that purpose. Finally, we show in Fig. 5.9 the comparison for C_2H_2 and C_2H_4 for the C_2 mechanism. The maximum is shifted between the starting mechanism and the reduced 4-step mechanism and is two orders of magnitude off for C_2H_2 which also is important for soot formation. This clearly indicates that a four step mechanism is not sufficient to calculate the combustion process if soot formation shall also be well predicted.

5.3 Truncation of steady state relations

In the previous analysis we have presented results from calculations where the non-linear steady state relations have been solved numerically by iterations. The most important step for an efficient use of a reduced mechanism is to introduce explicit expressions for the steady state species. This is not always easy and it often requires the truncation of the steady state relation by neglecting unimportant contributions from some of the reactions. Even then it may lead to higher-order non-linear algebraic equations. For the C_1-chain several such explicit formulations for the O and the OH radical concentration have been compared in [5.4]. It has turned out that a quadratic form for [O] that involves the coupling to [CH_3] through reaction 35, gives the best results. If now the C_2-chain is included, reaction 36 would lead to a third-order polynomial. Rather than to derive an explicit formulation for this case we have used a fix-point iteration for calculating the O concentration. It converges in a few iterations starting from the partial equilibrium expression of reaction 2 given below.

In the following we will compare three different cases:

Case 1: The OH radical concentration is calculated from the partial equilibrium of reaction 3 as

$$[\mathrm{OH}] = \frac{k_{3\mathrm{b}}\,[\mathrm{H_2O}]\,[\mathrm{H}]}{k_{3\mathrm{f}}\,[\mathrm{H_2}]}\,, \tag{5.10}$$

but the O-concentration is calculated from the quadratic form in the case of the C_1-mechanism or from a fix-point iteration in the case of the C_2-mechanism.

Case 2: In addition to the partial equilibrium approximation of reaction 3 for the OH-concentration, the O-concentration is calculated from partial equilibrium of reaction 2 as

$$[O] = \frac{k_{2b}\,[H]\,[OH]}{k_{2f}\,[H_2]} \tag{5.11}$$

where [OH] is calculated from (5.10).

Case 3: In addition to the partial equilibrium of reaction 3 for [OH], the O-concentration is calculated from a balance of reactions 1, 2 and 4, but excluding the effect of reaction 35 on the O-balance. This leads to

$$[O] = \frac{k_{1f}\,[H]\,[O_2] + k_{2b}\,[H]\,[OH] + k_{4f}\,[OH]^2}{k_{1b}\,[OH] + k_{2f}\,[H_2] + k_{4b}\,[H_2O]}\,. \tag{5.12}$$

For all these cases the remaining steady state relation can be expressed using their steady state relations in a similar way as shown in [5.8]. These formulas will not be repeated here.

Comparison of these three cases with the C_1-starting mechanism and the reduced four step mechanism in Fig. 5.10 shows the following: Cases 1 and 2 perform quite well, in particular on the lean side, while the case 3 yields much too large values for the burning velocity for stoichiometric and rich flames. This case should therefore not be considered any further.

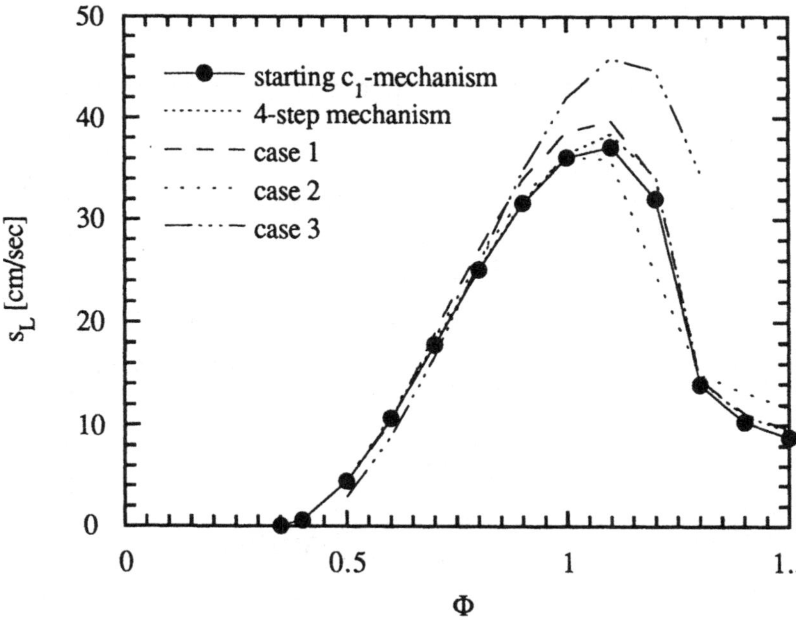

Fig. 5.10. Burning velocities of atmospheric methane-air flames with the starting C_1-mechanism, the reduced 4-step mechanism and three cases of former simplification. Case 1: partial equilibrium of reaction 3. Case 2: partial equilibrium for reactions 2 and 3. Case 3: partial equilibrium of reaction 3 and neglection of reaction 35 by calculating the concentration of O.

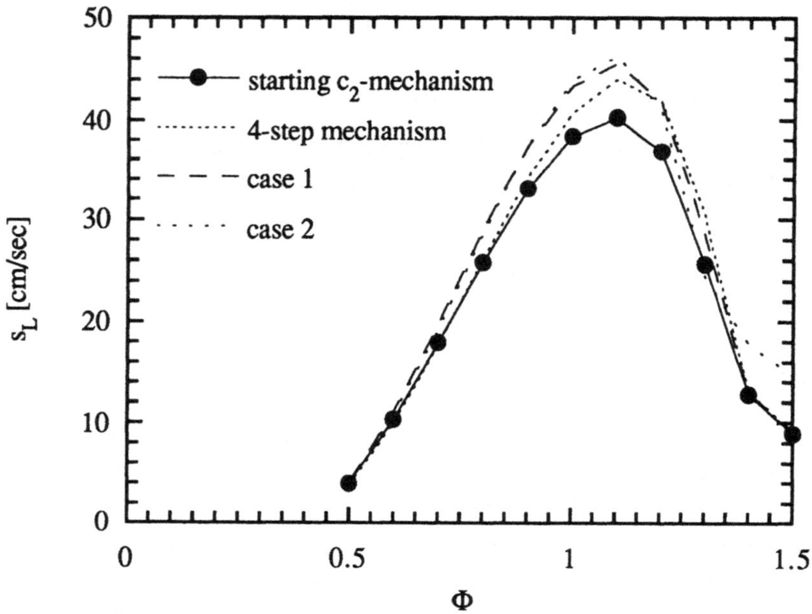

Fig. 5.11. Burning velocities of atmospheric methane-air flames with the starting C_2-mechanism and two cases of former simplification. Case 1: partial equilibrium of reaction 3. Case 2: partial equilibrium for reactions 2 and 3.

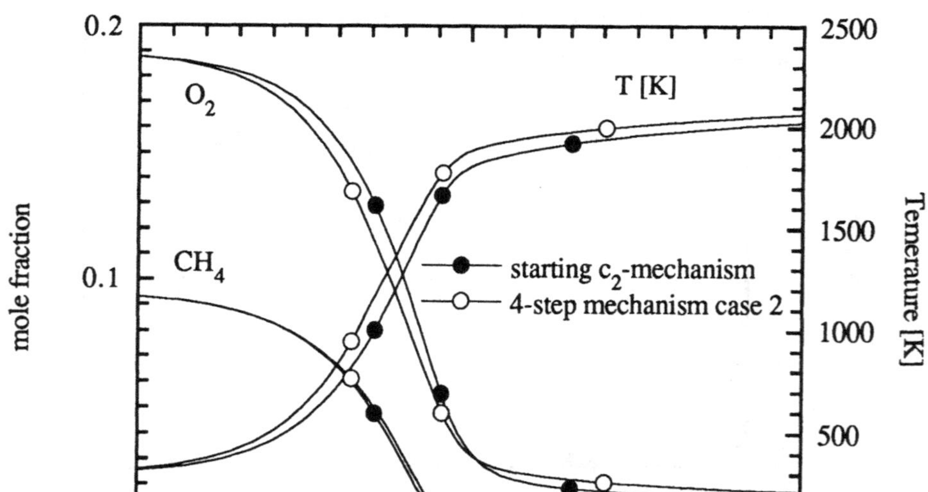

Fig. 5.12. Numerically calculated mole fractions of O_2 and CH_4 and the temperature profil with the starting C_2-mechanism and the simple 4-step mechanism (case 2) for atmospheric, stoichiometric methane-air flames

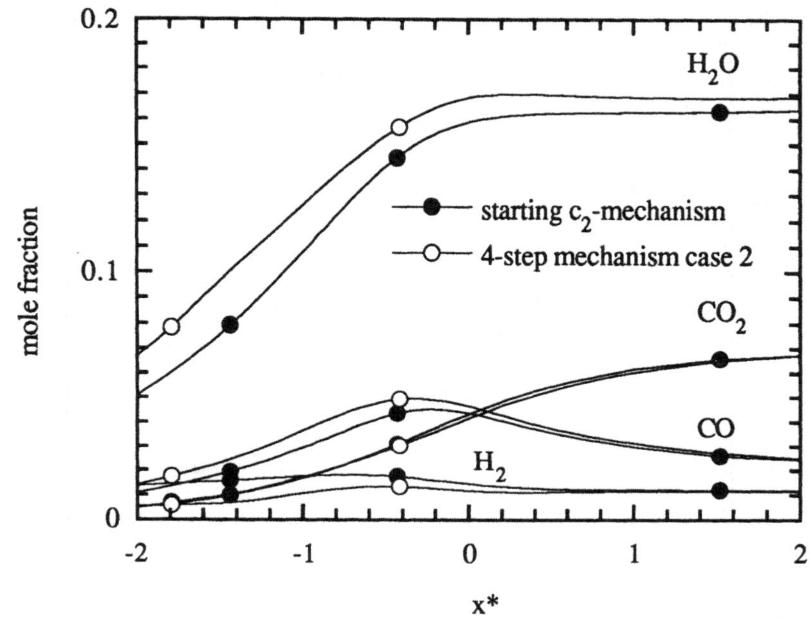

Fig. 5.13. Numerically calculated mole fractions of H_2, H_2O, CO and CO_2 with the starting C_2-mechanism and the simple 4-step mechanism (case 2) for atmospheric, stoichiometric methane-air flames

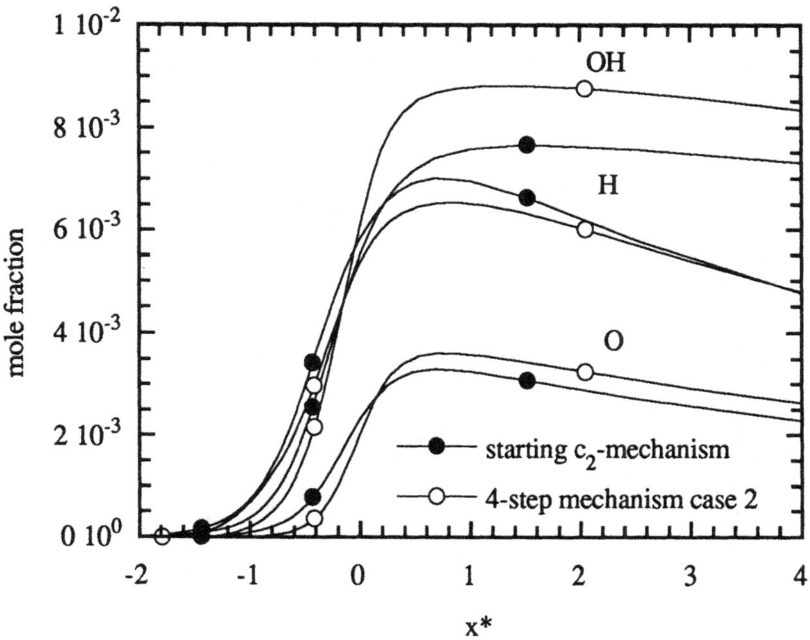

Fig. 5.14. Numerically calculated mole fractions of H, O and OH with the starting C_2-mechanism and the simple 4-step mechanism (case 2) for atmospheric, stoichiometric methane-air flames

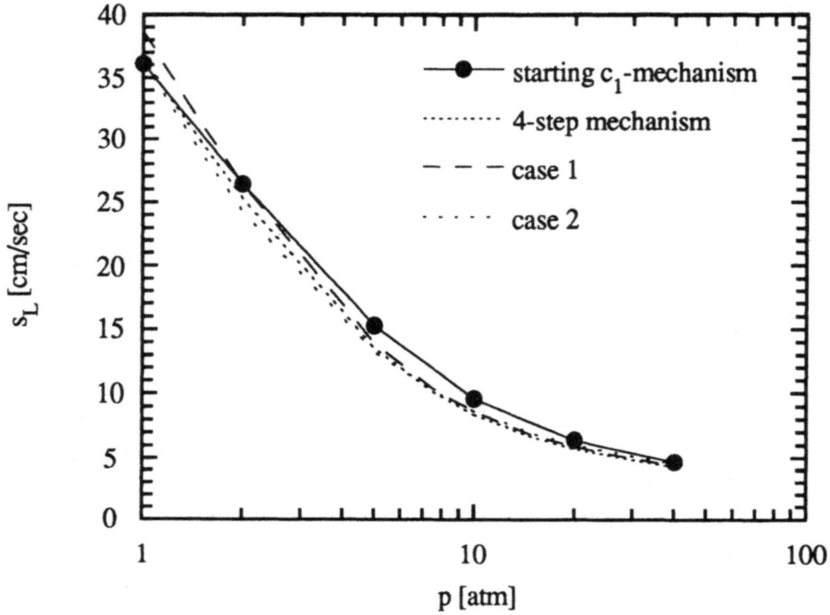

Fig. 5.15. Burning velocities of stoichiometric methane-air flames with the starting C_1-mechanism, the reduced 4-step mechanism and two cases of former simplification. Case 1: partial equilibrium of reaction 3. Case 2: partial equilibrium for reactions 2 and 3.

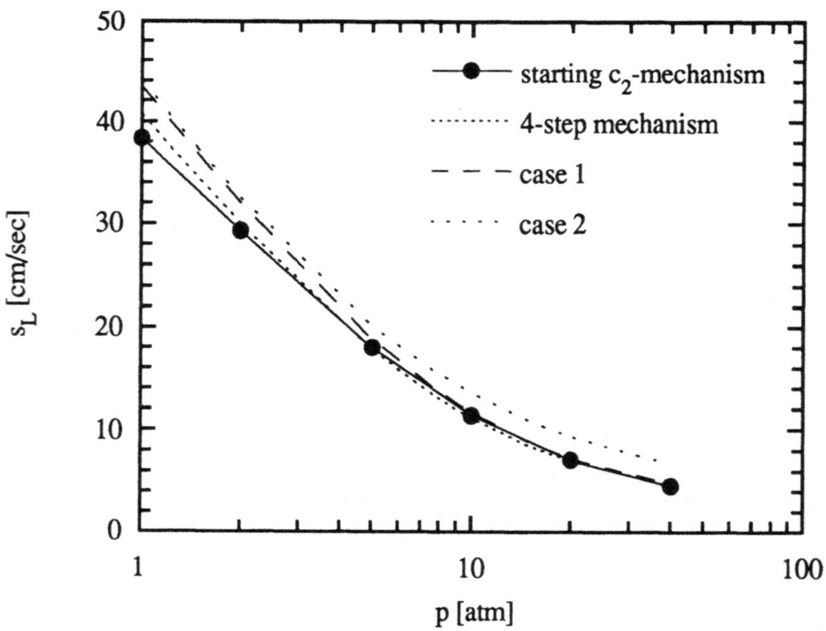

Fig. 5.16. Burning velocities of stoichiometric methane-air flames with the starting C_2-mechanism, the reduced 4-step mechanism and two cases of former simplification. Case 1: partial equilibrium of reaction 3. Case 2: partial equilibrium for reactions 2 and 3.

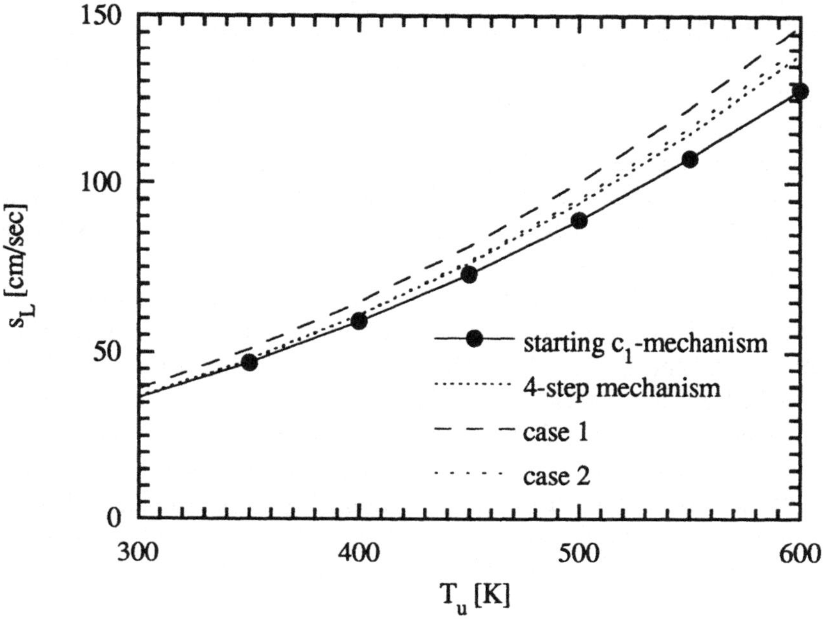

Fig. 5.17. Burning velocities of atmospheric, stoichiometric methane-air flames with the starting C_1-mechanism, the reduced 4-step mechanism and two cases of former simplification. Case 1: partial equilibrium of reaction 3. Case 2: partial equilibrium for reactions 2 and 3.

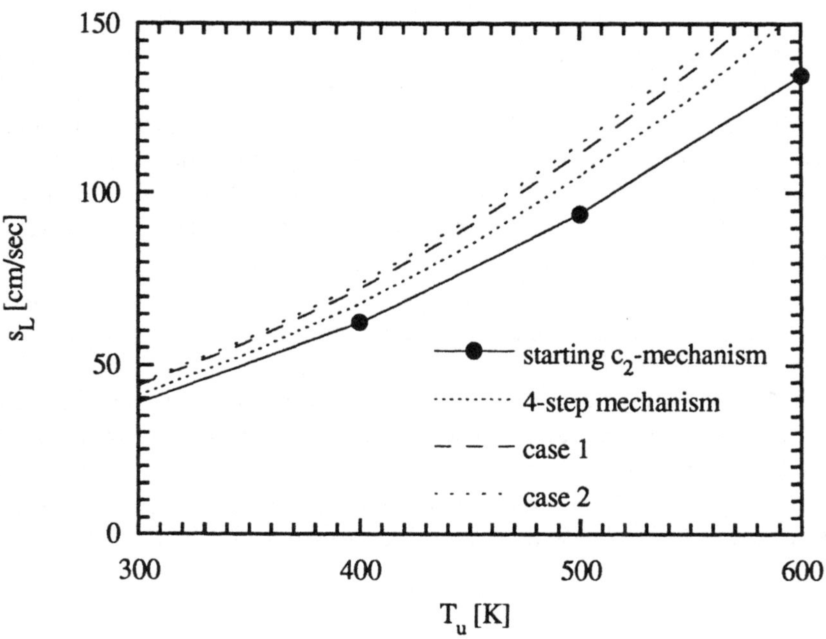

Fig. 5.18. Burning velocities of atmospheric, stoichiometric methane-air flames with the starting C_2-mechanism, the reduced 4-step mechanism and two cases of former simplification. Case 1: partial equilibrium of reaction 3. Case 2: partial equilibrium for reactions 2 and 3.

A comparison of cases 1 and 2 with the C_2-starting and the corresponding 4-step mechanism is shown in Fig. 5.11. Here the two truncated cases do not differ very much. They predict, however, burning velocities that fortuously are in better agreement with experimental values shown in Fig. 5.1 than the starting mechanism used here.

It can be seen from Figs. 5.12–5.14 that the simplest 4-step mechanism (case 2) predicts also the flame structure for a stoichiometric atmospheric-pressure flame accurately. The good results for this case are perhaps fortuous but will facilitate future asymptotic analysis.

In Figs. 5.15 and 5.16 burning velocities of stoichiometric flames at $T_u =$ 298 K are plotted as a function of pressure for the different cases considered above. It is seen that for the C_1-mechanism shown in Fig. 5.12, cases 1 and 2 perform well, but that for the C_2-mechanism the more complex formulation of case 1 rather than case 2 should be retained at higher pressures.

Finally, Figs. 5.17 and 5.18 show the burning velocities of stoichiometric, atmospheric-pressure flames for different unburnt gas temperatures. It can be seen that the relative error of the calculated burning velocities increases in all cases for C_1 and C_2-mechanisms with increasing temperature.

5.4 Conclusions

A 4-step mechanism has been derived for premixed methane flames, and validated for pressures between 1 and 40 atm and preheat temperatures up to 600 K.

It has been found that the influence of the C_2-chain is low for stoichiometric methane flames. For rich methane flames the chain propagating effect of reaction 36 introducing the C_2-chain has to be considered. Further simplification of the reduced mechanism by introducing the partial equilibrium approximation of reaction 3 for calculating the OH-concentration and of reaction 2 for calculating the O-concentration leads to a simple algebraic formulation of the 4-step mechanism, that can be used for further asymptotic analysis.

It could be shown, that the comparison of calculated burning velocities with experimental data is saticfactory for all stages of reducing the chemical mechanism.

The flame structure calculated with the 4-step mechanism and the simple 4-step mechanism has been found to be accurate in comparison with calculations using the starting mechanism.

Further reduction of the chemical mechanism by introducing the steady state assumption for H leads to unacceptabel errors and can not be recommended.

References

[5.1] Peters, N., "Numerical and asymptotic analysis of systematically reduced reaction schemes for hydrocarbon flames", Numerical Simulation of Combustion Phenomena, Lecture Notes in Physics **241**, pp. 90–109,1985.

[5.2] Paczko, G., Lefdal, P. M., Peters, N., "Reduced reaction schemes for methane, methanol and propane flames", Twenty-First Symposium (International) on Combustion, pp. 739–748, The Combustion Institute, 1988.

[5.3] Peters, N., Kee, R. J., "The computation of stretched laminar methane-air diffusion flames using a reduced four-step mechanism ", Comb. and Flame, **68**, pp. 17–30, 1987.

[5.4] Smooke, M. D., "Reduced Kinetic Mechanisms and Asymptotic Approximations for Methane-Air Flames", Springer 1991.

[5.5] Peters, N. and Williams, F. A., "The Asymptotic Structure of Stoichiometric Methane Air Flames", Comb. and Flame, **68**, p. 185, 1987.

[5.6] Law, C. K., "A compilation of Recent Experimental Data of Premixed Laminar Flames", Chapter 2 of this book.

[5.7] Warnatz, J., Eighteenth Symposium (International) on Combustion, pp. 369–384, The Combustion Institute, 1981.

[5.8] Peters, N., "Reducing Mechanisms", in: Reduced Kinetic Mechanisms and Asymptotic Approximations for Methane-Air Flames, (M. D. Smooke, Ed.), Springer 1991

6. Premixed Ethylene/Air and Ethane/Air Flames: Reduced Mechanisms Based on Inner Iteration

W. Wang
Department of Engineering, University of Cambridge, Trumpington Street, Cambridge CB2 1PZ, England

B. Rogg
Department of Engineering, University of Cambridge, Trumpington Street, Cambridge CB2 1PZ, England

6.1 Introduction

Various experimental and numerical studies of structures and burning velocities of laminar, premixed ethylene-air and ethane-air flames have been performed. Examples are the early experimental studies by Scholte and Vaags [6.1] and by Gibbs and Calcote [6.2], and the numerical study by Warnatz [6.3]. The results of recent experimental investigations are reported by Law in Chap. 2 of the present book [6.4]. These references should be consulted for additional literature on the subject.

In the present chapter, for ethylene-air and ethane-air flames, first we derive from the complete mechanism presented in Chap. 1 so-called "short" detailed mechanisms of elementary reactions. The short mechanisms comprise substantially fewer chemical species and elementary kinetic steps than the complete mechanism but yet allow realistic predictions of flame structures and burning velocities over a wide range of stoichiometries and pressures. Second, using systematical methods, we derive four-step reduced kinetic mechanisms for ethylene-air and ethane-air flames that from a computational point of view are more economical to use than the respective short mechanisms but yet give results for both flame structure and burning velocity that are in excellent to very good agreement with those predicted on the basis of the complete and short mechanisms. Furthermore, the reduced mechanisms derived herein are ideal for future research involving flame-structure analyses with asymptotic methods.

In the numerical implementation of the systematically reduced kinetic mechanisms we make use of the so-called "inner iteration". Although inner iteration has been used by various researchers in the area of reduced kinetic mechanisms, it has never been documented. Therefore, in the present chapter inner iteration is discussed in detail and is illustrated for the ethylene and ethane flames computed herein.

The structure of the remainder of this chapter is as follows. In Sect. 2 we specify and derive, respectively, the detailed mechanisms of elementary reactions used as the starting point for the derivation of the four-step mechanisms, subsequently termed the "short mechanisms", and the reduced mechanisms themselves. In Sect. 3 we describe "inner iteration". Section 4 gives a short overview over the numerical method used to solve the governing equations, and in Sect. 5 results validating the short and the reduced kinetic mechanisms are presented and discussed. A summary is given in Sect. 6.

6.2 Chemistry Models

6.2.1 The Short Mechanisms

The complete kinetic mechanism given in Chap. 1, which in this chapter subsequently is referred to as the "full mechanism", was taken as the point of departure for the derivation of a so-called "short mechanism" for both the ethylene and the ethane flame. Herein the terminology "short mechanism" is used for a detailed mechanism of elementary reactions that contains as few as possible elementary steps among as few as possible chemical species but yet allows an accurate representation of the flame structure and hence the burning velocity. In the literature various synonyms exist for "short mechanism" such as "starting mechanism" or "skeletal mechanism".

From detailed numerical sensitivity analysis similar to the one employed in [6.5], it was found that a short mechanism for the ethylene flame comprises the following elementary steps of the full mechanism, viz., 1f, 1b, 2f, 2b, 3f, 3b, 4f, 4b, 5f, 5b, 7, 15, 16, 18f, 18b, 21, 24f, 27, 31, 33b, 34, 35, 36, 39, 40f, 45, 49, 51f, 52f, 53, 54f, 56f, 58f, 58b and 59; a short mechanism for ethane flames was identified to comprise steps 1f, 1b, 2f, 2b, 3f, 3b, 4f, 4b, 5f, 7, 9, 15, 18f, 18b, 21, 24f, 27, 31, 33b, 34, 35, 36, 38f, 40f, 45, 49, 51f, 54f, 56f, 58f, 58b, 59 and 61. It will be seen in the figures presented and discussed below in Sect. 5 of this chapter that for a wide range of stoichiometries and pressures both short mechanisms are able to represent flame structures and burning velocities with an accuracy that is comparable to the accuracy achieved by means of the full mechanism.

6.2.2 Four-Step Kinetic Mechanisms and Global Reaction Rates

6.2.2.1 Four-Step Kinetic Mechanisms

The short mechanism of both the ethylene and the ethane flame comprise 19 reacting species plus inert nitrogen. Therefore, with 4 chemical elements and with 12 species assumed in steady state, four-step kinetic mechanisms exist for both flames. Detailed numerical analysis of the flame structure under various conditions shows that for both the ethylene and the ethane flame sensible choices for the steady-state species are OH, O, HO_2, CHO, CH_2O, CH_4, CH_3, CH_2, C_2H_5, C_2H_3 and C_2H_2. In addition, analysis has shown that for the ethylene flame ethane, C_2H_6, and for the ethane flame ethylene, C_2H_4, closely maintains a steady-state. With steady-state assumptions for these 12 species, using standard methods the four-step mechanisms

$$\text{(I)} \qquad C_2H_{2n} + O_2 \rightleftharpoons 2CO + n\,H_2\,,$$
$$\text{(II)} \qquad CO + H_2O \rightleftharpoons CO_2 + H_2\,,$$
$$\text{(III)} \qquad 2H + M \rightarrow H_2 + M\,,$$
$$\text{(IV)} \qquad O_2 + 3H_2 \rightleftharpoons 2H_2O + 2H$$

can be derived, where $n = 2$ for the ethylene flame and $n = 3$ for the ethane flame. In this mechanism, step 1 is the overall fuel-consumption step, and step II is the overall CO consumption step which neither creates nor destroys reaction intermediaries. Step III represents an overall recombination step, and step IV is an overall radical-production, oxygen-consumption step. From reactions (I)–(IV) it is seen that the mass rates of production of the seven species appearing explicitly in the reduced mechanism can be expressed in terms of the rates of the four global steps as

$$w_{C_2H_{2n}} = -\omega_I\,, \tag{6.1.a}$$
$$w_{O_2} = -\omega_I - \omega_{IV} \tag{6.1.b}$$
$$w_{CO} = 2\,\omega_I - \omega_{II}\,, \tag{6.1.c}$$
$$w_{H_2} = n\,\omega_I + \omega_{II} + \omega_{III} - 3\,\omega_{IV}\,, \tag{6.1.d}$$
$$w_{H_2O} = -\omega_{II} + 2\,\omega_{IV}\,, \tag{6.1.e}$$
$$w_{CO_2} = \omega_{II}\,, \tag{6.1.f}$$
$$w_H = -2\,\omega_{III} + 2\,\omega_{IV}\,. \tag{6.1.g}$$

Detailed numerical analysis has shown that it is advantageous to express the global rates of the four-step mechanism in terms of the rates of the elementary steps contained in the respective short mechanism as described in the following subsections.

6.2.2.2 Global Rates for Ethylene Flames

Sensitivity analyses and other numerical analyses have shown that for the ethylene flames the global rates are most advantageously expressed in terms of the rates of the elementary steps as

$$w_I = [w_{52f}] + w_{53} + w_{54f} - w_{58f} + w_{58b}\,, \tag{6.2}$$
$$w_{II} = w_{18f} - w_{18b}\,, \tag{6.3}$$
$$w_{III} = w_{5f} - w_{5b} + w_{15} + [w_{16}] + w_{21} + [w_{34}] + w_{36} + w_{49} - w_{58f} + w_{58b}\,, \tag{6.4}$$
$$w_{IV} = w_{1f} - w_{1b} - w_{33b} - w_{53} + w_{58f} - w_{58b}\,. \tag{6.5}$$

In deriving (6.2)–(6.5) use has been made of the steady-state relationships for the steady-state species identified above. On the r.h.s.'s of (6.2)–(6.5) rates of elementary steps in brackets, such as $[w_{52f}]$ in (6.2), have been neglected *in the computations* based on the reduced mechanism. It is important to note that the decisions to neglect certain rates of elementary steps in the rates of certain global steps have not been made in an arbitrary manner but are the result of a carefully conducted numerical sensitivity analysis which aimed to finely tune the reduced mechanism so as to give best results for flame structure and burning velocity for a wide range of stoichiometries and pressures.

In the elementary rates appearing on the right-hand-sides of (6.2)–(6.5), the concentrations of the steady-state species must be expressed in terms of the concentrations of the species appearing explicitly in the four-step mechanism. Therefore, the global rates are algebraically more or less complex expressions containing rate data of many of the elementary steps of the short mechanism; the degree of complexity of the global rates depends, of course, on the specific assumptions introduced in their derivation.

In the following we assume that the concentrations of the species appearing explicitly in the four-step mechanism, i.e., $[C_2H_4]$, $[H_2]$, $[O_2]$, $[CO]$, $[CO_2]$, $[H_2O]$ and $[H]$, are known by solving numerically the equations governing conservation of overall mass, species mass, momentum and energy for a specific problem in laminar premixed combustion. The concentrations of the steady-state species OH, O, HO_2, CHO, CH_2O, CH_4, CH_3, CH_2, C_2H_5, C_2H_3, C_2H_2 and C_2H_6, which do not appear explicitly in the reduced mechanism but are required to evaluate the global rates according to (6.2)–(6.5), are expressed in terms of the known species concentrations as follows.

The concentration of the OH radical is obtained by assuming partial equilibrium for reaction 3, i.e.,

$$[OH] = \frac{k_{3b}}{k_{3f}} \frac{[H_2O][H]}{[H_2]} . \tag{6.6}$$

With [OH] known from (6.6), the concentrations of the remaining steady-state species are determined using for some of them their steady-state relationship and for others their steady-state relationship with certain elementary rates neglected. The former steady-state relationships are usually referred to as "full steady-state relationships" or "full steady-states", the latter are usually termed "truncated steady-state relationships" or "truncated steady-states". Based on careful numerical analysis, herein full steady states are adopted for O, HO_2, CHO, CH_2O, CH_4, C_2H_5 and C_2H_6, and truncated steady states for the remaining steady-state species. Specifically, in the steady-state relationship for CH_2 the elementary step 27 is truncated, in that for CH_3 steps 39 and 40f, in that for C_2H_2 step 51f, and in that for C_2H_3 step 54f. The respective full and truncated steady-state relationships are

$$\begin{aligned} 0 = {} & k_{1f}[H][O_2] - k_{1b}[O][OH] - k_{2f}[H_2][O] + k_{2b}[H][OH] \\ & + k_{4f}[OH]^2 - k_{4b}[H_2O][O] - k_{35}[CH_3][O] \end{aligned}$$

$$- k_{39}[CH_4][O] - k_{45}[C_2H_2][O] - k_{53}[C_2H_4][O] , \tag{6.7}$$

$$0 = k_{5f}[H][M][O_2] - k_{5b}[HO_2][M] - k_7[H][HO_2] , \tag{6.8}$$

$$0 = k_{21}[CHO][H] + k_{24f}[CHO][M] - k_{31}[CH_2O][OH] , \tag{6.9}$$

$$0 = k_{31}[CH_2O][OH] - k_{35}[CH_3][O] , \tag{6.10}$$

$$0 = k_{33b}[CH_2][H_2] - k_{45}[C_2H_2][O] , \tag{6.11}$$

$$0 = k_{33b}[CH_2][H_2] - k_{34}[CH_3][H] - k_{35}[CH_3][O] - 2k_{36}[CH_3]^2 + k_{53}[C_2H_4][O] + 2k_{56f}[C_2H_5][H] , \tag{6.12}$$

$$0 = k_{34}[CH_3][H] - k_{39}[CH_4][O] - k_{40f}[CH_4][OH] , \tag{6.13}$$

$$0 = k_{45}[C_2H_2][O] - k_{49}[C_2H_3][H] , \tag{6.14}$$

$$0 = k_{49}[C_2H_3][H] + k_{51f}[C_2H_3] - k_{52f}[C_2H_4][H] , \tag{6.15}$$

$$0 = k_{56f}[C_2H_5][H] + k_{58f}[C_2H_5] - k_{58b}[C_2H_4][H] - k_{59}[C_2H_6][H] , \tag{6.16}$$

$$0 = k_{36}[CH_3]^2 - k_{59}[C_2H_6][H] , \tag{6.17}$$

Inspection of (6.7)–(6.17) shows that some of these equations are strongly coupled. However, it is possible to achieve some simplifications in them by suitable algebraic manipulations. The equations resulting from these manipulations are

$$[HO_2] = \frac{k_{5f}[H][M][O_2]}{k_7[H] + k_{5b}[M]} , \tag{6.18.a}$$

$$[C_2H_3] = \frac{k_{52f}[C_2H_4][H]}{k_{49}[H] + k_{51f}} , \tag{6.18.b}$$

$$[C_2H_2] = \frac{k_{49}[C_2H_3][H]}{k_{45}[O]} , \tag{6.18.c}$$

$$[O] = \frac{N}{D} \tag{6.18.d}$$

$$N = k_{1f}[H][O_2] + k_{2b}[H][OH] + k_{4f}[OH]^2$$

$$D = k_{1b}[OH] + k_{2f}[H_2] + k_{4b}[H_2O] + k_{35}[CH_3] + k_{39}[CH_4] + k_{45}[C_2H_2] + k_{53}[C_2H_4] ,$$

$$[CH_2] = \frac{k_{45}[C_2H_2][O]}{k_{33b}[H_2]} , \tag{6.18.e}$$

$$[C_2H_5] = \frac{k_{59}[C_2H_6][H] + k_{58b}[C_2H_4][H]}{k_{56f}[H] + k_{58f}} , \tag{6.18.f}$$

$$[CH_3] = \frac{-B + \sqrt{B^2 + 4AC}}{2A} , \tag{6.18.g}$$

$$A = 2k_{36} ,$$

$$B = k_{34}[H] + k_{35}[O] ,$$

$$C = k_{33b}[CH_2][H_2] + k_{53}[C_2H_4][O] + 2k_{56f}[C_2H_5][H] ,$$

$$[CH_4] = \frac{k_{34}[CH_3][H]}{k_{39}[O] + k_{40f}[OH]} , \tag{6.18.h}$$

$$[C_2H_6] = \frac{k_{36}[CH_3]^2}{k_{59}[H]}, \tag{6.18.i}$$

$$[CH_2O] = \frac{k_{35}[CH_3][O]}{k_{31}[OH]}, \tag{6.18.j}$$

$$[CHO] = \frac{k_{31}[CH_2O][OH]}{k_{21}[H] + k_{24f}[M]}. \tag{6.18.k}$$

Some of the manipulated equations can be solved explicitly for unknown concentrations of steady-state species, some of them are still strongly coupled and must be solved by means of iteration. Such an iteration is sometimes referred to as *inner iteration.* Although inner iteration is often used in the numerical solution of combustion problems with reduced kinetic mechanisms, it has never been documented in the literature. Therefore, we describe inner iteration in detail in Sect. 3.

6.2.2.3 Global Rates for Ethane Flames

Since many of the details in the derivation of the global rates for the ethane flame are similar or even identical to their ethylene-flame counterparts, this subsection is kept rather short. Similarly as for the ethylene flames, sensitivity analysis and other numerical analysis has shown that for the ethane flames the global rates are most advantageously expressed in terms of the rates of the elementary steps as

$$\omega_{I} = -w_{36} + [w_{59}] + w_{61}, \tag{6.19}$$

$$\omega_{II} = w_{18f} - w_{18b}, \tag{6.20}$$

$$\omega_{III} = w_{5f} + w_{15} + w_{21} + w_{34} + w_{49} + w_{56f}, \tag{6.21}$$

$$\omega_{IV} = w_{1f} - w_{1b} + w_{9} - w_{33b} + w_{36} - w_{56f}. \tag{6.22}$$

In deriving (6.19)–(6.22) use has been made of the steady-state relationships for the steady-state species identified above for ethane. Similarly as for the ethylene flame, in (6.19) the brackets around the rate of step 59 indicate that in the calculations with the reduced mechanism this step was neglected in order to finely tune the reduced mechanism to be able to predict reasonable flame structures and burning velocities for a wide range of conditions. Specifically, in global rate I step 59 has been neglected to reduce for rich flames the prediction of the flame velocity and the CO concentration.

In the elementary rates appearing on the right-hand-sides of (6.19)–(6.22), the concentrations of the steady-state species must be expressed in terms of the concentrations of the species appearing explicitly in the four-step mechanism. In the following we assume that the latter concentrations, i.e., $[C_2H_6]$, $[H_2]$, $[O_2]$, $[CO]$, $[CO_2]$, $[H_2O]$ and $[H]$, are known by solving numerically the equations governing conservation of overall mass, species mass, momentum and energy for a specific problem in laminar premixed combustion. The concentrations of the steady-state species OH, O, HO_2, CHO, CH_2O, CH_4, CH_3, CH_2, C_2H_5, C_2H_3, C_2H_2 and C_2H_4, which do not appear explicitly in the reduced mechanism but are required to evaluate the global rates according to

(6.19)–(6.22), are expressed in terms full and truncated steady-state relationships only, i.e., no partial-equilibrium assumptions are made. Based on careful numerical analysis, full steady states are adopted for HO_2, CH_2O, CH_2, CH_3, CH_4, C_2H_2, C_2H_3 and C_2H_4, and truncated steady states for the remaining steady-state species. Specifically, in the steady-state relationship for OH the elementary step 54f is truncated, in that for O step 35, in that for CHO step 24f, and in that for C_2H_5 steps 56f and 59. The respective full and truncated steady-state relationships are

$$\begin{aligned}0 = {} & k_{1f}[H][O_2] - k_{1b}[O][OH] + k_{2f}[H_2][O] - k_{2b}[H][OH] \\ & - k_{3f}[H_2][OH] + k_{3b}[H][H_2O] - 2k_{4f}[OH]^2 + 2k_{4b}[H_2O][O] \\ & - k_{18f}[CO][OH] + k_{18b}[CO_2][H] + k_{27}[CH_2][O_2] \\ & - k_{31}[CH_2O][OH] - k_{40f}[CH_4][OH] - k_{61}[C_2H_6][OH] \,,\end{aligned} \tag{6.23}$$

$$\begin{aligned}0 = {} & k_{1f}[H][O_2] - k_{1b}[O][OH] - k_{2f}[H_2][O] + k_{2b}[H][OH] + k_{4f}[OH]^2 \\ & - k_{4b}[H_2O][O] + k_9[H][HO_2] - k_{45}[C_2H_2][O] \,,\end{aligned} \tag{6.24}$$

$$0 = k_{5f}[H][M][O_2] - k_7[H][HO_2] - k_9[H][HO_2] \,, \tag{6.25}$$

$$0 = k_{21}[CHO][H] - k_{31}[CH_2O][OH] \,, \tag{6.26}$$

$$0 = k_{31}[CH_2O][OH] - k_{35}[CH_3][O] \,, \tag{6.27}$$

$$0 = k_{27}[CH_2][O_2] + k_{33b}[CH_2][H_2] - k_{45}[C_2H_2][O] \,, \tag{6.28}$$

$$\begin{aligned}0 = {} & k_{33b}[CH_2][H_2] - k_{34}[CH_3][H] - k_{35}[CH_3][O] - 2k_{36}[CH_3]^2 \\ & + k_{38f}[CH_4][H] + k_{40f}[CH_4][OH] + 2k_{56f}[C_2H_5][H] \,,\end{aligned} \tag{6.29}$$

$$0 = k_{34}[CH_3][H] - k_{38f}[CH_4][H] - k_{40f}[CH_4][OH] \,, \tag{6.30}$$

$$0 = k_{45}[C_2H_2][O] - k_{49}[C_2H_3][H] - k_{51f}[C_2H_3] \,, \tag{6.31}$$

$$0 = k_{49}[C_2H_3][H] + k_{51f}[C_2H_3] - k_{54f}[C_2H_4][OH] \,, \tag{6.32}$$

$$0 = k_{54f}[C_2H_4][OH] - k_{58f}[C_2H_5] + k_{58b}[C_2H_4][H] \,, \tag{6.33}$$

$$0 = k_{58f}[C_2H_5] - k_{58b}[C_2H_4][H] - k_{61}[C_2H_6][OH] \,, \tag{6.34}$$

Suitable algebraic manipulations of (6.23)–(6.34) yield

$$[HO_2] = \frac{k_{5f}[M][O_2]}{k_7 + k_9} \,, \tag{6.35.a}$$

$$[C_2H_4] = \frac{k_{61}[C_2H_6]}{k_{54f}} \,, \tag{6.35.b}$$

$$[C_2H_5] = \frac{k_{61}[C_2H_6](k_{54f}[OH] + k_{58b}[H])}{k_{54f}k_{58f}} \,, \tag{6.35.c}$$

$$[C_2H_3] = \frac{k_{54f}[C_2H_4][OH]}{k_{49}[H] + k_{51f}} \,, \tag{6.35.d}$$

$$[C_2H_2] = \frac{[C_2H_3](k_{49}[H] + k_{51f})}{k_{45}[O]} \,, \tag{6.35.e}$$

$$[CH_2] = \frac{k_{45}[C_2H_2][O]}{k_{27}[O_2] + k_{33b}[H_2]} \,, \tag{6.35.f}$$

$$[CH_3] = \frac{-B + \sqrt{B^2 + 4AC}}{2A}, \tag{6.35.g}$$
$$A = 2k_{36},$$
$$B = k_{34}[H] + k_{35}[O],$$
$$C = k_{33b}[CH_2][H_2] + k_{38f}[CH_4][H] + k_{40f}[CH_4][OH] + 2k_{56f}[C_2H_5][H],$$
$$[CH_4] = \frac{k_{34}[CH_3][H]}{k_{38f}[H] + k_{40f}[OH]}, \tag{6.35.h}$$
$$[CH_2O] = \frac{k_{35}[CH_3][O]}{k_{31}[OH]}, \tag{6.35.i}$$
$$[OH] = \frac{N}{D}, \tag{6.35.j}$$
$$N = k_{1f}[H][O_2] + k_{2f}[H_2][O] + k_{3b}[H][H_2O] + 2k_{4b}[H_2O][O] + k_{18b}[CO_2][H] + k_{27}[O_2][CH_2],$$
$$D = k_{1b}[O] + k_{2b}[H] + k_{3f}[H_2] + 2k_{4f}[OH] + k_{18f}[CO] + k_{31}[CH_2O] + k_{40f}[CH_4] + k_{61}[C_2H_6],$$
$$[O] = \frac{k_{1f}[H][O_2] + k_{2b}[H][OH] + k_{4f}[OH]^2 + k_9[H][HO_2]}{k_{1b}[OH] + k_{2f}[H_2] + k_{4b}[H_2O] + k_{45}[C_2H_2]}, \tag{6.35.k}$$
$$[CHO] = \frac{k_{31}[CH_2O][OH]}{k_{21}[H]}. \tag{6.35.l}$$

As for the ethylene flame, some of the manipulated equations can be solved explicitly for unknown concentrations of steady-state species, some of them have to be solved by inner iteration. Inner iteration is described in detail in the next section.

6.3 Inner Iteration

In this section we discuss the so-called "inner iteration". The terminology "inner iteration" refers to an iterative solution of the generally nonlinear, algebraic systems of equations in which the concentrations of the so-called steady-state species are the unknowns.

First, two points are worth mentioning. The first point is that by "steady-state species" we mean all species which do not appear explicitly in the systematically reduced mechanism. Thus, by our definition not only species whose concentrations are determined from full or truncated steady-state relationships are steady-state species, but also species for which the determination of their concentration involves one or more partial equilibrium assumptions.

The second point is the origin of the adjective "inner". Historically this adjective was introduced to distinguish between the iterative solution, say by Newton's method, of the equations governing the conservation of mass, momentum, energy and species mass from the iteration required to solve the afore-mentioned nonlinear, algebraic systems of equations for the concentrations of the steady-state species. Thus, within each iteration step of the Newton method, say, " inner iterations" were performed to obtain the concentrations of the steady-state species that are required to evaluate the global rates of a systematically reduced mechanism. Of course, a method for the numerical integration of the conservation equations of overall mass, species mass, momentum and energy needs not to be iterative; it may well be a non-iterative method such as the PISO method [6.6]. Nevertheless, if an iterative approach is required to solve the nonlinear system of equations governing the concentrations of the steady-state species, this approach is referred to as "inner" iteration.

Second, we shortly discuss under which conditions inner iteration is important. For instance, for the flames considered in this chapter it has been found that for lean to stoichiometric flames inner iteration involving steady-state relationships may be replaced by the simpler algebraic equations that arise if some of the steady-state relationships are replaced by partial equilibrium assumptions. Specifically, for the ethylene flame, for which only partial equilibrium of reaction 3 has been assumed, in addition partial equilibrium for step 2 could be assumed to replace one of the steady-state relationships. Similarly, for the ethane flame, for which no partial equilibrium has been assumed, partial equilibrium for steps 2 and 3 could be assumed to replace two of the steady-state relationships. For lean to stoichiometric flames results for flame structures and burning velocities obtained with these partial-equilibrium assumptions are in reasonable agreement with results obtained from both the full and the short mechanism. However, for stoichiometric to rich flames it has been found that for ethylene and ethane flames the partial equilibrium assumption for reaction 2 and for reactions 2 and 3, respectively, becomes poor, and that inner iteration on the more complicated algebraic system of equations presented in Sect. 2 of this chapter is required to predict reasonable flame structures and hence burning velocities. Since it is desirable to describe the combustion chemistry in a unique manner for a wide range of conditions rather than describing it differently for different conditions, inner iteration is employed regardless of these conditions, i.e., also under conditions where a less complicated formulation based on one or two additional partial-equilibrium assumptions would give results of comparable quality.

We now describe in detail inner iteration for the premixed ethane flames investigated in the present chapter. Shown in Fig. 6.1 is a flow chart illustrating inner iteration for the ethane flame. In this chart "input" means that the temperature and the concentrations of the species appearing explicitly in the reduced mechanism are considered as known quantities. Prior to entering the inner-iteration cycle, initial values for $[HO_2]$ and $[C_2H_4]$ are obtained from (6.35.a) and (6.35.b), respectively; the initial value for $[OH]$ is obtained from the partial-equilibrium assumption for reaction 3, $H_2 + OH \rightleftharpoons H_2O + H$, that for $[O]$ from partial equilibrium for step 2, $H_2 + O \rightleftharpoons OH + H$, and the initial value for $[CH_4]$ is taken as zero. Then the inner-iteration cycle is entered and the concentrations for $[C_2H_5]$, $[C_2H_3]$, $[C_2H_2]$, $[CH_2]$, $[CH_3]$, $[CH_4]$, $[CH_2O]$,

[OH], and [O] are obtained one after the other by evaluating successively (6.35.c)–(6.35.k). The inner-iteration cycle is repeated until the concentrations of the steady states evaluated in the cycle do not change any more, i.e., until convergence has been achieved. Once the inner iteration has converged, [CHO] is calculated from (6.35.l). It is worthwhile to note that in general convergence of the inner iteration is achieved in as few as 2 to 5 iteration steps so that in practice inner iteration can be replaced by a predictor step followed by one or two corrector steps. Inner iteration for the ethylene flame works similarly.

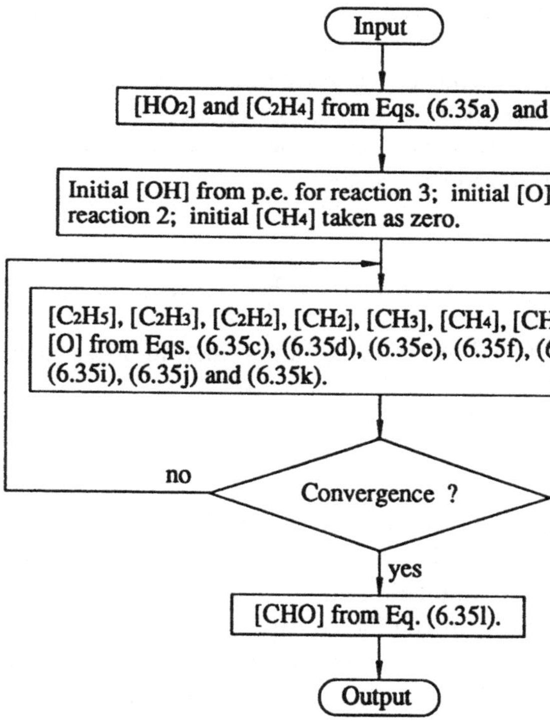

Fig. 6.1. Flow chart illustrating the inner iteration for ethane flames calculated with the reduced mechanism derived in Sect. 6.2.

6.4 Numerical Method

The calculations were performed with the Cambridge laminar-flame computer code RUN-1DL [6.7, 6.8] which has been developed for the numerical solution of premixed burner-stabilized and freely propagating flames, strained premixed, non-premixed and partially premixed flames subject to a variety of boundary conditions, and tubular flames. RUN-1DL employs fully self-adaptive gridding; it solves both transient and steady-state problems in physical space and for diffusion flames, alternatively, in mixture-fraction space; models for thermodynamics and molecular transport are implemented that range from trivially simple to very sophisticated; a model of thermal radiation is also implemented. Chemistry models that can be handled range from overall global one-step reactions over reduced kinetic mechanisms to detailed mechanisms comprising an arbitrary number of chemical species and elementary steps. Furthermore, RUN-1DL is able to simulate two-phase combustion problems, such as droplet and spray combustion. A copy of RUN-1DL is available from the author upon request [6.7, 6.8].

6.5 Validation of the Short and Reduced Mechanisms

In this section we present and discuss results obtained in order to validate the short and the reduced mechanisms derived in Sect. 2.

6.5.1 Ethylene Flames

Results pertaining to ethylene flames are shown in Figs. 6.2 to 6.8. Shown in Fig. 6.2 is the burning velocity as a function of the equivalence ratio for an atmospheric-pressure ethylene-air flame. In all figures symbols represent experimental data referenced in the figure caption, lines represent numerical results obtained herein. The solid line has been computed with the full mechanism of Chap. 1, the dashed line and the dashed-dotted line with the short and the reduced mechanism, respectively, derived herein. It is seen that from very lean to very rich flames the agreement between the burning velocities computed with the three different mechanisms is excellent. The agreement between numerical results and experimental data is excellent for lean to stoichiometric flames and good for rich flames. Specifically, for slightly rich flames the burning velocity is somewhat over-predicted, for very rich flames it is slightly under-predicted.

Shown in Fig. 6.3 is the burning velocity for a stoichiometric ethylene-air flame. It is seen that the results obtained with the full and the short mechanism are in excellent agreement for the entire range of pressure considered. With the reduced mechanism reasonable results are predicted. Specifically,ffor sub-atmospheric pressures the reduced mechanism leads to a slight underprediction of the burning velocity, for pressures between 5 and 10 bars to a slight overprediction.

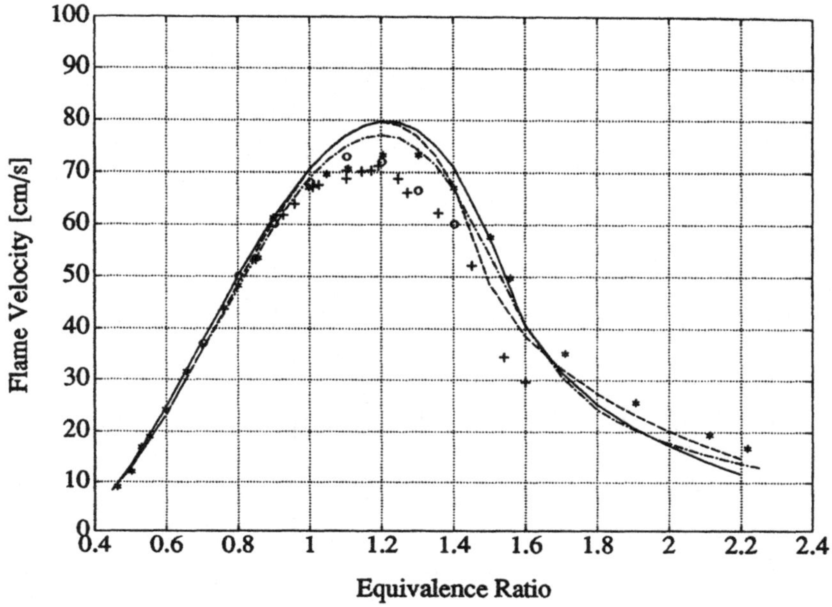

Fig. 6.2. Burning velocity of an atmospheric-pressure ethylene-air flame as a function of equivalence ratio. Solid line: full mechanism; dashed line: short mechanism; dash-dotted line: reduced mechanism; circles: data by Gibbs [6.2]; stars: data by Law [6.4]; pluses: data by Scholte [6.1].

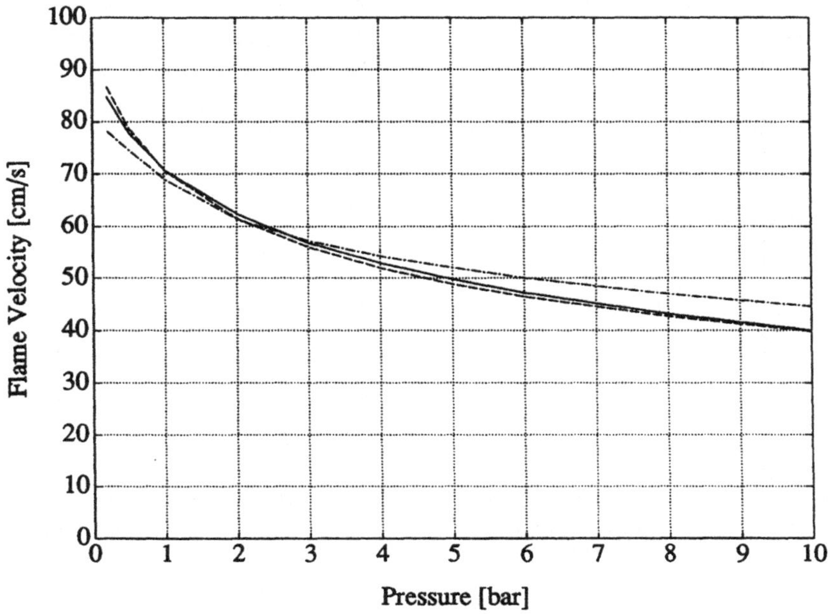

Fig. 6.3. Burning velocity of a stoichiometric ethane-air as a function of pressure. Lines as in Fig. 6.3.

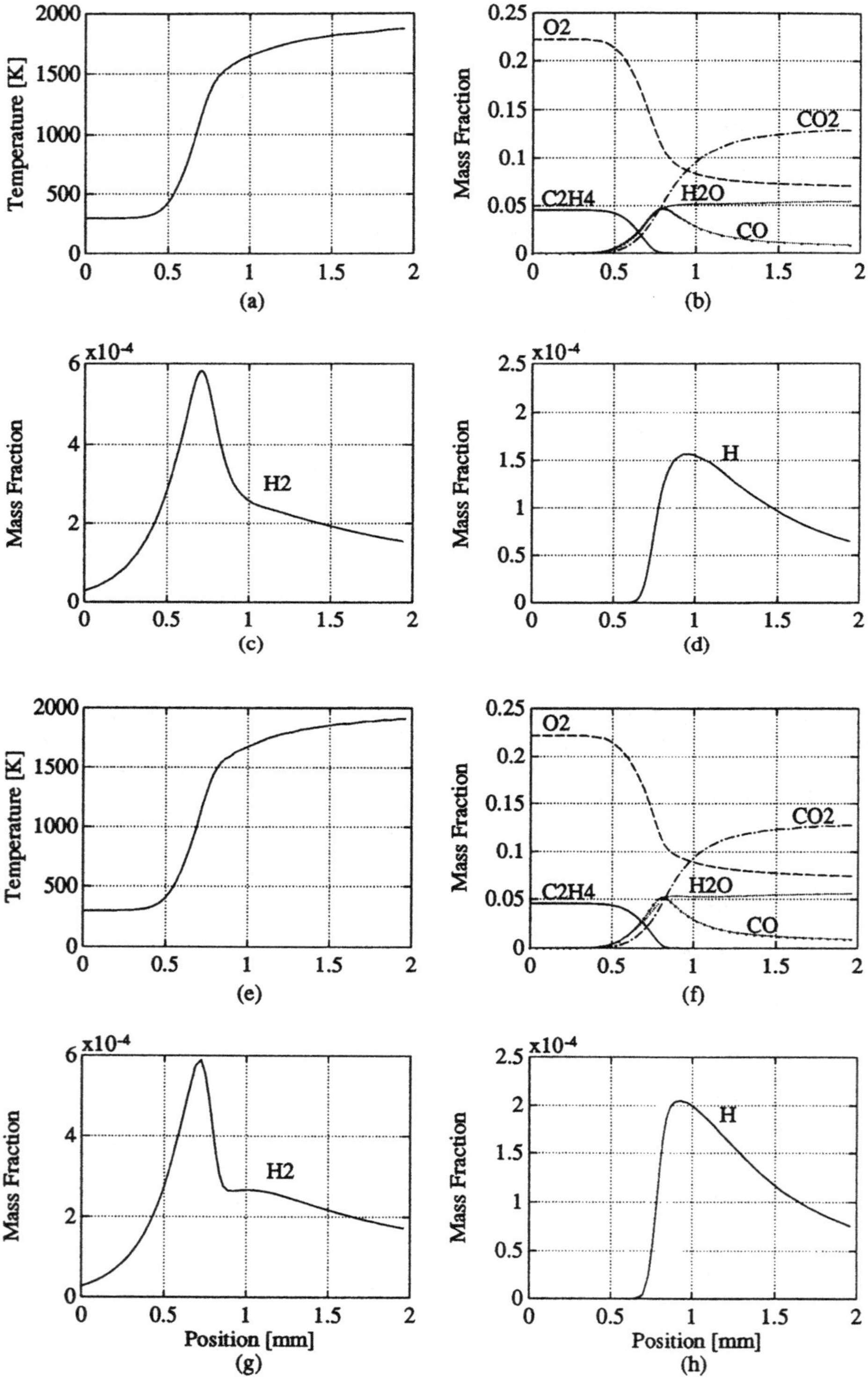

Fig. 6.4. Temperature and species mass-fraction profiles of a lean atmospheric-pressure ethylene-air flame (equivalence ratio 0.7) calculated with the short mechanism (top four pictures) and the reduced mechanism (bottom four pictures).

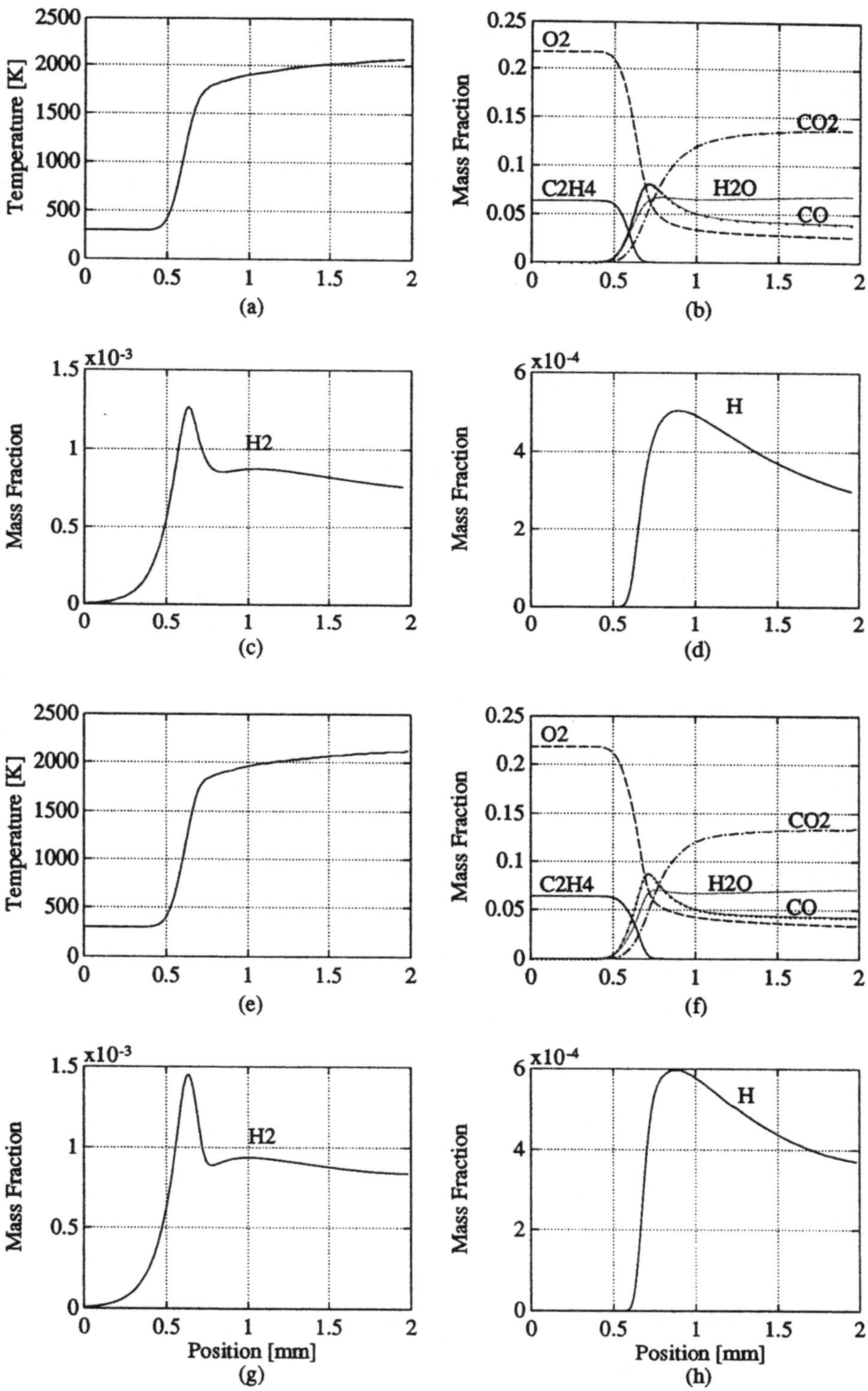

Fig. 6.5. As Fig. 6.4, but for a stoichiometric ethylene-air flame.

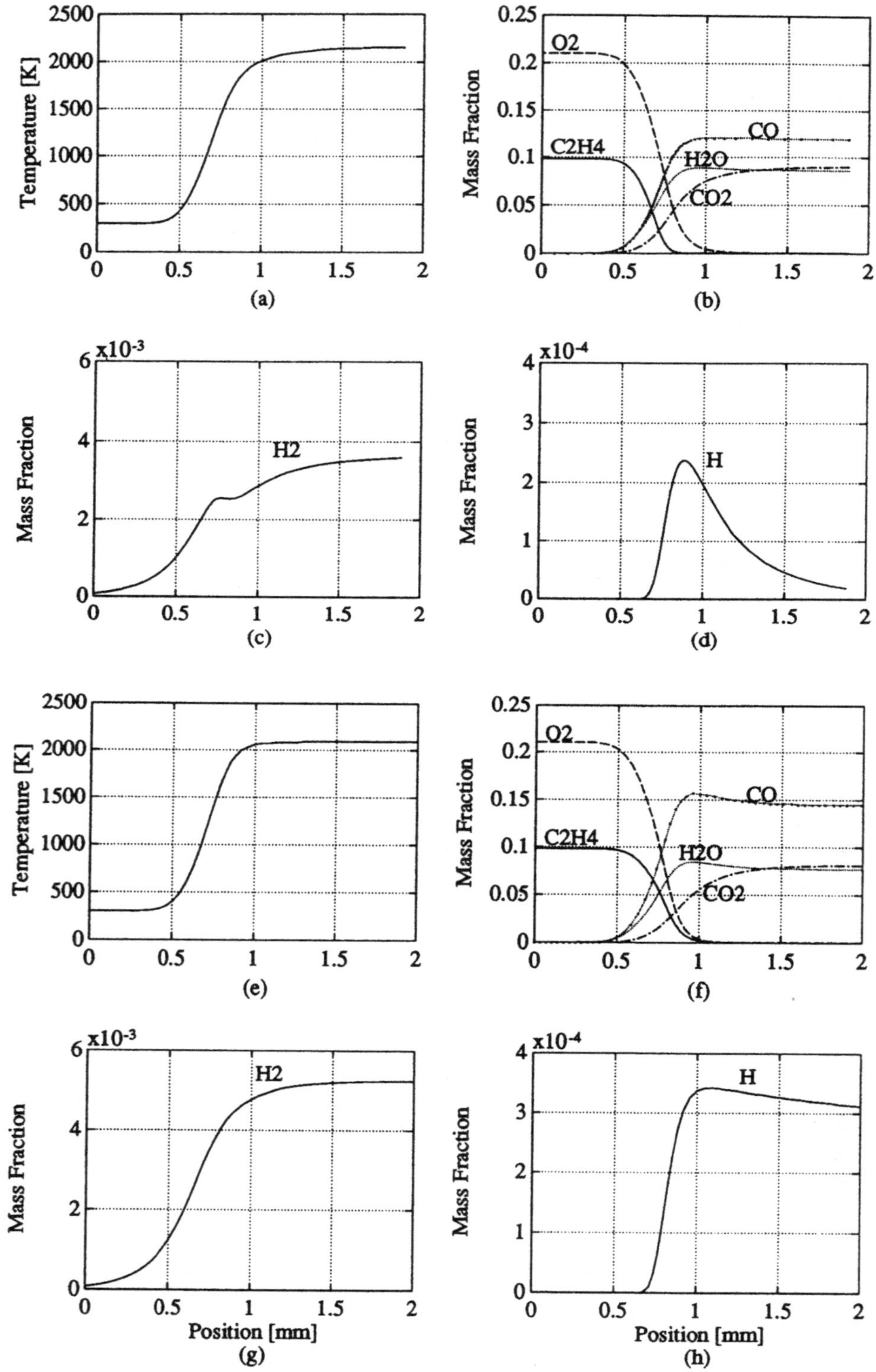

Fig. 6.6. As Fig. 6.4, but for a rich ethylene-air flame (equivalence ratio 1.6).

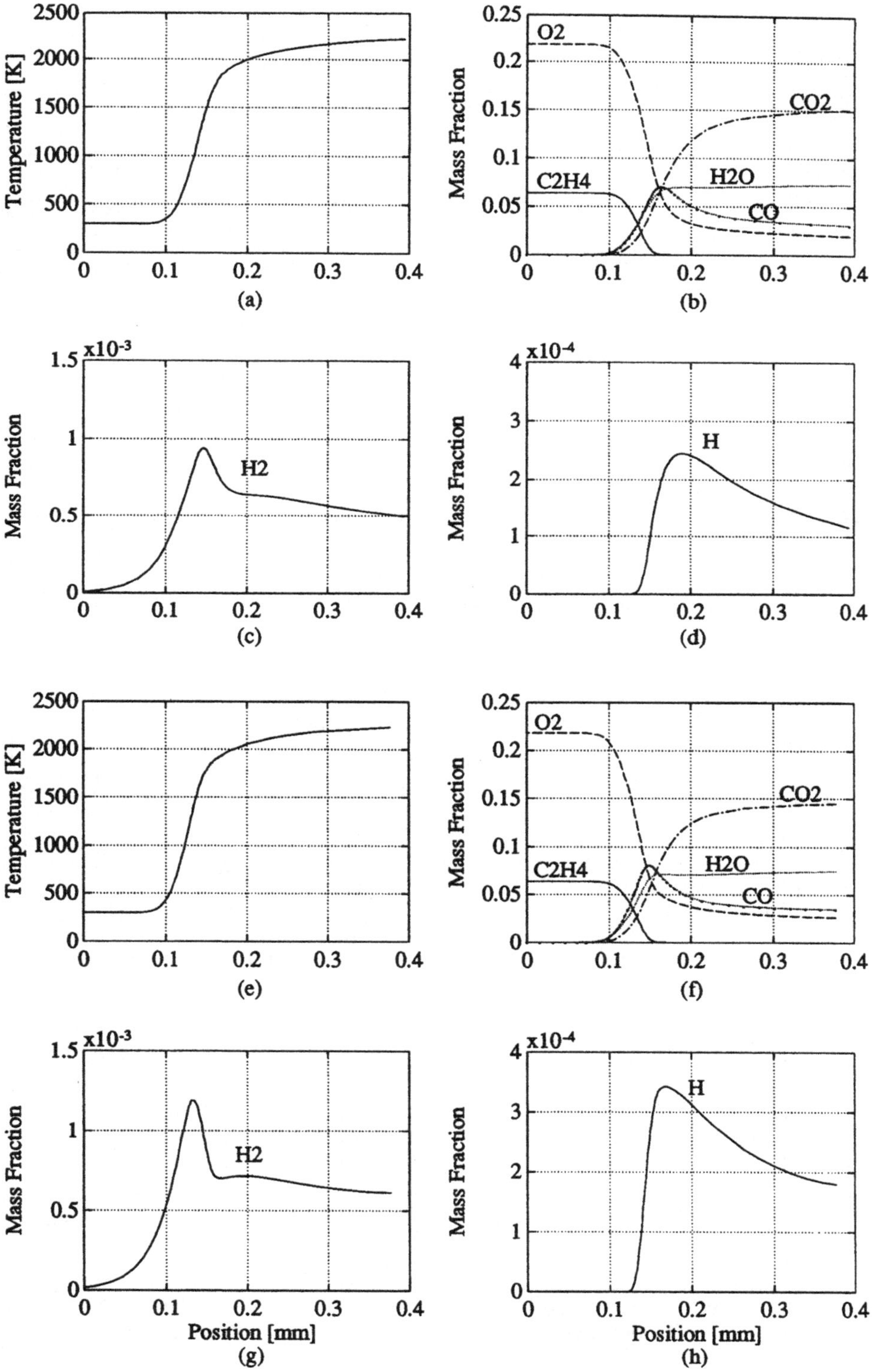

Fig. 6.7. As Fig. 6.4, but for a stoichiometric ethylene-air flame at a pressure of 5 bars.

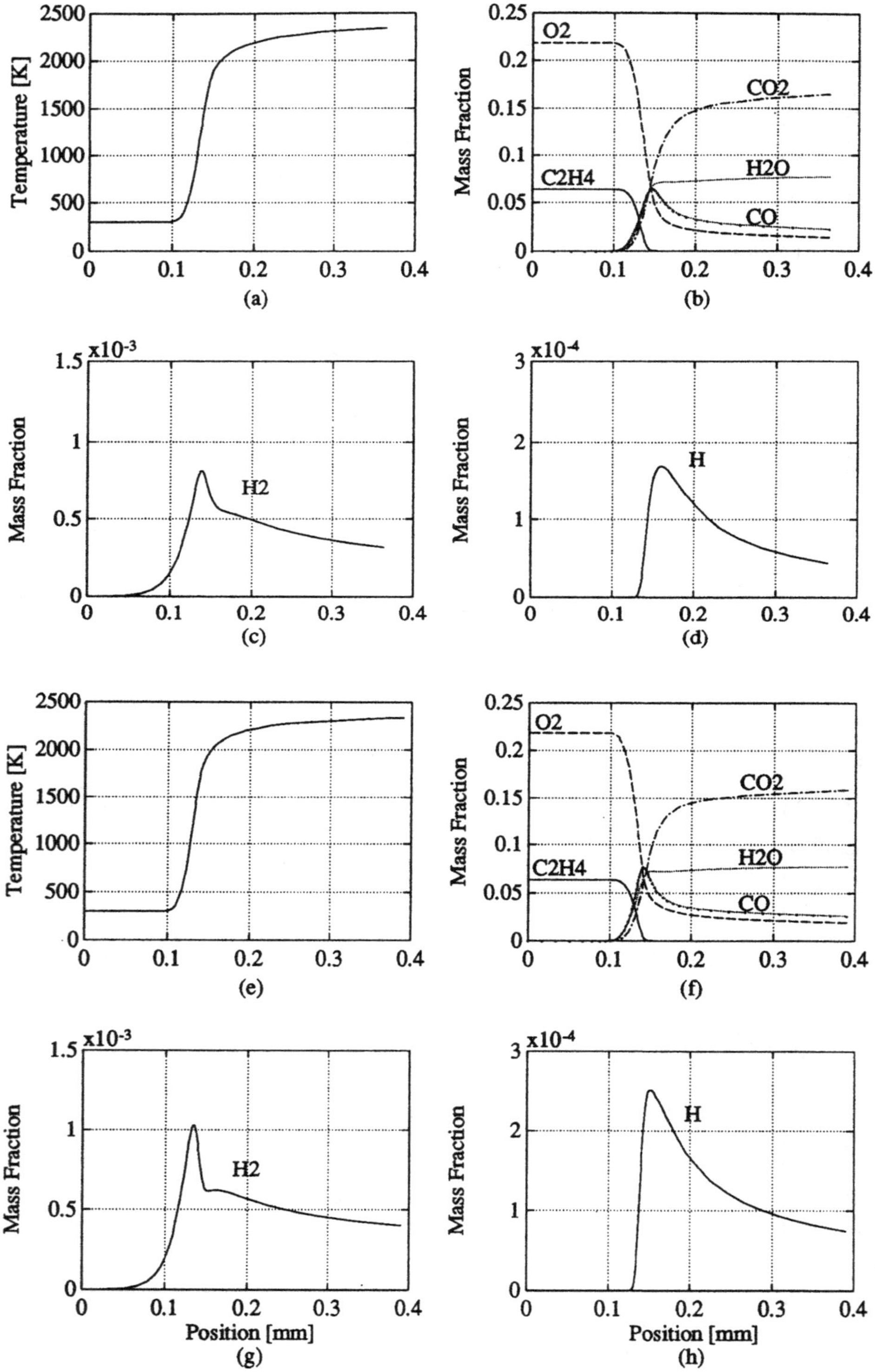

Fig. 6.8. As Fig. 6.4, but for a stoichiometric ethylene-air flame at a pressure of 10 bars.

Shown in Figs. 6.4 to 6.8 are numerically predicted flame structures. In each of these figures the top four pictures show the profiles of temperature and species-mass fractions of the species appearing explicitly in the reduced mechanism as obtained with the short mechanism; the bottom four pictures show the same quantities as obtained with the reduced mechanism. Profiles obtained with the full mechanism are not shown because for the conditions considered herein (stoichiometric flames between 0.2 and 10 bars and lean to rich atmospheric-pressure flames), the profiles obtained with the full and short mechanism are in excellent agreement.

Shown in Figs. 6.4 and 6.5 is the flame structure of a lean flame with a fuel-air equivalence ratio of 0.7 and a stoichiometric flame, respectively. It is seen that for both flames for all profiles, except for that of H-atoms, the agreement is very good; with the reduced mechanism the H-atom mass fraction is slightly overpredicted.

Shown in Fig. 6.6 is the flame structure of a rich flame with a fuel-air equivalence ratio of 1.6. It is seen that generally the flame structures predicted with the short and the reduced mechanism are in good agreement. With the reduced mechanism the maximum values of the CO and H-atom mass fractions are slightly overpredicted.

Shown in Figs. 6.7 and 6.8 is the flame structure of a stoichiometric ethylene-air flame for a pressure of 5 and 10 bars, respectively. It is seen that for both flames the results obtained with the short and the reduced mechanism agree very well. Again, with the reduced mechanism for both flames there is a slight overprediction of the maximum H-atom mass fraction.

6.5.2 Ethane Flames

Although the derivation of the reduced mechanism for the ethane flame is different from the derivation of the reduced mechanism for the ethylene flame, there is considerable similarity in the quality of the flame structures predicted for these two flames. Therefore, the discussion of the results obtained for the ethane flame is kept rather short. The meaning and the symbols in Figs. 6.9 to 6.15 are the same as for the ethylene flame.

Figures 6.9 and 6.10 are the ethane-flame counterparts of Figs. 6.2 and 6.3. By comparing Fig. 6.9 with 6.2 and Fig. 6.10 with 6.3 it is seen that for very rich atmospheric-pressure flames slightly better results are obtained for the ethylene flame than for the ethane flame, but that the quality of the agreement between full-mechanism and reduced-mechanism results is vice versa with respect to the pressure dependence of the burning velocity.

Figures 6.11 to 6.15 are the ethane-flame counterparts of Figs. 6.4 to 6.8. Similarly as for the ethylene flame, it is seen that flame structures obtained with the short mechanism and the reduced mechanism agree very well for lean to stoichiometric atmospheric pressure flames and for stoichiometric flames at 5 and 10 bars. For very rich flames the reduced mechanism slightly underpredicts the maximum H-atom mass fraction in the flame; it also predicts a

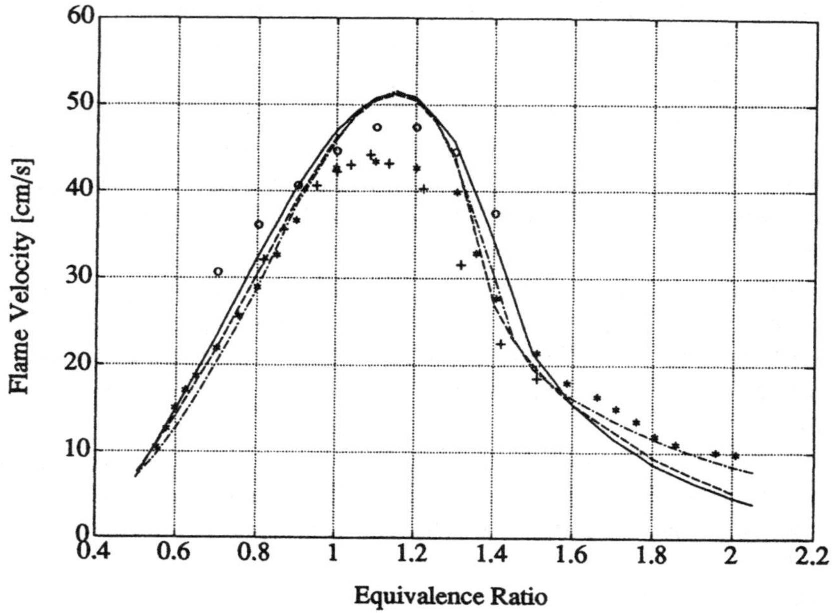

Fig. 6.9. Burning velocity of an atmospheric-pressure ethane-air flame as a function of equivalence ratio. Solid line: full mechanism; dashed line: short mechanism; dash-dotted line: reduced mechanism; circles: data by Gibbs [6.2]; stars: data by Law [6.4]; pluses: data by Scholte [6.1].

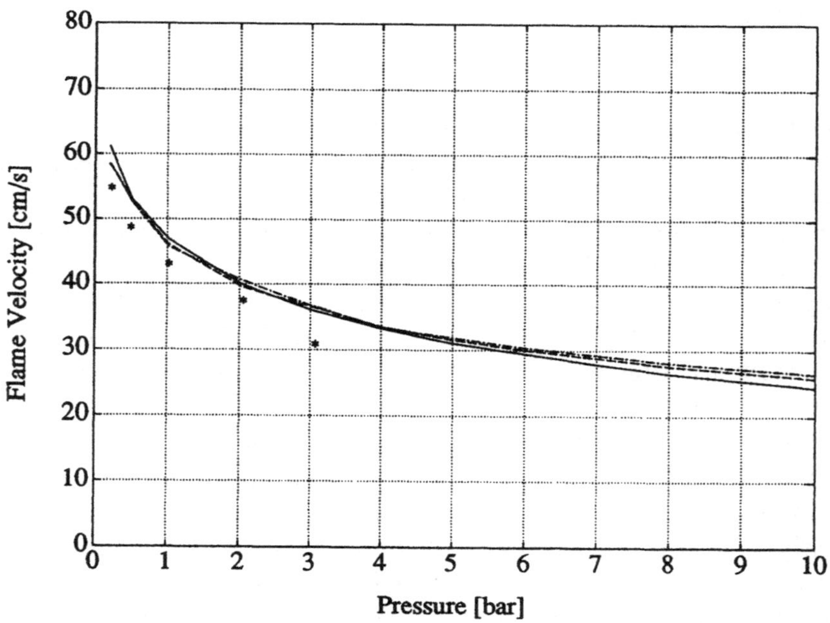

Fig. 6.10. Burning velocity of a stoichiometric ethane-air flame as a function of pressure. Lines and symbol as in Fig. 6.10.

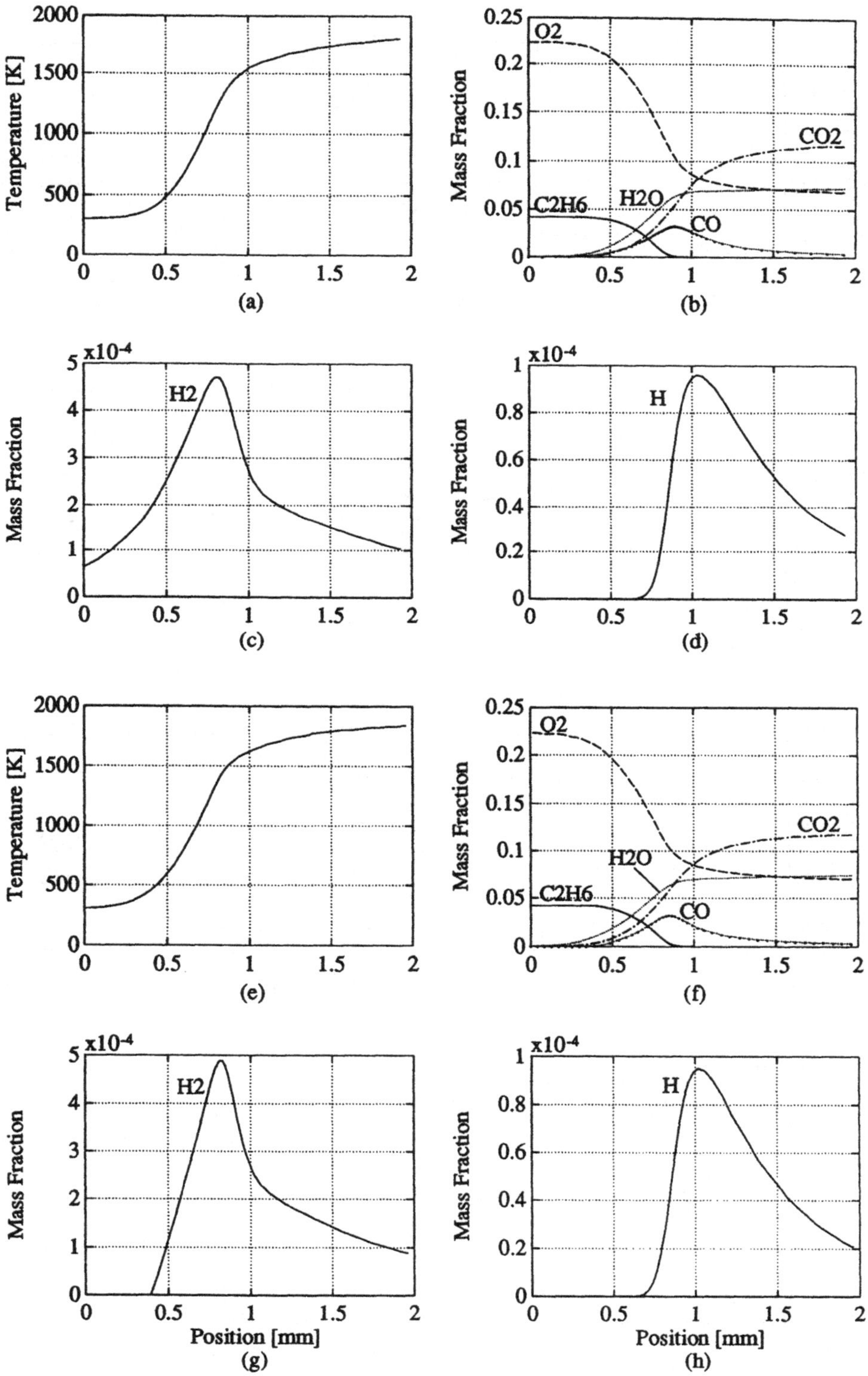

Fig. 6.11. Temperature and species mass-fraction profiles of a lean atmospheric-pressure ethane-air flame (equivalence ratio 0.7) calculated with the short mechanism (top four pictures) and the reduced mechanism (bottom four pictures).

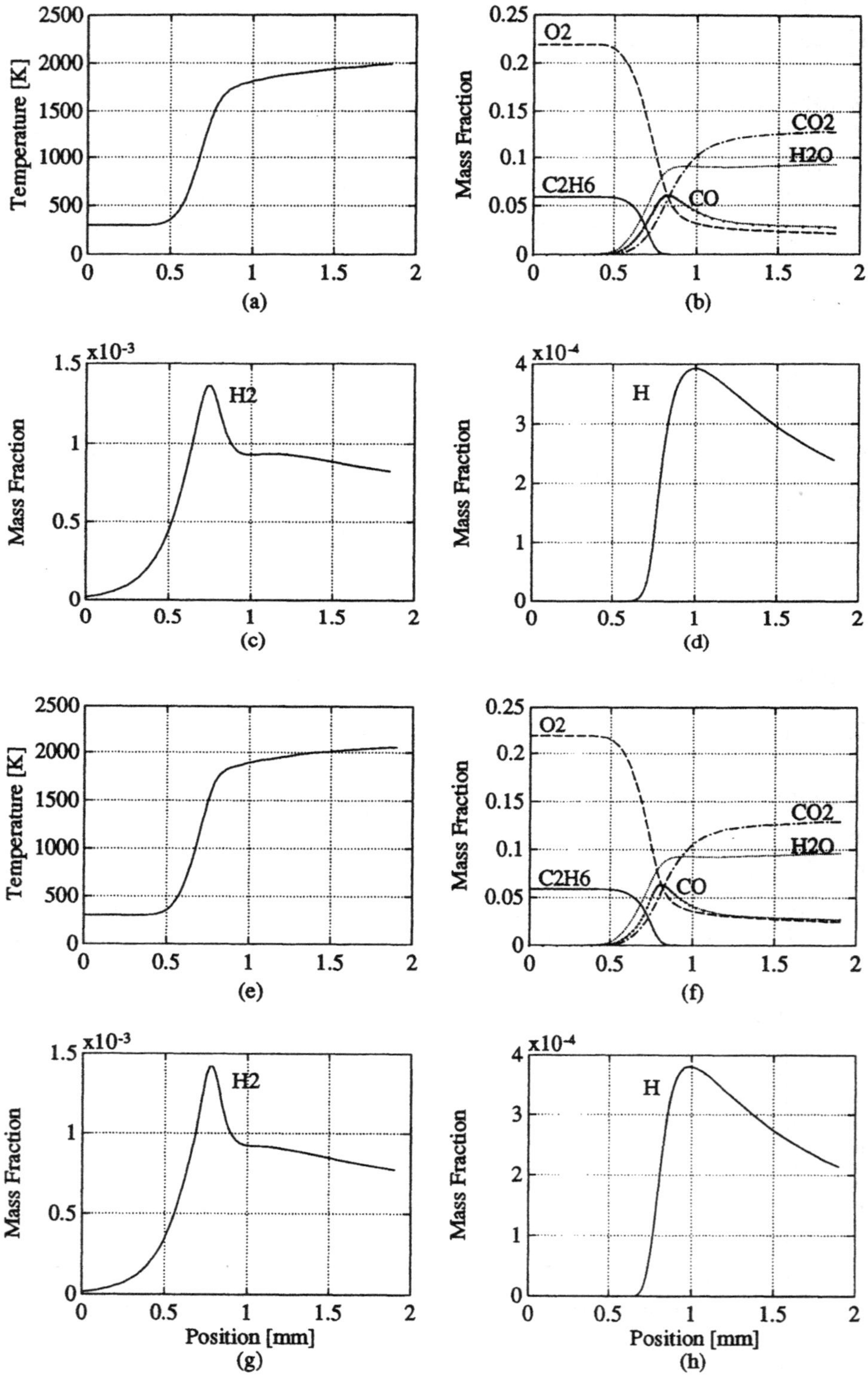

Fig. 6.12. As Fig. 6.11, but for a stoichiometric ethane-air flame.

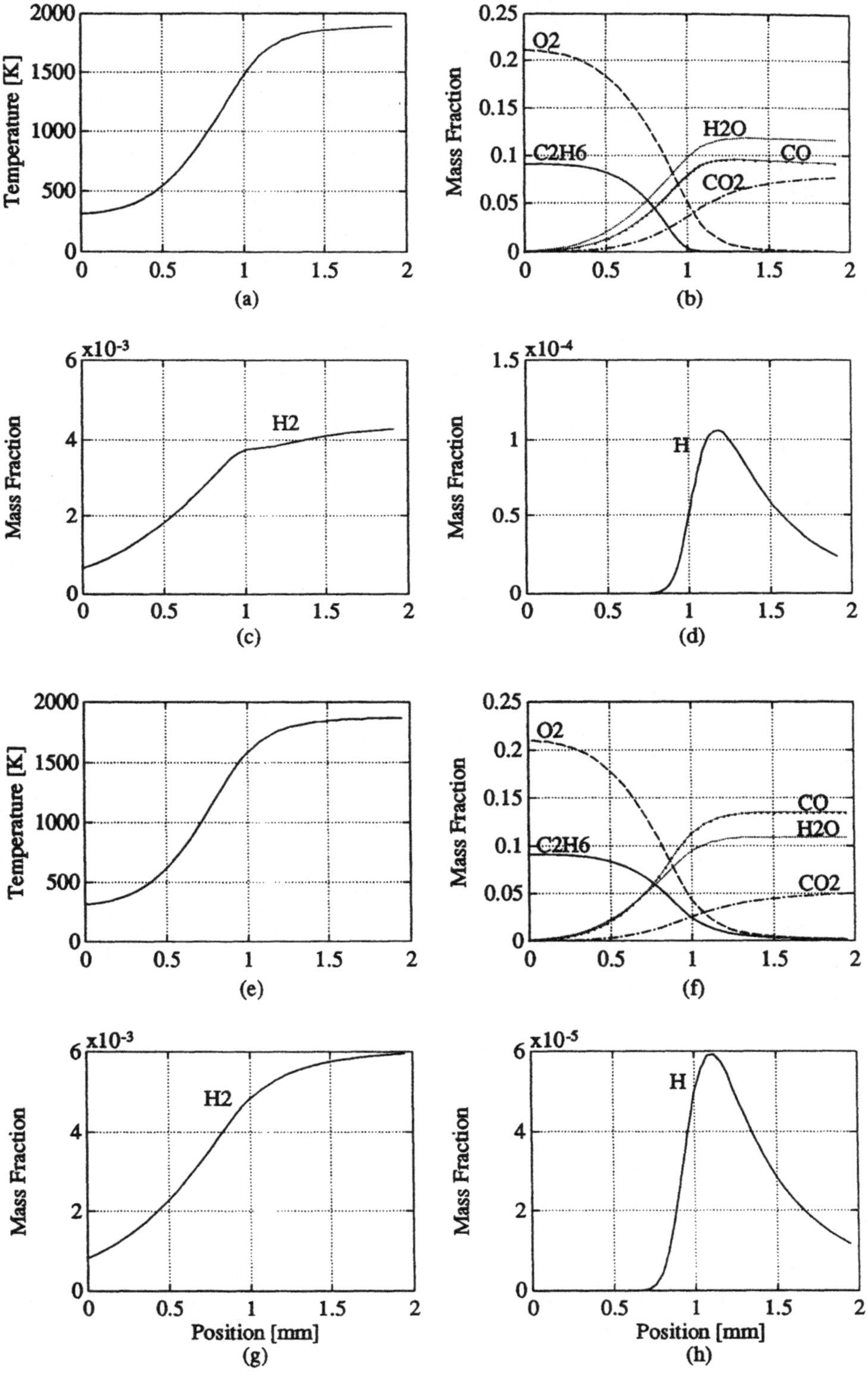

Fig. 6.13. As Fig. 6.11, but for a rich ethane-air flame (equivalence ratio 1.6).

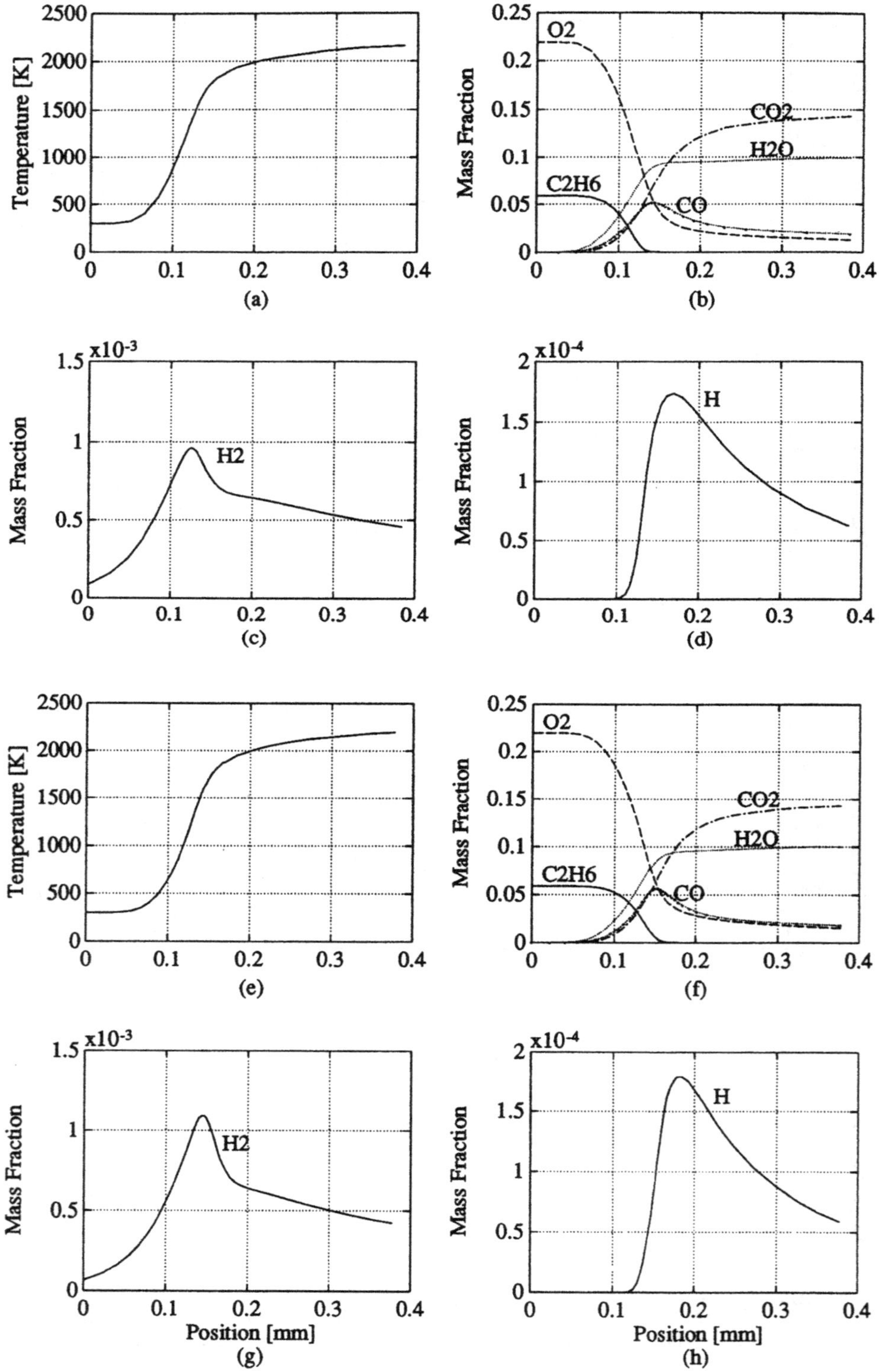

Fig. 6.14. As Fig. 6.11, but for a stoichiometric ethane-air flame at a pressure of 5 bars.

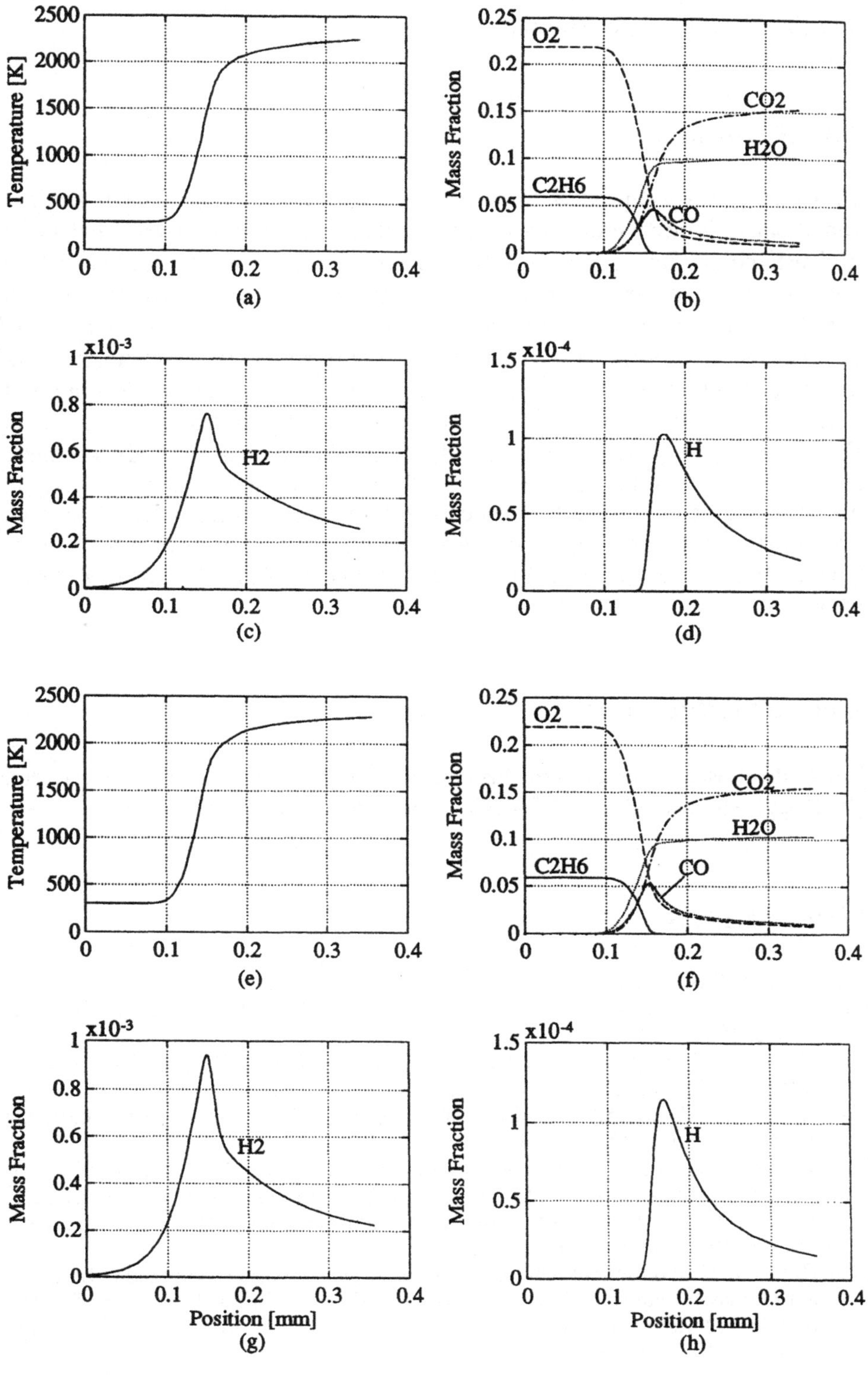

Fig. 6.15. As Fig. 6.11, but for a stoichiometric ethane-air flame at a pressure of 10 bars.

somewhat slower formation of CO_2 and a corresponding faster formation of CO than the short mechanism.

6.6 Summary

In Sect. 2 of this chapter we have derived short detailed mechanisms and systematically reduced four-step kinetic mechanisms for premixed ethylene-air and ethane-air flames, respectively. In Sect. 5 these mechanisms have been validated by comparison of results obtained with them to results obtained with the full mechanism of Chap. 1. For a wide range of stoichiometries and pressures it has been found that results obtained for flame structures and burning velocities on the basis of the short, detailed mechanisms are in excellent agreement with the results obtained on the basis of the full mechanism. For the same range of conditions it has been found that the reduced mechanisms predict results that are in excellent to good agreement with results obtained with the full and the respective short mechanism. In Sect. 3 we have presented a short overview over the numerical method used to integrate the governing equations. In Sect. 4 we have discussed in detail the so-called "inner iteration".

Acknowledgement

Much of the work presented in this chapter was done whilst the second author (B. Rogg) was hosted by CERMICS at Sophia-Antipolis, France.

References

[6.1] Scholte, T. G., Vaags, P. B., Combust. Flame, **3**, p. 495, 1959.
[6.2] Gibbs, G. J., Calcote, H. F., Journal of Chemical and Engineering Data, **4**, p. 226, 1959.
[6.3] Warnatz, J., Eighteenth Symposium (International) on Combustion, p. 369–384, The Combustion Institute, 1981.
[6.4] Law, C. K., A Compilation of Recent Experimental Data of Premixed Laminar Flames. Chap. 2 of this book.
[6.5] Rogg, B., Sensitivity Analysis of Laminar Premixed CH_4-Air Flames Using Full and Reduced Kinetic Mechanisms, in M. D. Smooke, editor, "Reduced Kinetic Mechanisms and Asymptotic Approximations for Methane-Air Flames", pp. 159–192, Springer, Berlin, 1991.
[6.6] Rogg, B., On Numerical Analysis of Two-Dimensional, Axisymmetric, Laminar Jet Diffusion Flames. in Brauner, C.-M. and Schmidt-Lainé, C., editors, "Mathematical Modeling in Combustion and Related Topics", pp. 551–560, Dortrecht, Martinus Nijhoff, 1988.

[6.7] Rogg, B., RUN–1DL: A Computer Program for the Simulation of One-Dimensional Chemically Reacting Flows. Report CUED/A-THERMO/TR39, Cambridge University Engineering Department, April 1991.
[6.8] Rogg, B., RUN-1DL: The Cambridge Laminar-Flamelet Code, Appendix C of this book.

7. Reduced Kinetic Mechanisms for Premixed Acetylene–Air Flames

Fabian Mauss
Institut für Technische Mechanik, RWTH Aachen, W-5100 Aachen, Germany

R. P. Lindstedt
Department of Mechanical Engineering, Imperial College, Exhibition Road, London SW7 2BX, England

7.1 Introduction

It has been shown in a number of publications [7.1], [7.2] that the formation of acetylene is a precondition of the formation of soot in fuel rich hydrocarbon flames. Accordingly, studies of acetylene flames are closely connected with the study of PAH growth and soot formation [7.3].

In this paper we present a reduced 7-step mechanism that predicts burning velocities and species concentrations as accurately as the starting mechanism (Table 1 in Chap. 1) for pressures between 1 atm and 40 atm and initial gas temperatures between 298 K and 600 K and over a wide range of the fuel-air-equivalence ratio. This mechanism will be a basis for future calculations of soot formation with the help of a fast polymerisation assumption [7.4].

Furthermore a simple 4-step mechanism which can be used for future asymptotic analysis of acetylene-air flames will be derived from the 7-step mechanism. In order to derive the four-step mechanism we introduce steady-state assumptions for the O and OH radicals and for the stable species C_2H_4.

Burning velocities of acetylene air flames are, in contrast to other hydrocarbon flames, nearly as high as the burning velocities of hydrogen-air flames [7.5]. Also the concentrations of the major radicals H, O and OH are found to be as high as in hydrogen flames. This is due to the fact, that in contrast to methane flames [7.6] (Chap. 5) and propane flames (Chap. 8) [7.7] the oxidation of acetylene via the main chain

(45) $$C_2H_2 + O = CH_2 + CO$$

(25*f*) $$CH_2 + H = CH + H_2$$

(19) $$CH + O_2 = CHO + O$$

(24*f*) $$CHO + M' = CO + H + M'$$

$$\overline{}$$

$$C_2H_2 + O_2 \rightarrow 2CO + H_2$$

is neutral in terms of radical consumption and production.

Additional chain breaking side paths that are relevant to the oxidation of other C_2-hydrocarbons [7.8] (Chap. 6) are only of minor importance. The main chain breaking reactions are reactions producing HO_2.

This allows for treatment similar to that applied to hydrogen-air flames [7.9], for which explicit formulations for the O and OH radicals were derived.

7.2 A Reduced Seven Step Mechanism

In the following discussion a subset of the mechanism shown in Table 1 of Chap. 1 is used as a starting mechanism. In detail, the starting mechanism includes all reactions of the C_1- and C_2-chain (1–61), not considering C_2H_5 and C_2H_6 (36 and 56–61). In addition we include the consumption of C_3H_3 (62 and 63) and the recombination of C_3H_6 and n-C_3H_7 (71, 73 and 74). The entire starting mechanism consists of 59 reactions and 24 chemical species.

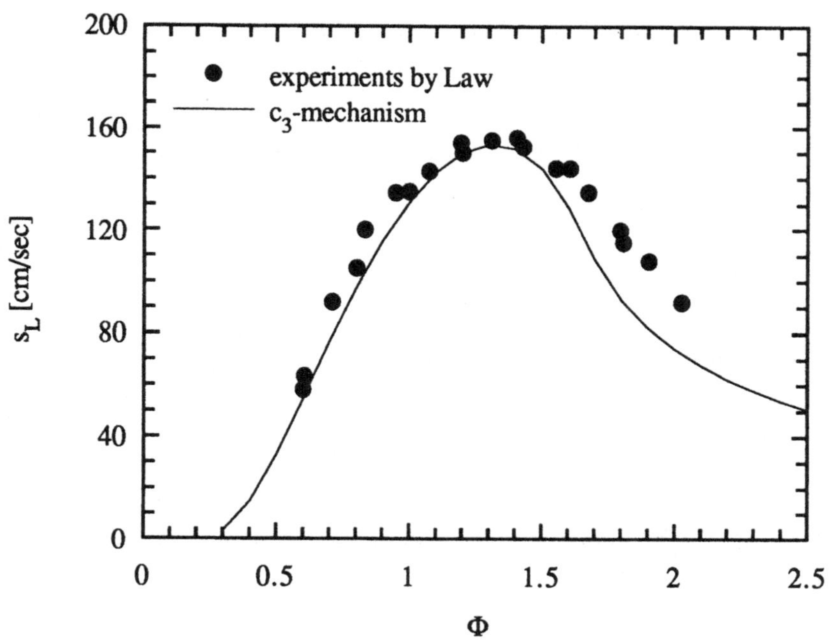

Fig. 7.1. The numerical solution with the starting C_3-mechanism and recent data by Law [7.10] for atmospheric methan-air flames

In Fig. 7.1 burning velocities calculated with the starting mechanism for different fuel-air-equivalence ratios are compared in Fig. 7.1 with recent experimental data compiled by Law [7.10] (Chap. 2). The calculated data are found to always be on the lower side of the experimental data. For rich flames the difference increases up to 10%.

The starting mechanism leads to the following system of balance equations

$$\begin{aligned} L([\mathrm{H}]) = & -w_1 + w_2 + w_3 - w_5 - w_6 - w_7 - w_9 \\ & - w_{13} - 2w_{15} - w_{16} + w_{18} - w_{21} + w_{24} \\ & - w_{25} + 2w_{26} + w_{27} + 2w_{28} - w_{29} + w_{32} \\ & - w_{33} - w_{34} + w_{35} - w_{38} + w_{41} - w_{43} \\ & + w_{44} + w_{46} - w_{49} + w_{51} - w_{52} + w_{53} \\ & + w_{74} \end{aligned}$$

$$\begin{aligned} L([\mathrm{O}]) = & \; w_1 - w_2 + w_4 + w_9 - w_{10} - 2w_{17} + w_{19} \\ & - w_{26} - w_{30} - w_{35} - w_{39} + w_{42} - w_{44} \\ & - w_{45} - w_{46} - w_{53} - w_{63} \end{aligned}$$

$$\begin{aligned} L([\mathrm{OH}]) = & \; w_1 + w_2 - w_3 - 2w_4 + 2w_6 - w_8 + w_{10} \\ & - 2w_{12} + w_{13} - w_{14} - w_{16} - w_{18} - w_{22} \\ & + w_{27} + w_{30} - w_{31} + w_{37} + w_{39} - w_{40} \\ & - w_{47} - w_{54} \end{aligned}$$

$$\begin{aligned} L([\mathrm{HO_2}]) = & \; w_5 - w_6 - w_7 - w_8 - w_9 - w_{10} - 2w_{11} \\ & + w_{14} + w_{23} + w_{50} \end{aligned}$$

$$L([\mathrm{H_2O_2}]) = w_{11} + w_{12} - w_{13} - w_{14}$$

$$\begin{aligned} L([\mathrm{O_2}]) = & -w_1 - w_5 + w_7 + w_8 + w_{10} + w_{11} + w_{17} \\ & - w_{19} - w_{23} - w_{27} - w_{28} - w_{37} - w_{42} \\ & - w_{50} - w_{62} \end{aligned}$$

$$\begin{aligned} L([\mathrm{H_2}]) = & -w_2 - w_3 + w_7 + w_{15} + w_{21} + w_{25} + w_{29} \\ & + w_{33} + w_{38} - w_{41} + w_{49} + w_{52} + w_{55} \end{aligned}$$

$$\begin{aligned} L([\mathrm{H_2O}]) = & \; w_3 + w_4 + w_8 + w_9 + w_{13} + w_{14} + w_{16} \\ & + w_{22} + w_{31} + w_{40} + w_{47} + w_{54} \end{aligned}$$

$$
\begin{aligned}
L([\mathrm{CO}]) &= -w_{18} + w_{20} + w_{21} + w_{22} + w_{23} + w_{24} + w_{26} \\
&\quad + w_{27} + w_{43} + 2w_{44} + w_{45} + w_{53} + w_{63} \\
L([\mathrm{CO_2}]) &= w_{18} - w_{20} + w_{28} \qquad (7.1) \\
L([\mathrm{CH_4}]) &= w_{34} - w_{38} - w_{39} - w_{40} \\
L([\mathrm{CH_3}]) &= -w_{33} - w_{34} - w_{35} - w_{37} + w_{38} + w_{39} + w_{40} \\
&\quad + w_{53} + w_{71} + w_{73} \\
L([\mathrm{CH_2}]) &= -w_{25} - w_{26} - w_{27} - w_{28} + w_{33} + w_{43} + w_{45} \\
L([\mathrm{CH_2O}]) &= -w_{29} - w_{30} - w_{31} - w_{32} + w_{35} + w_{37} + w_{62} \\
L([\mathrm{CH}]) &= -w_{19} - w_{20} + w_{25} - w_{48} \\
L([\mathrm{CHO}]) &= w_{19} + w_{20} - w_{21} - w_{22} - w_{23} - w_{24} + w_{29} \\
&\quad + w_{30} + w_{31} + w_{32} \\
L([\mathrm{C_2H_4}]) &= -w_{52} - w_{53} - w_{54} - w_{55} + w_{73} \\
L([\mathrm{C_2H_3}]) &= -w_{49} - w_{50} - w_{51} + w_{52} + w_{54} + w_{63} + w_{71} \\
L([\mathrm{C_2H_2}]) &= w_{41} - w_{45} - w_{46} - w_{47} - w_{48} + w_{49} + w_{50} \\
&\quad + w_{51} + w_{55} \\
L([\mathrm{C_2H}]) &= -w_{41} - w_{42} + w_{47} \\
L([\mathrm{CHCO}]) &= w_{42} - w_{43} - w_{44} + w_{46} + w_{62} \\
L([n\text{-}\mathrm{C_3H_7}]) &= -w_{73} - w_{74} \\
L([\mathrm{C_3H_6}]) &= -w_{71} + w_{74} \\
L([\mathrm{C_3H_3}]) &= w_{48} - w_{62} - w_{63}
\end{aligned}
$$

$L([X_i])$ is the convective-diffusive operator in the balance equation. The species HO_2, H_2O_2, CH_4, CH_3, CH_2, CH_2O, CH, CHO, C_2H_3, C_2H, CHCO, n-C_3H_7, C_3H_6 and C_3H_3 are assumed to be in steady state with $L([X_i])= 0$. This leads tothe elimination of the following reaction rates: 8, 12, 38, 35, 25, 31, 19, 24, 51, 42, 44, 73, 74 and 63. A simple Gaussian eliminiation of these rates in (7.1) leads to the following 7-step mechanism.

$$\begin{array}{ll} \mathrm{I} & C_2H_2 + O_2 = 2CO + H_2\,, \\ \mathrm{II} & CO + H_2O = CO_2 + H_2\,, \\ \mathrm{III} & H + H + M' = H_2 + M'\,, \\ \mathrm{IV} & O_2 + 3H_2 = 2H_2O + 2H\,, \\ \mathrm{V} & 2H_2 + O = H_2O + 2H\,, \\ \mathrm{VI} & H_2 + OH = H_2O + H\,, \\ \mathrm{VII} & C_2H_4 = C_2H_2 + H_2\,. \end{array} \tag{7.2}$$

The corresponding global reaction rates are:

$$\begin{aligned}
w_{\mathrm{I}} &= -w_{41} + w_{45} + w_{46} + w_{47} + w_{53} + w_{62} \\
w_{\mathrm{II}} &= w_{18} - w_{20} + w_{28} \\
w_{\mathrm{III}} &= w_5 + w_{15} + w_{16} + w_{17} + w_{21} + w_{22} + w_{23} \\
&\quad - w_{32} + w_{34} + w_{48} + w_{49} + w_{50} - w_{71} \\
w_{\mathrm{IV}} &= w_1 + w_6 + w_9 - w_{17} - w_{20} - w_{26} + w_{33} + w_{37} \\
&\quad + w_{43} - w_{46} - w_{48} - w_{53} \\
w_{\mathrm{V}} &= -w_1 + w_2 - w_4 - w_9 + w_{10} + 2w_{17} + w_{20} + 2w_{26} \\
&\quad + w_{27} + w_{28} + w_{30} - 2w_{33} - w_{37} + w_{39} - 2w_{43} \\
&\quad + 2w_{46} + 2w_{48} + 2w_{53} \\
w_{\mathrm{VI}} &= -w_1 - w_2 + w_3 + 2w_4 + w_5 - 3w_6 - w_7 - w_9 \\
&\quad - 2w_{10} + w_{16} + w_{18} + w_{22} + w_{23} - w_{27} - w_{29} \\
&\quad - 2w_{30} - w_{32} - w_{33} - w_{37} - w_{39} + w_{40} + w_{47} \\
&\quad + w_{50} + w_{53} + w_{54} + w_{62} \\
w_{\mathrm{VII}} &= w_{52} + w_{53} + w_{54} + w_{55} + w_{71}
\end{aligned} \tag{7.3}$$

It should be noted here, that the choice of the main chain reactions has no influence on the global reaction rates of the reduced reaction sheme. This is true as long as truncations have not been introduced in the algebraic system.

The elementary reactions found in the global reaction rate of reaction I contain the important attack of C_2H_2 by the O-radical

$$\text{(45)} \qquad C_2H_2 + O \rightarrow CH_2 + CO$$

$$\text{(46)} \qquad C_2H_2 + O \rightarrow CHCO + H$$

and the less important reactions with H, OH and O_2 to C_2H and CHCO

$$\text{(41}b\text{)} \qquad C_2H_2 + H \rightarrow C_2H + H_2$$

$$\text{(47}f\text{)} \qquad C_2H_2 + OH \rightarrow C_2H + H_2O$$

$$\text{(42)} \qquad C_2H + O_2 \rightarrow CHCO + O.$$

There is also a path via C_2H_3 to C_2H_4

$$\text{(51}b\text{)} \qquad C_2H_2 + H + M \rightarrow C_2H_3 + M$$

$$\text{(52}b\text{)} \qquad C_2H_3 + H_2 \rightarrow C_2H_4 + H$$

$$\text{(53)} \qquad C_2H_4 + O \rightarrow CH_3 + CO + H$$

and via C_3H_3 to CHCO and CH_2O

$$\text{(48)} \qquad C_2H_2 + CH \rightarrow C_3H_3$$

$$\text{(62)} \qquad C_3H_3 + O_2 \rightarrow CHCO + CH_2O.$$

The reactions 46 and 53 appear with a minus sign in the global reaction IV and are chain breaking. While the reaction path via reaction 53 is only of minor importance, the chain breaking effect via reaction 46 is reduced by the chain branching effect of reaction 43, the main chain of CHCO that is formed by reaction 46.

The most important chain breaking path leading to HO_2 as an intermediate is the sequence of

$$51b \qquad C_2H_2 + H = C_2H_3$$

$$50 \qquad C_2H_3 + O_2 = C_2H_2 + HO_2$$

$$8 \qquad HO_2 + OH = H_2O + O_2$$

The influence of this chain on the hydrogen-oxygen mechanism will be further discussed in section 7.3 below concerns the Propagyl radical.

There are two other side paths involving C_3-species. The first path

$$48 \qquad C_2H_2 + CH \rightarrow C_3H_3$$

$$63 \qquad C_3H_3 + O \rightarrow C_2H_3 + CO$$

$$51f \qquad C_2H_3 \rightarrow C_2H_2 + H$$

is chain breaking. Propagyl has been found [7.11] to recombine to benzene

$$C_3H_3 + C_3H_3 = c\text{-}C_6H_6.$$

Due to the absence of this reaction from the starting mechanism, the chain breaking effect of reaction 48 is overpredicted. This may explain the disagreement between calculated and measured fuel rich acetylene flames as shown in Fig. 7.1. The second path via

71*b*	$C_2H_3 + CH_3 = C_3H_6$
74*b*	$C_3H_6 + H = n\text{-}C_3H_7$
73	$n\text{-}C_3H_7 = C_2H_4 + CH_3$

is also chain breaking and involves an other path to form C_3H_3 via C_3H_5 and C_3H_4. This latter path is not considered in the paper. The system of algebraic equations for the steady state species would appear at first to involve a set of non linear algebraic equations due to reaction 71b. But the same amount of CH_3 is consumed by reaction 71b as is produced by reaction 73. Accordingly, reaction 71b can be eliminated from the algebraic equation for $[CH_3]$ with the help of the steady state assumption for C_3H_6 and n-C_3H_7. The following equations for the steady state species result from such an analysis:

$$[C_2H] = \frac{[C_2H_2] \cdot ([H]k_{41b} + [OH]k_{47f})}{[H_2]k_{41f} + [O_2]k_{42} + [H_2O]k_{47b}} \tag{7.4}$$

$$\begin{aligned} Z_{CH_2} &= [C_2H_2] \cdot ([O]k_{45} + (([H]k_{41b} + [OH]k_{47f})F_{42} + [O]k_{46})F_{43f}) \\ &\quad + [C_2H_4][O]k_{53}F_{33f} \\ N_{CH_2} &= [H]k_{25f}(1 - F_{25b} - F_{48,a}F_{62}) + [O]k_{26} + [O_2](k_{27} + k_{28}) \\ &\quad + [H_2]k_{33b}(1 - F_{33f}) + [CO]k_{43b}(1 - F_{43f}) \\ [CH_2] &= \frac{Z_{CH_2}}{N_{CH_2}} \end{aligned} \tag{7.5}$$

$$\begin{aligned} Z_{CH_3} &= [C_2H_2] \cdot ([O]k_{45} + (([H]k_{41b} + [OH]k_{47f})F_{42} + [O]k_{46})F_{43f})F_{33b} \\ &\quad + [C_2H_4][O]k_{53} \\ [CH_3] &= \frac{Z_{CH_3}}{N_{CH_3}} \end{aligned} \tag{7.6}$$

$$\begin{aligned} Z_{CH} &= [C_2H_2] \cdot ((([H]k_{41b} + [OH]k_{47f})F_{42} + [O]k_{46})F_{43f} \\ &\quad + [O]k_{45})F_{25f,a} + [C_2H_4][O]k_{53}F_{33f}F_{25f,a} \\ N_{CH} &= [H_2]k_{25b}(1 - F_{25f,a}) + [O_2]k_{19} + [CO_2]k_{20} \\ &\quad + [C_2H_2]k_{48}(1 - F_{62}F_{43f}F_{25f,a}) \\ [CH] &= \frac{Z_{CH}}{N_{CH}} \end{aligned} \tag{7.7}$$

$$\begin{aligned} Z_{\mathrm{CHCO}} &= [\mathrm{C_2H_2}] \cdot \big(([\mathrm{H}]k_{41b} + [\mathrm{OH}]k_{47f})F_{42} + [\mathrm{O}]k_{46} \\ &\qquad + [\mathrm{O}]k_{45}(F_{43b} + F_{25f,b}F_{48,c}F_{62})\big) \\ &\quad + [\mathrm{C_2H_4}][\mathrm{O}]k_{53}F_{33f}(F_{43b} + F_{25f,b}F_{48,c}F_{62}) \\ N_{\mathrm{CHCO}} &= [\mathrm{H}]k_{43f}(1 - F_{43b} - F_{25f,b}F_{48,c}F_{62}) + [\mathrm{O}]k_{44} \\ [\mathrm{CHCO}] &= \frac{Z_{\mathrm{CHCO}}}{N_{\mathrm{CHCO}}} \end{aligned} \tag{7.8}$$

$$\begin{aligned} Z_{\mathrm{C_3H_3}} &= [\mathrm{C_2H_2}] \cdot \big(([\mathrm{H}]k_{41b} + [\mathrm{OH}]k_{47f})\, F_{42} + [\mathrm{O}]k_{46}\big)\, F_{43f} \\ &\qquad + [\mathrm{O}]k_{45} + [\mathrm{C_2H_4}][\mathrm{O}]k_{53}F_{33f})F_{25f,a}F_{48,b} \\ [\mathrm{C_3H_3}] &= \frac{ZC_3H_3}{[\mathrm{O_2}]k_{62}(1 - F_{43f}F_{25f,a}F_{48,b}) + [\mathrm{O}]k_{63}} \end{aligned} \tag{7.9}$$

The following definitions were used in deriving (7.4)–(7.9):

$$\begin{aligned} Z_{25f} &= [\mathrm{H}]k_{25f} + [\mathrm{O}]k_{26} + [\mathrm{O_2}](k_{27} + k_{28}) + [\mathrm{H_2}]k_{33b} + [\mathrm{CO}]k_{43b} \\ F_{25f,a} &= \frac{[\mathrm{H}]k_{25f}}{Z_{25f} - [\mathrm{H_2}]k_{33b}F_{33f} - [\mathrm{CO}]k_{43b}F_{43f}} \\ F_{25f,b} &= \frac{[\mathrm{H}]k_{25f}}{Z_{25f} - [\mathrm{H_2}]k_{33b}F_{33f}} \\ F_{25b} &= \frac{[\mathrm{H_2}]k_{25b}}{[\mathrm{H_2}]k_{25b} + [\mathrm{O_2}]k_{19} + [\mathrm{CO_2}]k_{20} + [\mathrm{C_2H_2}]k_{48}} \\ F_{33f} &= \frac{[\mathrm{H}]k_{33f}}{[\mathrm{H}]k_{33f} + [\mathrm{O}]k_{35} + [\mathrm{O_2}]k_{37}} \\ F_{33b} &= \frac{[\mathrm{H_2}]k_{33b}}{Z_{25f} - [\mathrm{H}]k_{25f}(F_{25b} + F_{48,a}F_{62}) - [\mathrm{CO}]k_{43b}F_{43f}} \\ F_{42} &= \frac{[\mathrm{O_2}]k_{42}}{[\mathrm{H_2}]k_{41f} + [\mathrm{O_2}]k_{42} + [\mathrm{H_2O}]k_{47b}} \\ F_{43f} &= \frac{[\mathrm{H}]k_{43f}}{[\mathrm{H}]k_{43f} + [\mathrm{O}]k_{44}} \\ F_{43b} &= \frac{[\mathrm{CO}]k_{43b}}{Z_{25f} - [\mathrm{H}]k_{25f}F_{25b} - [\mathrm{H_2}]k_{33b}F_{33f}} \\ F_{48,a} &= \frac{[\mathrm{C_2H_2}]k_{48}}{[\mathrm{H_2}]k_{25b} + [\mathrm{O_2}]k_{19} + [\mathrm{CO_2}]k_{20} + [\mathrm{C_2H_2}]k_{48}} \\ F_{48,b} &= \frac{[\mathrm{C_2H_2}]k_{48}}{[\mathrm{H_2}]k_{25b}(1 - F_{25f,a}) + [\mathrm{O_2}]k_{19} + [\mathrm{CO_2}]k_{20} + [\mathrm{C_2H_2}]k_{48}} \\ F_{48,c} &= \frac{[\mathrm{C_2H_2}]k_{48}}{[\mathrm{H_2}]k_{25b}(1 - F_{25f,b}) + [\mathrm{O_2}]k_{19} + [\mathrm{CO_2}]k_{20} + [\mathrm{C_2H_2}]k_{48}} \\ F_{62} &= \frac{[\mathrm{O_2}]k_{62}}{[\mathrm{O_2}]k_{62} + [\mathrm{O}]k_{63}} \end{aligned} \tag{7.10}$$

The numerators of (7.4)–(7.9) contain all possible paths of any non-steady state species toward the respective steady-state species. The denominators represent the brutto consumption of the steady-state species. For example the brutto consumption of CHCO via reaction 43f is given by

$$[\mathrm{CHCO}][\mathrm{H}]k_{43f}(1 - F_{43b} - F_{25f,b}F_{48,c}F_{62}),$$

where F_{43b} gives the amount of CHCO that is recreated directly by the backward reaction 43b

$$\begin{array}{ll}(43f) & \mathrm{CHCO + H = CH_2 + CO} \\ (43b) & \mathrm{CH_2 + CO = CHCO + H,}\end{array}$$

and the product $F_{25f,b}F_{48,c}F_{62}$ gives the amount of CHCO that is recreated via the path

$$\begin{array}{ll}(43f) & \mathrm{CHCO + H = CH_2 + CO} \\ (25f) & \mathrm{CH_2 + H = CH + H_2} \\ (48) & \mathrm{C_2H_2 + CH = C_3H_3} \\ (62) & \mathrm{C_3H_3 + O_2 = CHCO + CH_2O.}\end{array}$$

We note that all factors F in (7.10) satisfy the relationship

$$0 \le F \le 1. \tag{7.11}$$

This feature is particulary useful for the introduction of possible truncations. Steady state species not given by (7.2)–(7.9) can be calculated as follows:

$$\begin{aligned} Z_{\mathrm{C_2H_3}} &= [\mathrm{C_2H_2}][\mathrm{H}][\mathrm{M}]k_{51b} + [\mathrm{C_2H_4}]([\mathrm{H}]k_{52f} + [\mathrm{OH}]k_{54f}) \\ &\quad + [\mathrm{C_3H_3}][\mathrm{O}]k_{63} \\ N_{\mathrm{C_2H_3}} &= [\mathrm{H}]k_{49} + [\mathrm{O_2}]k_{50} + [\mathrm{M}]k_{51f} + [\mathrm{H_2}]k_{52b} \\ &\quad + [\mathrm{H_2O}]k_{54b} + [\mathrm{CH_3}]k_{71b}(1 - F_{71f}) \\ [\mathrm{C_2H_3}] &= \frac{Z_{\mathrm{C_2H_3}}}{N_{\mathrm{C_2H_3}}} \end{aligned} \tag{7.12}$$

$$[\mathrm{C_3H_6}] = \frac{[\mathrm{C_2H_3}][\mathrm{CH_3}]k_{71b}}{k_{71f} + [\mathrm{H}]k_{74b}(1 - F_{74f})} \tag{7.13}$$

$$[n\text{-}\mathrm{C_3H_7}] = \frac{[\mathrm{C_3H_6}][\mathrm{H}]k_{74b}}{k_{73} + k_{74f}} \tag{7.14}$$

$$[\mathrm{CH_2O}] = \frac{[\mathrm{CH_3}]([\mathrm{O}]k_{35} + [\mathrm{O_2}]k_{37}) + [\mathrm{C_3H_3}][\mathrm{O_2}]k_{62}}{[\mathrm{H}]k_{29} + [\mathrm{O}]k_{30} + [\mathrm{OH}]k_{31} + [\mathrm{M}]k_{32}} \tag{7.15}$$

$$\begin{aligned} Z_{\mathrm{CHO}} &= [\mathrm{CH_3}] \cdot ([\mathrm{O}]k_{35} + [\mathrm{O_2}]k_{37}) + [\mathrm{C_3H_3}][\mathrm{O_2}]k_{62} \\ &\quad + [\mathrm{CH}]([\mathrm{O_2}]k_{19} + [\mathrm{CO_2}]k_{20}) + [\mathrm{CO}][\mathrm{H}][\mathrm{M}]k_{24b} \\ N_{\mathrm{CHO}} &= [\mathrm{H}]k_{21} + [\mathrm{OH}]k_{22} + [\mathrm{O_2}]k_{23} + [\mathrm{M}]k_{24f} \\ [\mathrm{CHO}] &= \frac{Z_{\mathrm{CHO}}}{N_{\mathrm{CHO}}} \end{aligned} \tag{7.16}$$

$$[CH_4] = \frac{[CH_3]([M]k_{34} + [H_2]k_{38b} + [H_2O]k_{40b})}{[H]k_{38f} + [O]k_{39} + [OH]k_{40f}} \tag{7.17}$$

with

$$F_{74f} = \frac{k_{74f}}{k_{73} + k_{74f}} \tag{7.18}$$

The introduction of (7.4)–(7.9) in (7.12)–(7.17) also leads to algebraic equations, only depend on non-steady state species. The expanded equations can subsequently be used in automatically written subroutines for the calculation of reduced mechanisms.

A comparison of the profiles calculated with the starting mechanism and with the seven-step mechanism is shown for temperature and the reactants C_2H_2 and O_2 in Fig. 7.2, for the products CO, CO_2, H_2 and H_2O in Fig. 7.3, and for the radicals H, O and OH in Fig. 7.4. The only difference that can be seen is a downstream shift of the profiles calculated with the seven-step-mechanism. It can be concluded that the seven-step-mechanism is able to predict not only the structure of acetylene flames but also important intermediate species which influence the formation of pollutants in acetylene flames.

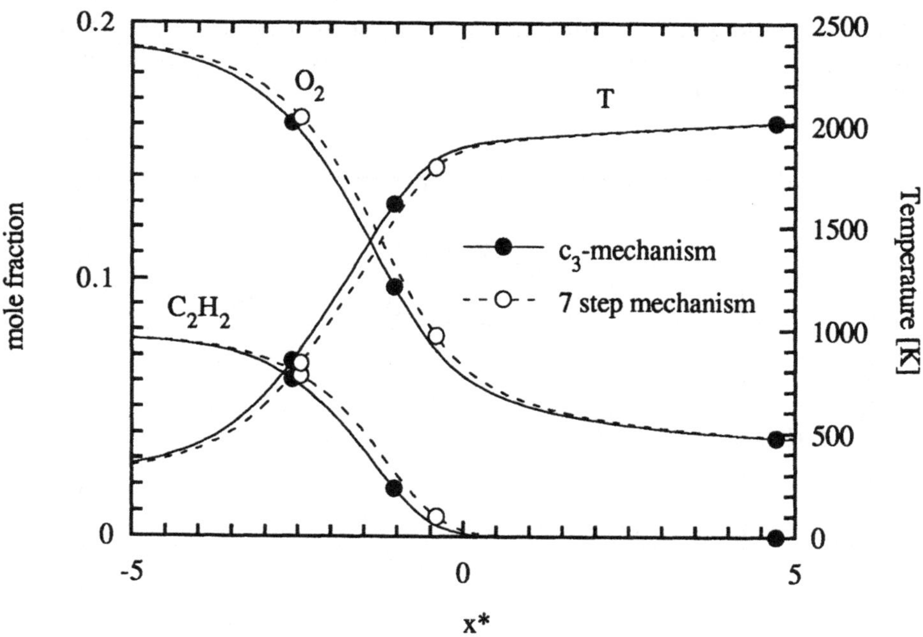

Fig. 7.2. Numerically calculated mole fractions of O_2 and C_2H_2 and the temperature profile with the starting C_3-mechanism and the reduced 7 step mechanism for atmospheric, stoichiometric acetylene-air flames

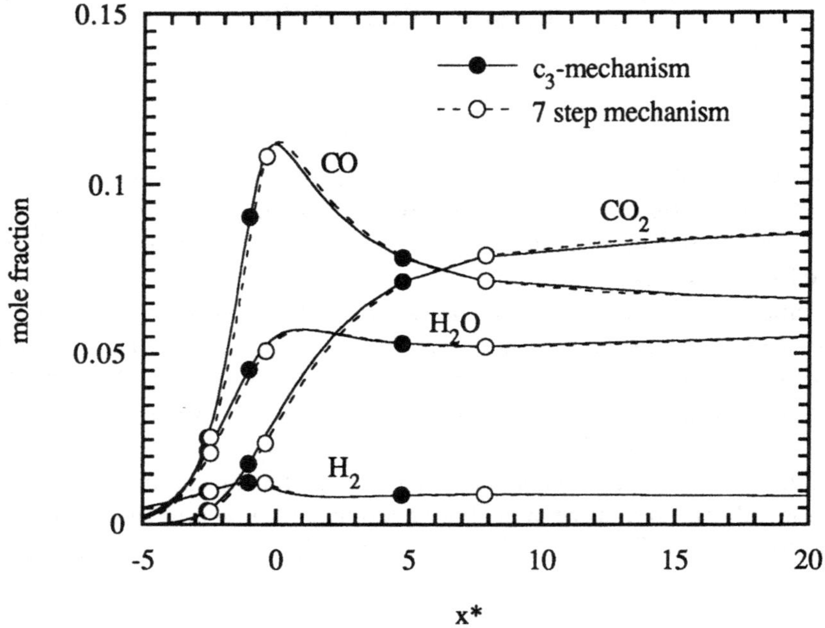

Fig. 7.3. Numerically calculated mole fractions of H_2, H_2O, CO, and CO_2 with the starting C_3-mechanism and the reduced 7 step mechanism for atmospheric, stoichiometric acetylene-air flames

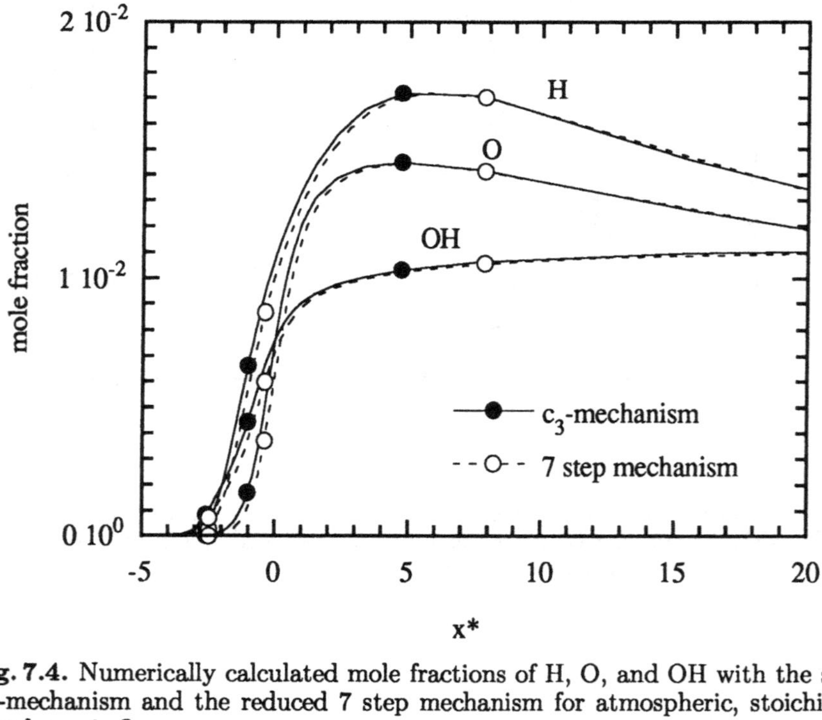

Fig. 7.4. Numerically calculated mole fractions of H, O, and OH with the starting C_3-mechanism and the reduced 7 step mechanism for atmospheric, stoichiometric acetylene-air flames

7.3 The derivation of a four-step mechanism

The first step towards the derivation of a simple four-step mechanism is the introduction of a steady state assumption for OH. While it can be shown that the convective diffusive operator of OH is much smaller than the chemical source term of OH in (7.1), the mass fraction of OH is not neglible in the mass balance. This leads to errors in the species profiles which are of the same order of magnitude as the errors calculated for the hydrogen flames in [7.9].

With the help of the steady state assumption for OH, reaction VI in (7.2) can be eliminated. The remaining global reaction rates in (7.3) are not changed by this operation.

Profiles calculated with the resulting six-step-mechanism are shown in Figs. 7.5–7.7. It can be seen that O_2 (Fig. 7.5), H_2O (Fig. 7.6) and the radicals H and O (Fig. 7.7) conserve the mass of the eliminated OH-radical. The latter leads to an increase of the calculated burning velocities of about 10 percent.

Due to the steady state assumption of OH, the algebraic equations in (7.1) are no longer linear. We calculated [OH] with the help of a simple fixpoint iteration starting with [OH] calculated from the partial equilibrium of reaction 3, viz.,

$$[\mathrm{OH}] = \frac{[\mathrm{H_2O}][\mathrm{H}]k_{3b}}{[\mathrm{H_2}]k_{3f}}. \tag{7.19}$$

The next step of the reduction is the introduction of the steady state assumption for C_2H_4. The elimination of the global reaction VII leads to a five step mechanism with the global reaction rates given by (7.3). Since C_2H_4 is a stable species, this steady state assumption is not good. However in acetylene-air flames the influence of C_2H_4 is small. Therefore the resulting errors of up to two orders of magnitude in the C_2H_4 profile leads only to small errors in the calculated flame structure (Figs. 7.8–7.10). To avoid further non-linearities in the system of equations (7.1), all terms concerning $[C_2H_4]$ were truncated in (7.5)–(7.9).

The last step in deducing the 4-step mechanism is the steady state assumption of the O-radical. In other hydrocarbon flames the fuel is mostly consumed by H-abstraction either due to radical attack or due to thermal decomposition. In contrast to other hydrocarbons acetylene is consumed by oxidation reactions, namely

(45) $\quad C_2H_2 + O = CH_2 + CO$

(46) $\quad C_2H_2 + O = CHCO + H.$

Therefore acetylene-air flames are very sensitive to errors in the O-radical. In analogy to the OH-radical it is found that the mass fraction of the O-radical is not negligibly small, although the convective-diffusive operator in (7.1) is small in comparison with the chemical source terms. With the help of a steady state assumption for O the global reaction V in (7.2) is eliminated. The global reaction rates of the resulting 4-step mechanism are given in (7.3).

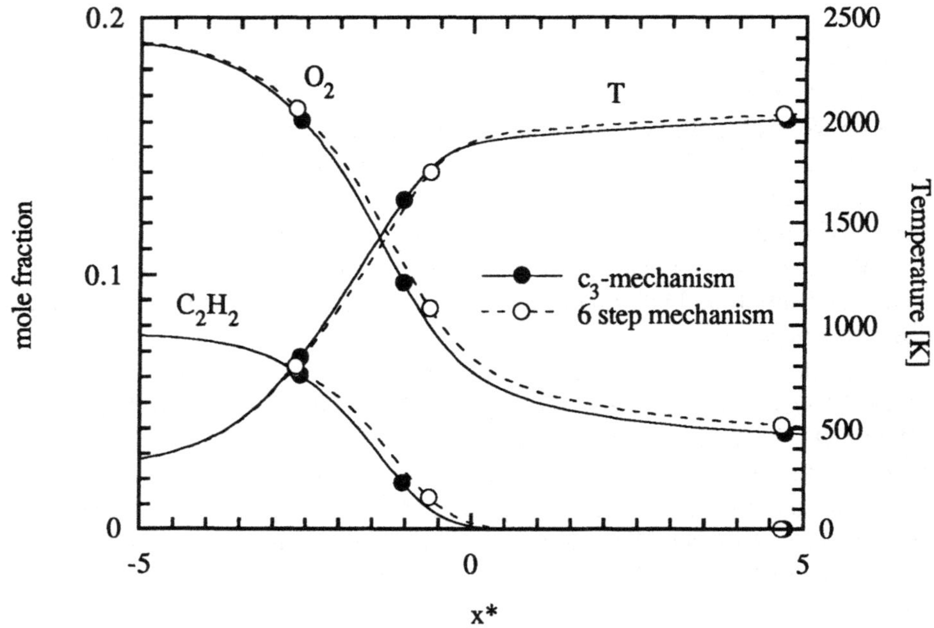

Fig. 7.5. Numerically calculated mole fractions of O_2 and C_2H_2 and the temperature profile with the starting C_3-mechanism and the reduced 6 step mechanism for atmospheric, stoichiometric acetylene-air flames

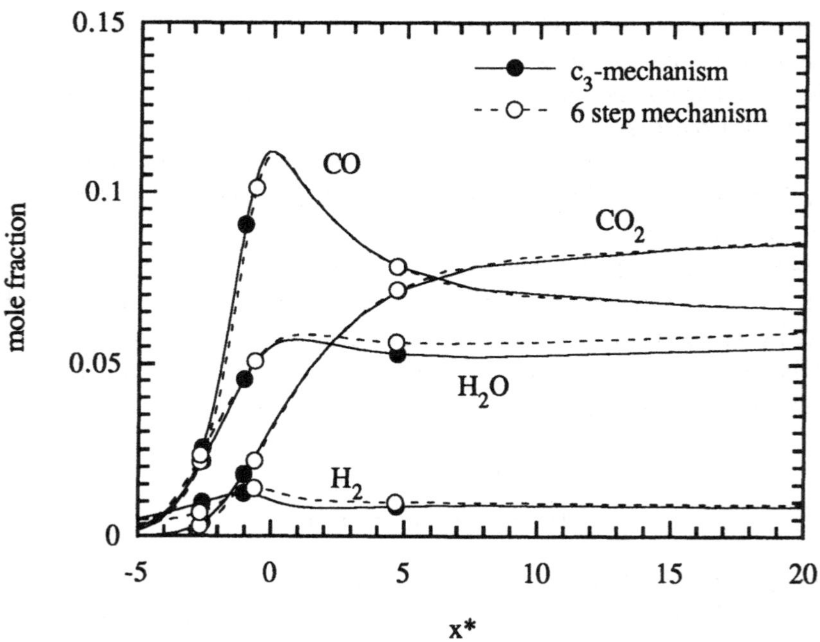

Fig. 7.6. Numerically calculated mole fractions of H_2, H_2O, CO, and CO_2 with the starting C_3-mechanism and the reduced 6 step mechanism for atmospheric, stoichiometric acetylene-air flames

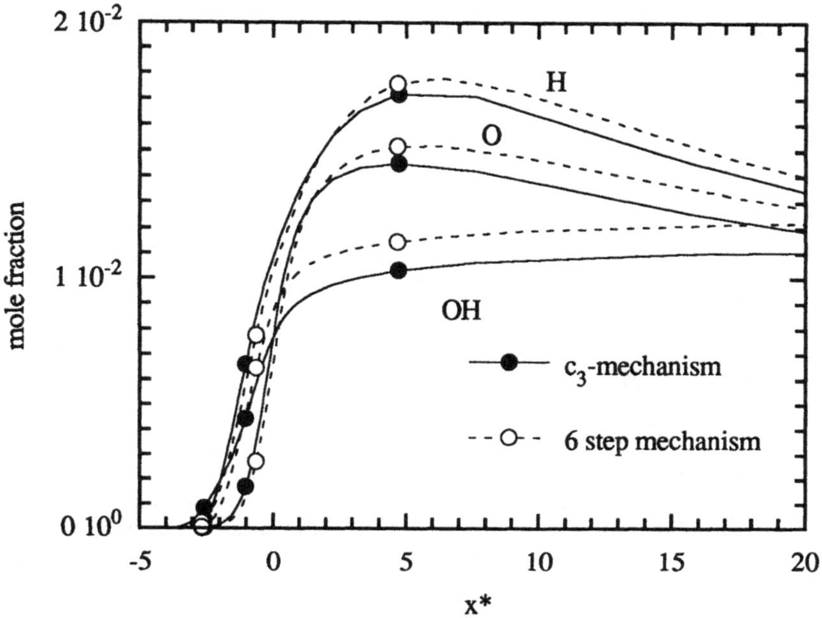

Fig. 7.7. Numerically calculated mole fractions of H, O, and OH with the starting C_3-mechanism and the reduced 6 step mechanism for atmospheric, stoichiometric acetylene-air flames

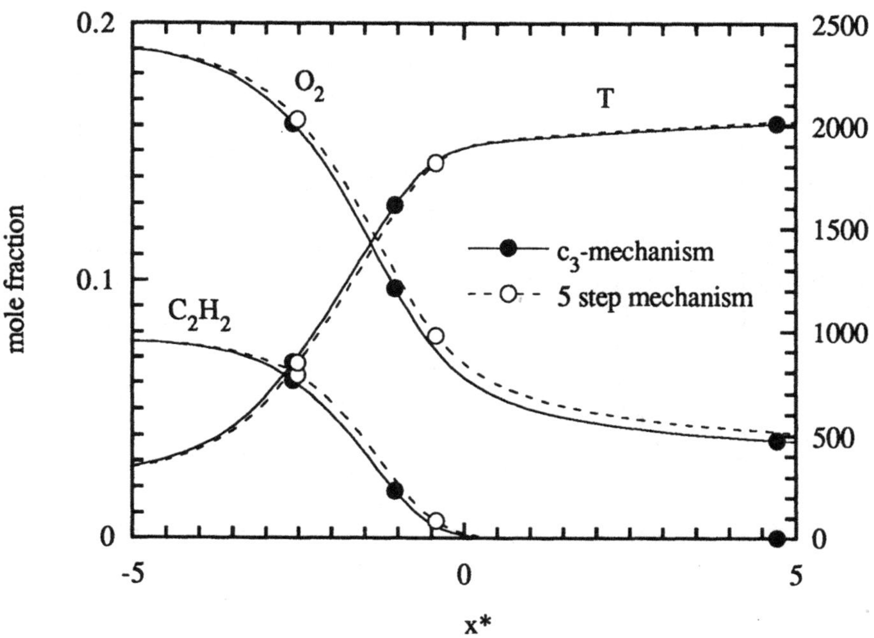

Fig. 7.8. Numerically calculated mole fractions of O_2 and C_2H_2 and the temperature profile with the starting C_3-mechanism and the reduced 5 step mechanism for atmospheric, stoichiometric acetylene-air flames

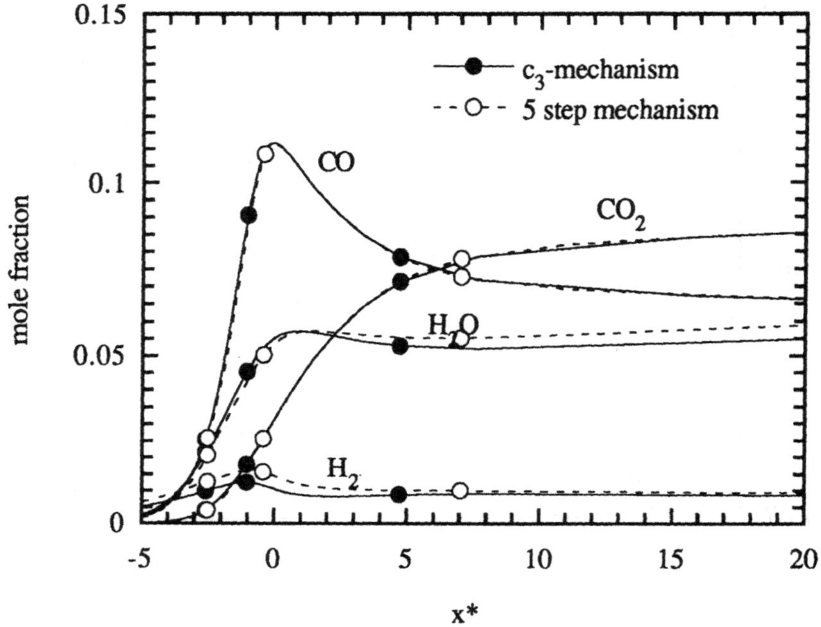

Fig. 7.9. Numerically calculated mole fractions of H_2, H_2O, CO, and CO_2 with the starting C_3-mechanism and the reduced 5 step mechanism for atmospheric, stoichiometric acetylene-air flames

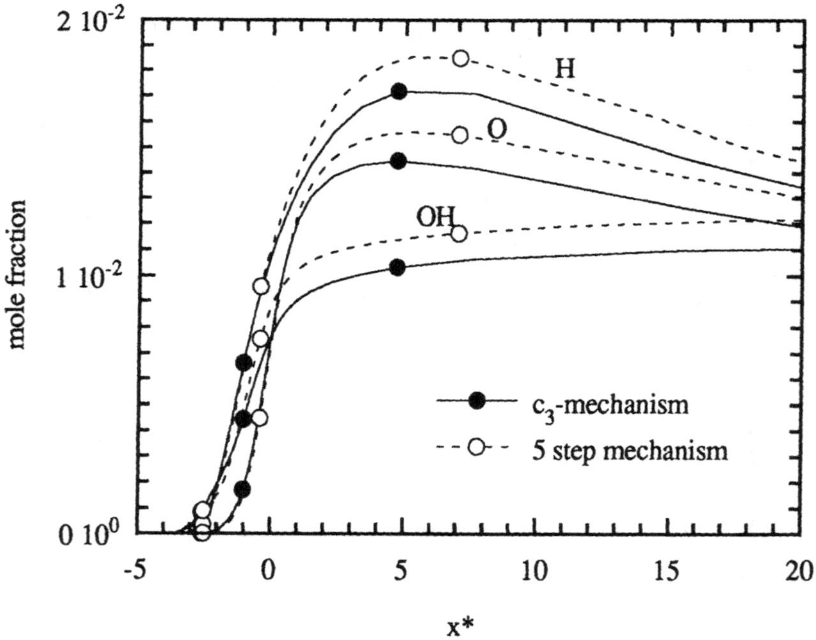

Fig. 7.10. Numerically calculated mole fractions of H, O, and OH with the starting C_3-mechanism and the reduced 5 step mechanism for atmospheric, stoichiometric acetylene-air flames

Since the 4-step mechanism will not maintain the accuracy needed for numerical purpose, we have introduced additional truncations of the algebraic equations in order to prepare a basis for future asymptotic analysis of the flame structure of acetylene air flames.

In order to retain the chain breaking reaction 50 is calculated $[C_2H_3]$ from

$$[C_2H_3] = \frac{[C_2H_2][H][M]k_{51b}}{[M]k_{51f} + [O_2]k_{50}} \tag{7.20}$$

and $[HO_2]$ from

$$[HO_2] = \frac{[O_2][H][M]k_{5f} + F_{50}[C_2H_2][H][M]k_{51b}}{[H](k_6 + k_7 + k_9) + [OH]k_8} \tag{7.21}$$

with

$$F_{50} = \frac{[O_2]k_{50}}{[M]k_{51f} + [O_2]k_{50}} \tag{7.22}$$

In analogy to [7.9] the algebraic equations for O and OH in (7.1) are reduced to

$$L([O]) = w_1 - w_2 - w_{45} - w_{46} \tag{7.23}$$

and

$$L([OH]) = w_1 + w_2 - w_3 + 2w_6 = 0 \tag{7.24}$$

In addition to [7.9] the fuel consuming reactions 45 and 46 appear in the balance of [O]. Using the steady state relation for HO_2 and neglecting w_8 we write

$$[OH] = \frac{w_{1f} + w_{2f} + w_{3b} + 2z_6(w_5 + F_{50}w_{51b})}{k_{1b}[O] + k_{2b}[H] + k_{3f}[H_2]} \tag{7.25}$$

where

$$z_6 = \frac{k_6}{k_6 + k_7 + k_9} \tag{7.26}$$

Still following [7.9] the steady state relation of O can be written as

$$[O] = [H]\frac{k_{1f}[O_2] + k_{2b}[OH]^*}{k_{1b}[OH]^* + k_{2f}[H_2] + (k_{45} + k_{46})[C_2H_2]} \tag{7.27}$$

with

$$[OH]^* = \frac{w_{3b} + 2z_6(w_5 + F_{50}w_{51b})}{k_{3f}[H_2]} \tag{7.28}$$

Equation (7.28) is found to be accurate upstream in the flame, where radical consumption takes place and downstream, where reaction 3 is in partial equilibrium. In the main reaction zone reactions 2b and 1b are negligibly small in the numerator and the denominator of (7.27) and the formulation for [O] is still accurate.

Finally, we calculate $[H_2O]$ in (7.21) by using only the second term in the nominator of (7.28) due to the fact that reaction w_8 is important only in the upstream region of the flame. This leads to

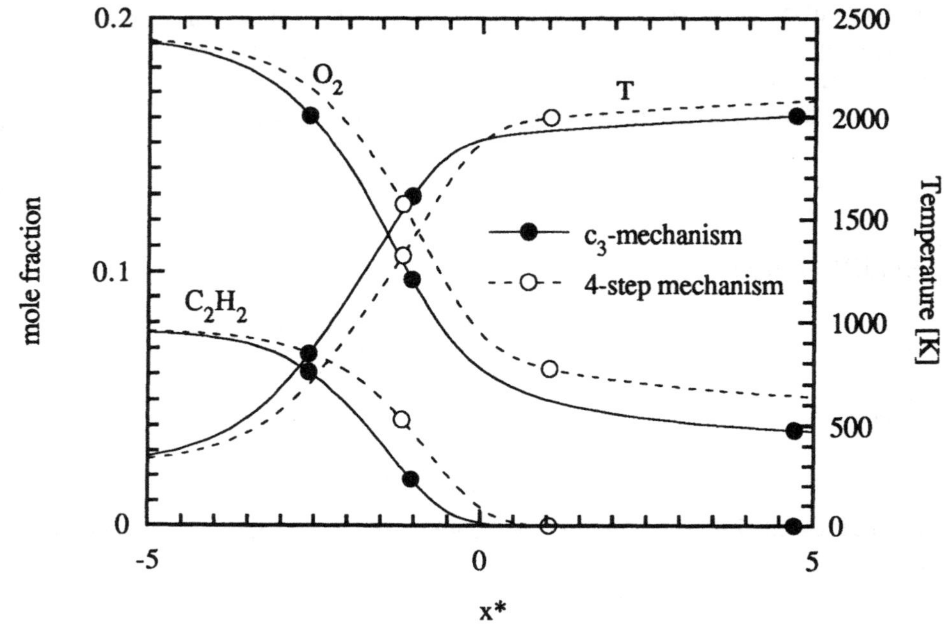

Fig. 7.11. Numerically calculated mole fractions of O_2 and C_2H_2 and the temperature profile with the starting C_3-mechanism and the reduced 4 step mechanism for atmospheric, stoichiometric acetylene-air flames

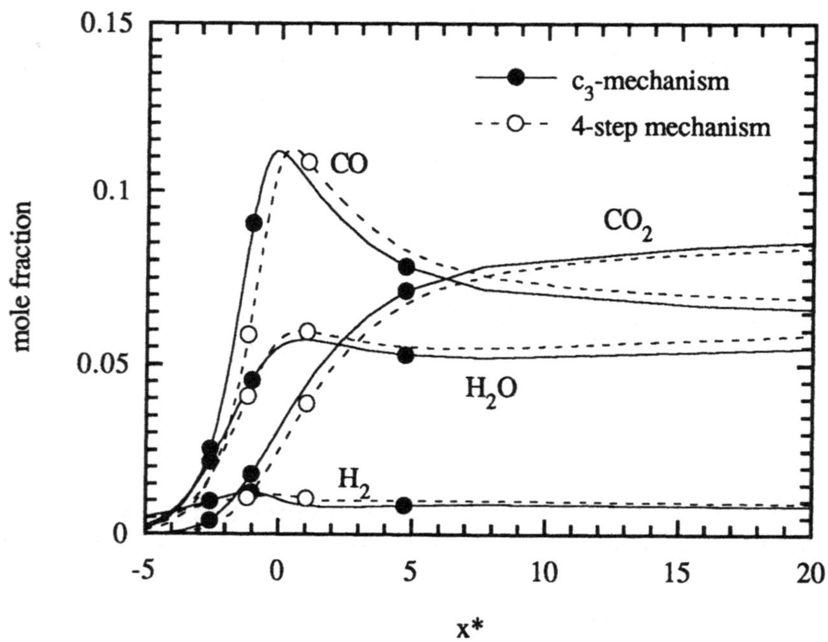

Fig. 7.12. Numerically calculated mole fractions of H_2, H_2O, CO, and CO_2 with the starting C_3-mechanism and the reduced 4 step mechanism for atmospheric, stoichiometric acetylene-air flames

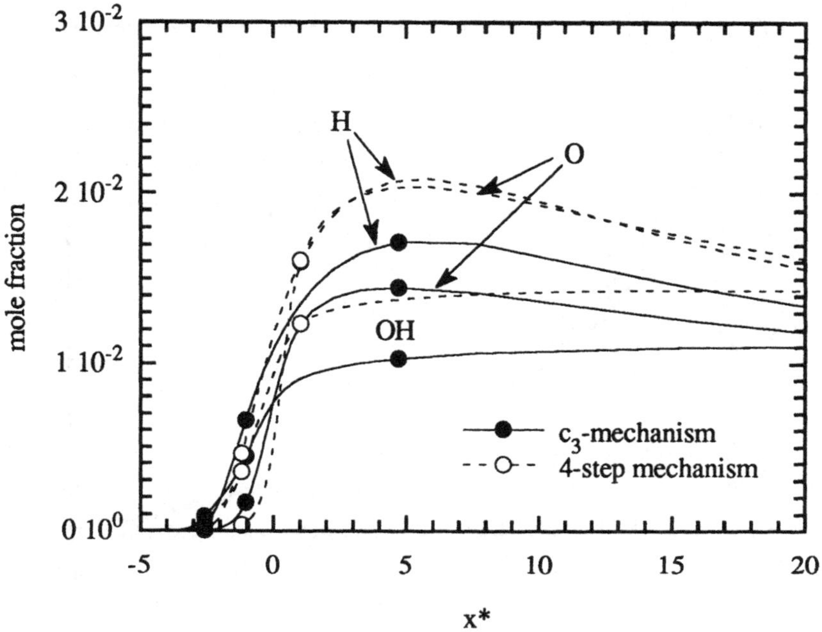

Fig. 7.13. Numerically calculated mole fractions of H, O, and OH with the starting C3-mechanism and the reduced 4 step mechanism for atmospheric, stoichiometric acetylene-air flames

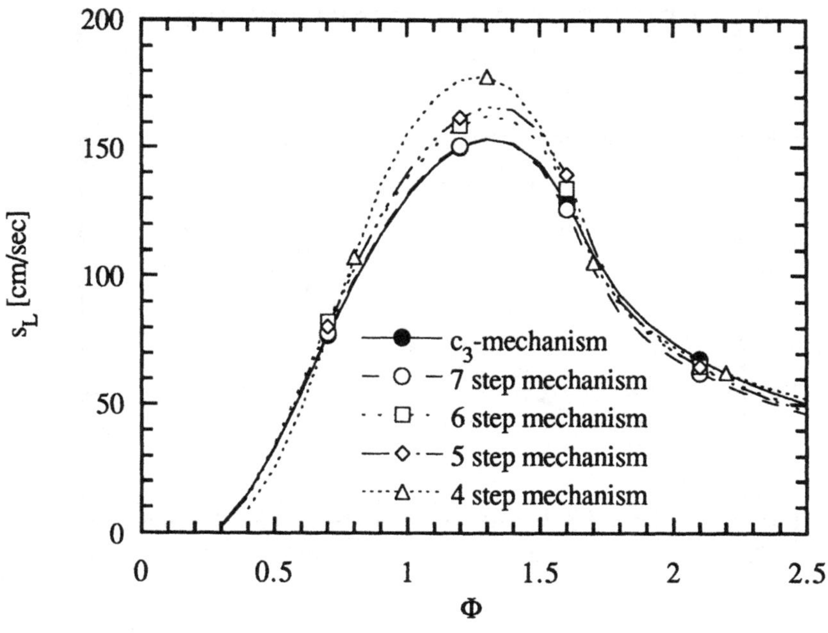

Fig. 7.14. Burning velocities of atmospheric acetylene-air flames with the starting C3-mechanism and the different reduced mechanisms

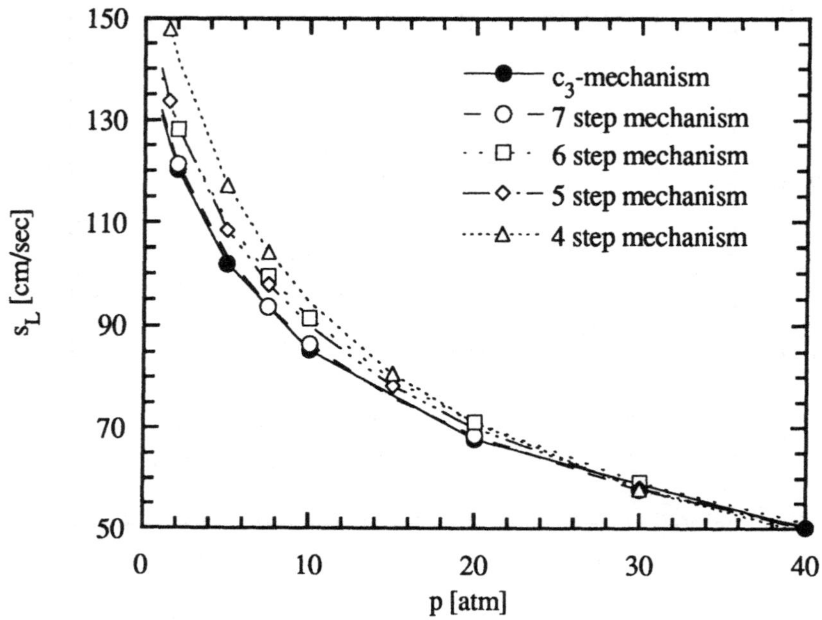

Fig. 7.15. Burning velocities of stoichiometric acetylene-air flames with the starting C_3-mechanism and the different reduced mechanisms

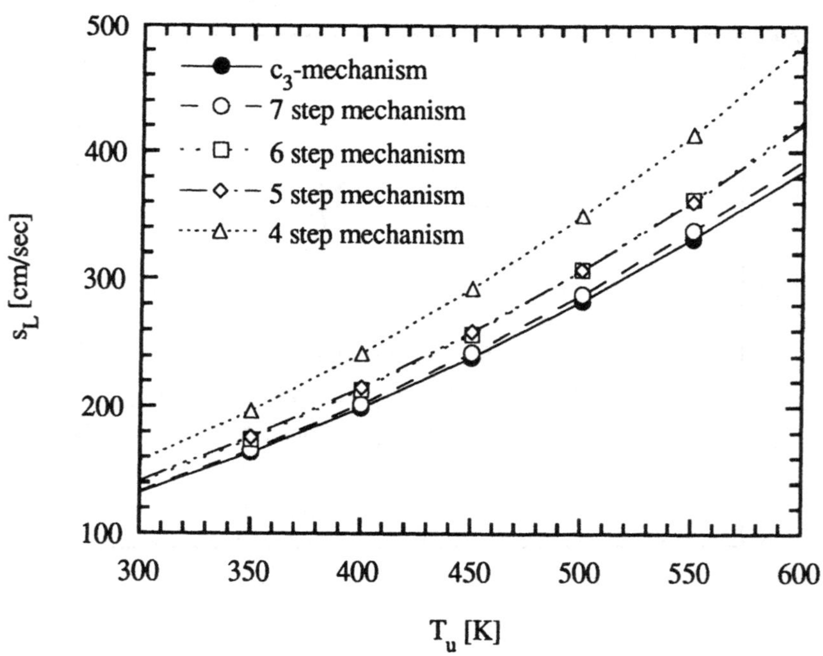

Fig. 7.16. Burning velocities of atmospheric, stoichiometric acetylene-air flames with the starting C_3-mechanism and the different reduced mechanisms

$$[HO_2] = \frac{k_5[H][O_2][M] + k_{51b}F_{50}[C_2H_2][H][M]}{[H](k_6 + k_7 + k_9) + 2k_8 z_6(w_5 + F_{50}w_{51b})/k_{3f}[H_2]} . \qquad (7.29)$$

The calculated profiles of temperature, reactants and products are shown in Figs. 7.11 and 7.12. It can be seen that the mass of O and OH is mainly conserved by O_2. The shift of the profiles in X^* can be influenced by the definition $X^* = 0$. The definition used in this paper is $X^* = 0$ in the position of maximal radical production. We point out, that despite the simplification the structure of the flame is well described by the short 4-step mechanism.

7.4 Conclusion

A seven step mechanism has been derived for premixed acetylene flames that is sufficiently accurate over a wide range of fuel-air-equivalence ratios (Fig. 7.14), pressures (Fig. 7.15) and initial gas temperatures (Fig. 7.16). It has been shown that this mechanism predicts accurately the profiles of temperature, reactants, and products as well as the profiles of the major radicals. This mechanism is recommended for numerical calculations of acetylene flames and their important intermediate species which influence the formation of pollutants.

We have further reduced the seven step mechanism further to derive a simple four step mechanism for use of further asymptotic calculations. The loss of accuracy of each step has been documented. It was found that while for O and OH the convective-diffusive operator is negligible small the elimination of these radicals leads to errors in the overall mass balance. These errors does not affect the flame structure, but the overall flame velocities. For lean ($\Phi < 0.8$) and rich ($\Phi > 1.6$) acetylene flames (Fig. 7.14) as well as for high pressure flames ($p > 20$ atm) (Fig. 7.15) the radical concentrations are low and as a result the four step mechanism becomes more accurate for both flame structure and burning velocities.

The most important chain breaking reaction in acetylene flames involves HO_2, and the description of the H_2-O_2 mechanism is different from other hydrocarbons and is more closely related to that of hydrogen flames.

References

[7.1] Warnatz, J., Bochhorn, H., Möser, A., Wenz, H. W.: "Experimental Investigation and Computational Simulation of Acetylene-Oxygen Flames From Near Stoichiometric to Sooting Conditions", Nineteenth Symposium (International) on Combustion, The Combustion Institute, 1982, pp. 197–209.

[7.2] Leung, K. M., Lindstedt, R. P., Jones, W. P.: "A Simplified Reaction Mechanism for Soot Formation in Nonpremixed Flames", Combustion and Flame **87**, 1991, pp. 289–305.

[7.3] Frenklach, M., Warnatz, J.: "Detailed Modelling of PAH Profiles in a Sooting Low-Pressure Acetylene Flame Combustion", Science and Technology **51**, 1987, pp. 265–283.

[7.4] Mauss, F., Trilken, B., Peters, N.: Soot formation in partially premixed diffusion flames at elevated pressure, International Workshop "Mechanisms and Models of Soot Formation", Heidelberg, 1991.
[7.5] Warnatz, J.: "The Structure of Laminar Alkane-, Alkene-, and Acetylene Flames", Eighteenth Symposium (International) on Combustion, The Combustion Institute, 1981, pp. 369–384.
[7.6] Mauss, F., Peters, N.: "Reduced Kinetic Mechanisms for Premixed Methane-Air Flames", Chapter 5 of this book.
[7.7] Kennel, C., Mauss, F., Peters, N.: "Reduced Kinetic Mechanisms for Premixed Propane-Air Flames", Chapter 8 of this book.
[7.8] Wang, W., Rogg, B.: "Premixed Ethylene/Air and Ethane/Air Flames: Reduced Mechanisms Based on Inner Iteration", Chapter 6 of this book.
[7.9] Mauss, F., Peters, N., Rogg, B., Williams, F. A.: "Reduced Kinetic Mechanisms for Premixed Methane-Air Flames", Chapter 3 of this book.
[7.10] Law, C. K.: "A Compilation of Experimental Data on Laminar Burning Velocities, Chapter 2 of this book.
[7.11] Alkemade, V., Homann, K. H.: Zeitschrift für Physikalische Chemie, Neue Folge **16**, 1989, pp. 19–34.

8. Reduced Kinetic Mechanisms for Premixed Propane–Air Flames

C. Kennel
P. S. A. études et recherches, F 92252 La Garenne Colombes

F. Mauss and N. Peters
Institut für Technische Mechanik, RWTH Aachen, W-5100 Aachen,
West Germany

8.1 Introduction

Because of its low price, propane is often used as a reference fuel for hydrocarbon combustion. The combustion of most of the complex hydrocarbon fuels proceeds through a rapid decomposition to small C_1-C_3 hydrocarbon fragments and their subsequent oxidation in a complex chain mechanism [8.1]. Therefore the oxidation of lower hydrocarbons such as methane and propane provides most of the kinetic information needed to understand the combustion of all hydrocarbon. While methane is particular in the sense that C_2 and C_3 species have to be formed by a slow recombination of CH_3, propane readily forms all the relevant C_1-C_3 intermediates by decomposition.

A full mechanism for popane flame contains about 200 elementary reactions or more. Reduced mechanisms for propane have been proposed previously by Paczko et al. [8.2] and by Djavdan et al. [8.3]. One of the difficulty that was encountered in these studies was related to the fact that the elementary kinetic mechanism was not satisfactory. Meanwhile, new kinetic data have been proposed in the literature. A sensitivity analysis for flames at pressures between 1 and 40 atm and for equivalence ratios ranging from 0.5 to 1.4 has shown that the first 82 reactions in Table 1, Chap. 1, are sufficient to describe the structure of propane flames with sufficient accuracy. This mechanism will therefore be used as the starting mechanism for the reduction procedure. It is by far superior to that of the 68 reactions that had been used in a first attempt to provide a reduced mechanism for propane flames [8.2]. The previous mechanism was found to be insufficient in particular for rich flames and for high pressure flames because it did not take important reaction routes via species such as C_3H_6 into account. Rather than to modify the previous mechanism, we decided to entirely redo the reduction procedure based on the most recent mechanism.

Most of the C_3H_8 is consumed by H and OH radicals to form the normal- and *iso*-propyl radical

(77) $$C_3H_8 + H \rightarrow n\text{-}C_3H_7 + H_2$$
(78) $$C_3H_8 + H \rightarrow i\text{-}C_3H_7 + H_2$$
(81) $$C_3H_8 + OH \rightarrow n\text{-}C_3H_7 + H_2O$$
(82) $$C_3H_8 + OH \rightarrow i\text{-}C_3H_7 + H_2O \quad .$$

Both chains are approximately of the same importance. The propyl radicals decompose to C_2H_4 and CH_3 as

(73) $$n\text{-}C_3H_7 \rightarrow C_2H_4 + CH_3$$
(75) $$i\text{-}C_3H_7 \rightarrow C_2H_4 + CH_3 \quad .$$

Ethene C_2H_4 follows the C_2-chain to C_2H_3 mainly via the reactions

(52*f*) $$C_2H_4 + H \rightarrow C_2H_3 + H_2$$
(54*f*) $$C_2H_4 + OH \rightarrow C_2H_3 + H_2O \quad .$$

Propene C_3H_6 is mainly formed by reaction 71b

(71*b*) $$C_2H_3 + CH_3 \rightarrow C_3H_6$$

and forms after H-addition via reaction (74b) n-C_3H_7

(74*b*) $$C_3H_6 + H \rightarrow n\text{-}C_3H_7 \quad .$$

The C_3-chain via C_3H_5 and C_3H_4 leading to the C_2-chain is only of minor importance. These consumption reactions for C_3H_6 and C_2H_4 are relatively slow compared to the four reactions of formation, which explains the relatively large maximum concentrations of C_3H_6 and C_2H_4 found in rich flames. Therfore these species cannot be assumed in steady-state for rich flames. The C_2-chain continues to C_2H_2 which is rapidely formed by the very fast reaction $C_2H_3 \rightleftharpoons C_2H_2 + H$. The oxidation of acetylene occurs mainly by atomic oxygen attach

(45) $$C_2H_2 + O \rightarrow CH_2 + CO.$$
(46) $$C_2H_2 + O \rightarrow CHCO + H \quad .$$

For rich flames, due to the lack of O radicals these steps are relatively slow such that C_2H_2 also may not be assumed in steady state.

8.2 The Nine-step Reduced Mechanism

From the discussion above it becomes apparent that all hydrocarbon species except C_3H_8, C_3H_6, C_2H_4 and C_2H_2 may be assumed in steady state. To simplify the numerical procedure, it sometimes proves to be convenient to retain the radicals O and OH as non-steady state species even though their steady state assumption is well justified, as will be seen below. Furthermore, as in previous reduced mechanisms for hydrocarbon flames, the H radical and the stable species O_2, CO, H_2, H_2O and CO_2 must be retained. With these 12 non-steady state species containing the three elements C, H and O, a nine step mechanism can be constructed. The fastest reactions responsible for rapid consumption of the remaining steady-state species are for n-C_3H_7, i-C_3H_7, C_3H_5, C_3H_4, C_3H_3, C_2H_6, C_2H_5, C_2H_3, CHCO, C_2H, CH_4, CH_3, CH_2O, CH_2, CH, CHO, H_2O_2, HO_2, the reactions 73, 75, 70, 68, 62, 59, 58, 51, 44, 42, 38, 35, 31, 25, 19, 24, 12, 7, respectively. This leads to the following global mechanism

I	$C_3H_8 + O + OH \rightarrow C_2H_4 + CO + H_2O + H_2 + H$
II	$CO + OH \rightarrow CO_2 + H$
III	$H + H + M' \rightarrow H_2 + M'$
IV	$O_2 + H \rightarrow OH + O$
V	$H_2 + OH \rightarrow H_2O + H$
VI	$H_2 + O \rightarrow OH + H$
VII	$C_2H_4 + OH \rightarrow C_2H_2 + H_2O + H$
VIII	$C_2H_2 + O_2 \rightarrow 2CO + H_2$
IX	$C_3H_6 + O + OH \rightarrow C_2H_4 + CO + H_2O + H$

with the rates

$$
\begin{aligned}
w_{\mathrm{I}} &= w_{77} + w_{78} + w_{79} + w_{80} + w_{81} + w_{82} \\
w_{\mathrm{II}} &= w_{18} - w_{20} + w_{28} \\
w_{\mathrm{III}} &= w_{5} - w_{11} + w_{13} + w_{14} + w_{15} + w_{16} + w_{17} + w_{21} + w_{22} \\
&\quad + w_{23} - w_{32} + w_{34} + w_{48} + w_{49} + w_{50} + w_{56} + w_{57} + w_{65} \\
&\quad - w_{69} - w_{71} + w_{76} \\
w_{\mathrm{IV}} &= w_{1} + w_{6} + w_{9} + w_{11} - w_{14} - w_{17} - w_{20} - w_{26} + w_{33} \\
&\quad + w_{36} + w_{37} + w_{43} - w_{46} - w_{48} - w_{53} - w_{56} \\
w_{\mathrm{V}} &= w_{3} + w_{4} + w_{8} + w_{9} + w_{13} + w_{14} + w_{16} + w_{22} - w_{29} \\
&\quad - w_{30} - w_{32} - w_{33} - w_{36} + w_{40} + w_{47} + w_{48} - w_{52} - w_{55} \qquad (8.2) \\
&\quad + w_{56} + w_{61} - w_{63} - w_{66} - w_{71} - w_{72} + w_{81} + w_{82} \\
w_{\mathrm{VI}} &= w_{2} - w_{4} + w_{6} + w_{10} + w_{11} - w_{14} + w_{17} + w_{26} + w_{27} \\
&\quad + w_{28} + w_{30} - w_{33} - w_{36} + w_{39} - w_{43} + w_{46} + w_{48} + w_{53} \\
&\quad + w_{56} + w_{60} + w_{64} + w_{65} + w_{66} - w_{72} + w_{79} + w_{80}
\end{aligned}
$$

$$
\begin{aligned}
w_{\mathrm{VII}} = & -w_{36} + w_{52} + w_{53} + w_{54} + w_{55} + w_{56} + w_{64} \\
& + w_{65} + w_{66} + w_{67} + w_{71} \\
w_{\mathrm{VIII}} = & -w_{36} - w_{41} + w_{45} + w_{46} + w_{47} \\
& + w_{48} + w_{53} + w_{56} - w_{63} + w_{64} \\
w_{\mathrm{IX}} = & \, w_{71} + w_{72} - w_{74} - w_{76} \,.
\end{aligned}
$$

The first rates appearing in (8.2) are the principle rates. The algebraic equations for the steady state species may be obtained directly from the balance equations for the full mechanism by setting the convective-diffusive operator equal to zero. These equations are explicit expressions for all steady state species except C_2H_3. An iteration procedure is employed to calculate C_2H_3 starting from a first guess based on the balance of reactions 49–52 and 54 only. This iteration converges within 3 to 4 steps. Quadratic equations are solved for CH_3 and HO_2. The 9-step mechanism thereby defined does not involve any truncation of steady state equations.

A comparison of species profiles between the full mechanism and the 9-step mechanism for a stoichiometric propane-air flame at 1 atm is shown in Fig. 8.1–8.5.

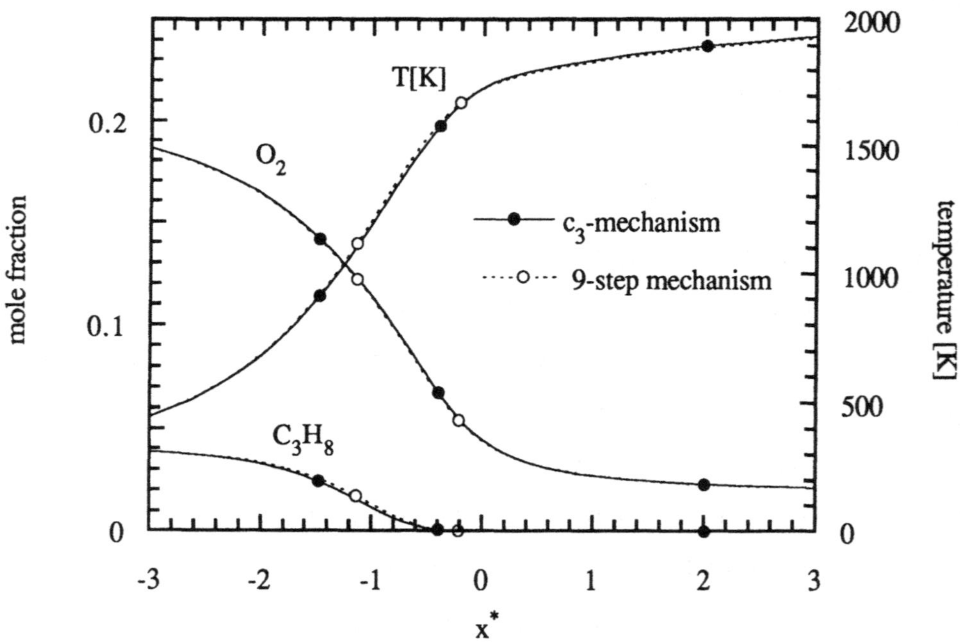

Fig. 8.1. Comparison of full kinetics with the 9-step mechanism for the temperature and the reactants.

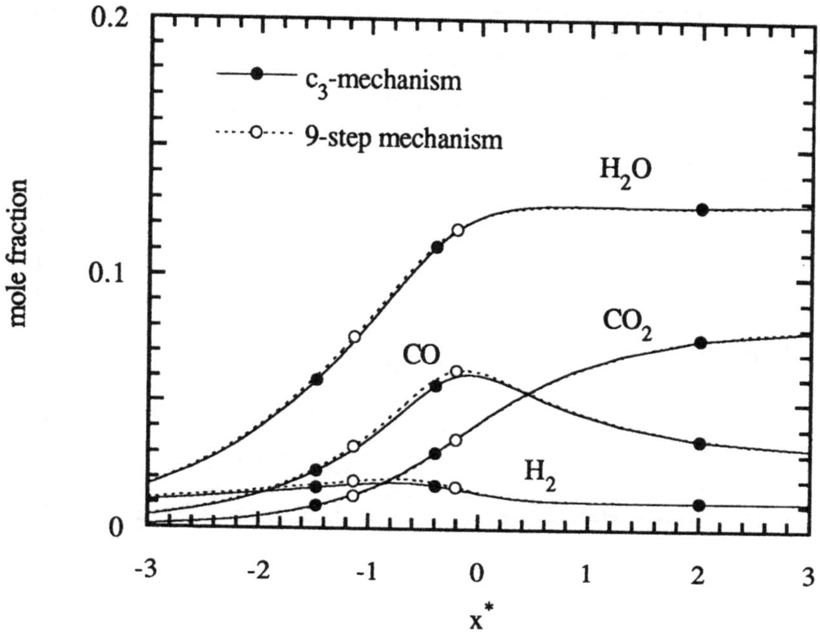

Fig. 8.2. Comparison of full kinetics with the 9-step mechanism for stable species.

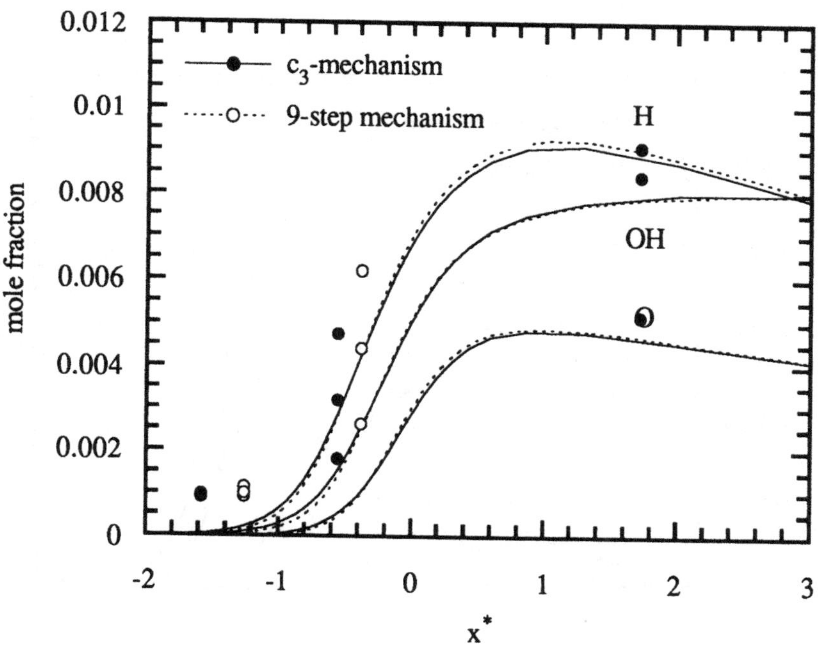

Fig. 8.3. Comparison of full kinetics with the 9-step mechanism for radicals.

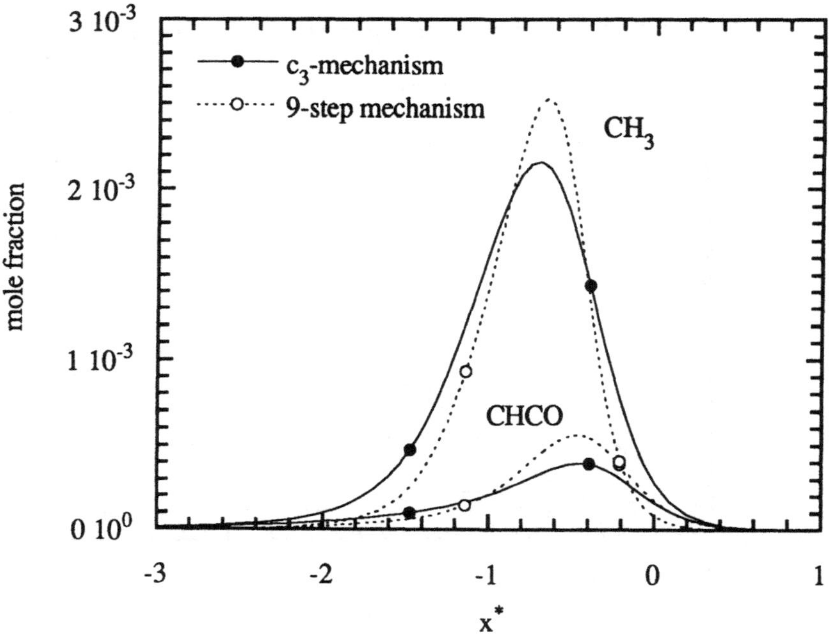

Fig. 8.4. Comparison of full kinetics with the 9-step mechanism for CH_3 and CHCO.

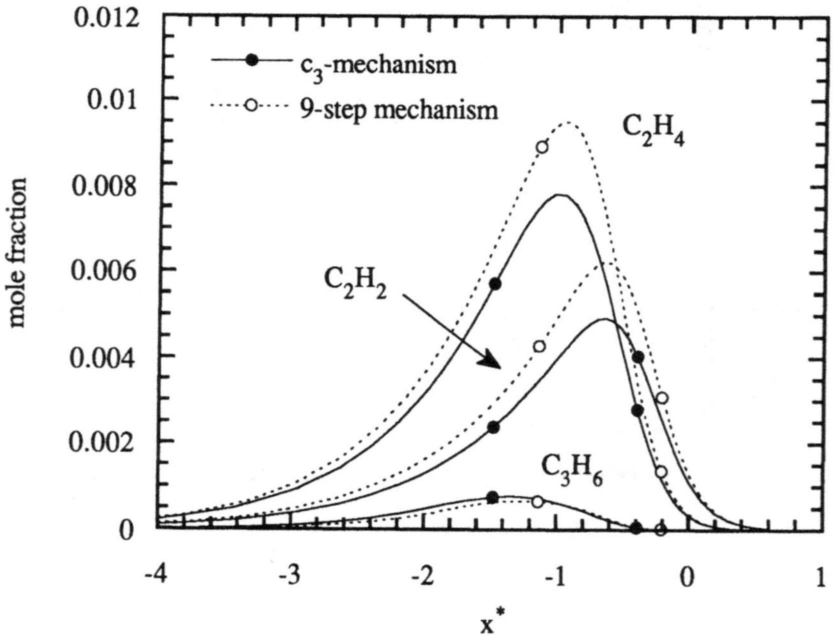

Fig. 8.5. Comparison of full kinetics with the 9-step mechanism for C_2H_4, C_2H_2 and C_3H_6.

They are plotted over the non-dimensional coordinate

$$x^* = (\rho_u \, s_L) \int_0^x \frac{c_p}{\lambda} \, dx \,. \tag{8.3}$$

The origin $x^* = 0$ was defined at the location where the sum of the production rates of the radicals, defined as $w_s = w_\mathrm{H} + w_\mathrm{OH} + 2w_\mathrm{O}$ has a maximum. The rate of O was taken twice because one O radical produces two radicals H or OH in the fast shuffling reactions 2–4.

There is very little difference for the temperature and the stable species C_3H_8, O_2, CO, H_2, CO_2 and H_2O as shown in Figs. 8.1 and 8.2. Even the concentrations of the radicals H, O and OH, shown in Fig. 8.3, compare very well. Up to twenty percent differences appear for the steady state species CH_3 and CHCO shown in Fig. 8.4 and the non-steady state species C_2H_4 and C_2H_2 shown in Fig. 8.5, while C_3H_6 differs somewhat less. Since the mole fractions of these species are smaller than one percent, the element mass fraction deficit introduced by the steady state assumptions of the many hydrocarbon species affects the percent accuracy of these species more than that of the species occuring in larger concentrations. The balance of convection, diffusion, chemical production and consumption for H, O, OH, C_2H_4, C_2H_2 and C_3H_6 is shown in Figs. 8.6–8.11, respectively. From these figures it is clear that the steady state assumption is very well justified for O and OH and somewhat less for C_3H_6, while the other species must be retained as non-steady state species.

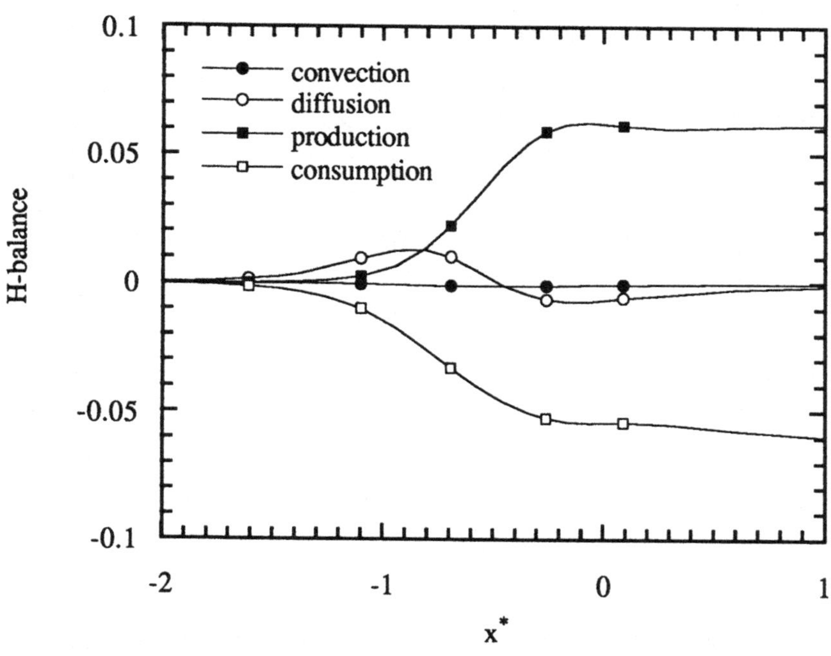

Fig. 8.6. The different terms in the balance equation for H-radical.

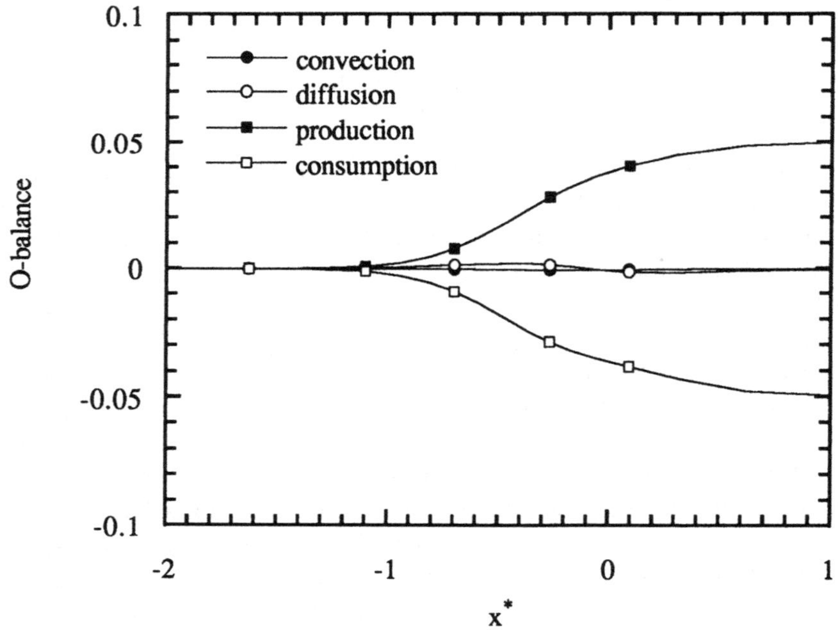

Fig. 8.7. The different terms in the balance equation for O-radical.

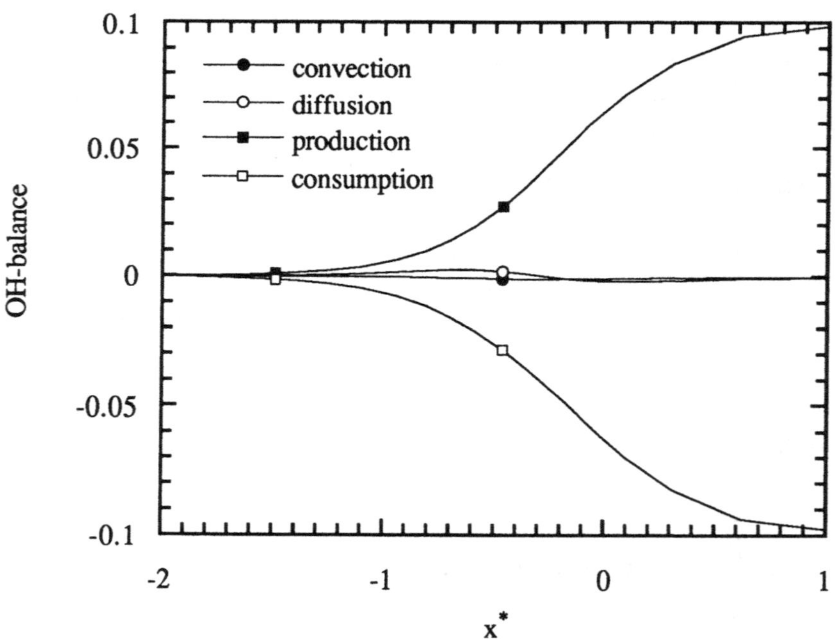

Fig. 8.8. The different terms in the balance equation for OH-radical.

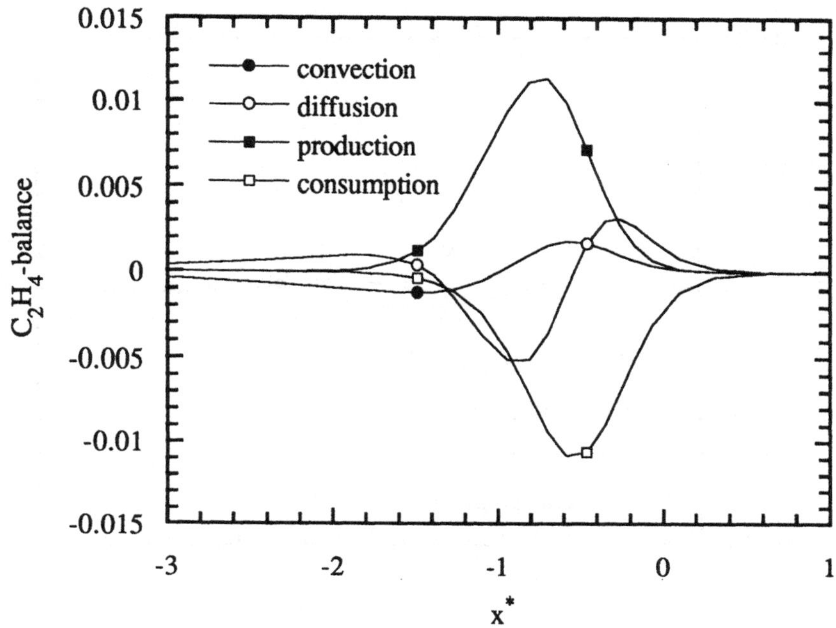

Fig. 8.9. The different terms in the balance equation for C_2H_4.

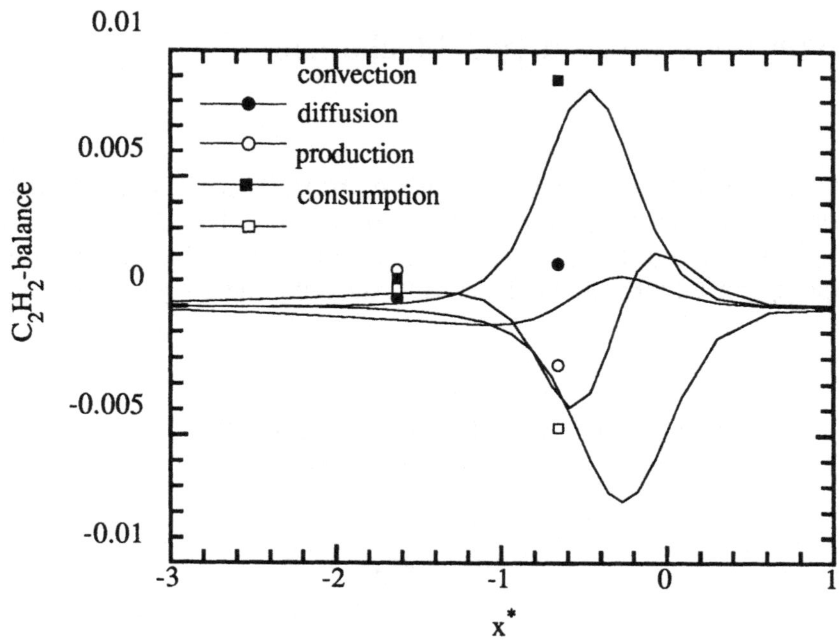

Fig. 8.10. The different terms in the balance equation for C_2H_2.

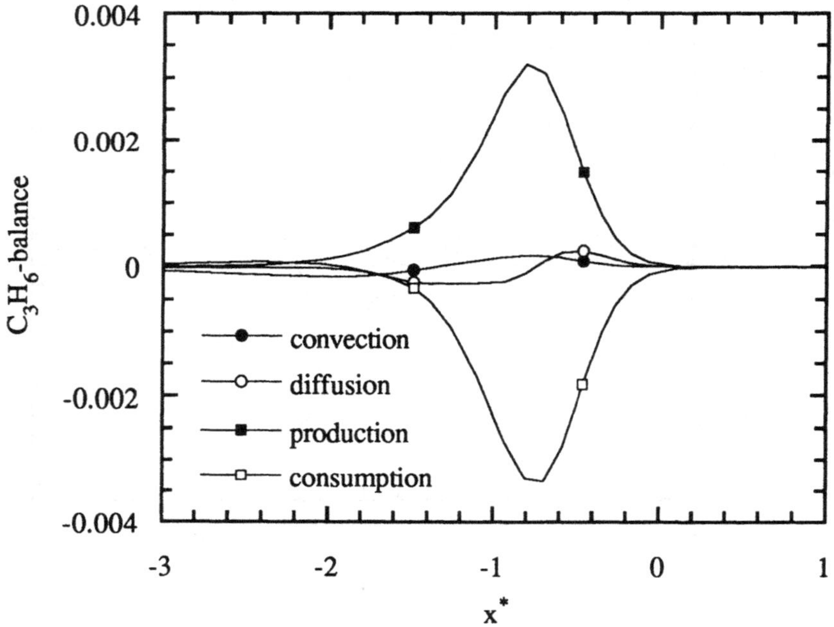

Fig. 8.11. The different terms in the balance equation for C_3H_6.

8.3 The Six-step Reduced Mechanism

While the nine-step mechanism does not require any truncations, but involves numerical iterations, the following reduction steps will be formulated such that explicit equations for all species are to be solved. Introducing steady state assumptions for O, OH and C_3H_6 leads to the following 6-step mechanism

I	$C_3H_8 + H_2O + 2H \rightarrow C_2H_4 + CO + 4H_2$
II	$CO + H_2O \rightarrow CO_2 + H_2$
III	$H + H + M' \rightarrow H_2 + M'$
IV	$O_2 + 3H_2 \rightarrow 2H_2O + 2H$
V	$C_2H_4 \rightarrow C_2H_2 + H_2$
VI	$C_2H_2 + O_2 \rightarrow CO + CO + H_2$

with carefully truncated rates

$$
\begin{aligned}
w_{\mathrm{I}} &= w_{77} + w_{78} + w_{79} + w_{80} + w_{81} + w_{82} \\
w_{\mathrm{II}} &= w_{18} - w_{20} + w_{28} \\
w_{\mathrm{III}} &= w_{5} + w_{15} + w_{16} + w_{21} + w_{22} + w_{23} + w_{34} + w_{48} + w_{49} \\
&\quad + w_{50} + w_{56} + w_{57} - w_{71} + w_{76} \\
w_{\mathrm{IV}} &= w_{1} + w_{6} + w_{9} - w_{20} - w_{26} + w_{36} - w_{46} - w_{48} - w_{56} \\
w_{\mathrm{V}} &= w_{54} - w_{36} + w_{52} + w_{71} \\
w_{\mathrm{VI}} &= - w_{36} + w_{45} + w_{46}
\end{aligned}
$$

Adding the steady state relations of CH_3, CH_4, n-C_3H_7, i-C_3H_7 and C_3H_6 leads to the following quadratic form for CH_3

$$
\begin{aligned}
2k_{36}\,[CH_3]^2 + (k_{35}\,[O] + k_{37}\,[O_2])\,[CH_3] &= [C_3H_8] \\
&\times ((k_{77}+k_{78})\,[H] + (k_{79}+k_{80})\,[O] + (k_{81}+k_{82})\,[OH])
\end{aligned}
\tag{8.6}
$$

The remaining steady state species can be calculated from the following explicit expressions

$$
[C_2H_3] = \frac{k_{51b}\,[H][C_2H_2] + (k_{52f}\,[H] + k_{54f}\,[OH])\,[C_2H_4]}{k_{49}\,[H] + k_{50}\,[O_2] + k_{51f} + k_{52b}\,[H_2] + k_{54b}\,[H_2O] + k_{71b}\,[CH_3]}
\tag{8.7}
$$

$$
[n\text{-}C_3H_7] = \frac{k_{74b}\,[C_3H_6][H] + (k_{77}\,[H] + k_{79}\,[O] + k_{81}\,[OH])\,[C_3H_8]}{k_{73} + k_{74f}}
\tag{8.8}
$$

$$
[i\text{-}C_3H_7] = \frac{(k_{78}\,[H] + k_{80}\,[O] + k_{82}\,[OH])\,[C_3H_8]}{k_{75} + k_{76}\,[O_2]}
\tag{8.9}
$$

$$
[C_3H_6] = \frac{k_{71b}\,[C_2H_3][CH_3]}{k_{71f} + k_{74b}\,[H]}
\tag{8.10}
$$

where w_{74f} is neglected for simplification. Furthermore we have

$$
[C_2H_6] = \frac{k_{36}\,[CH_3][CH_3]}{k_{59}\,[H] + k_{60}\,[O] + k_{61}\,[OH]}
\tag{8.11}
$$

$$
[C_2H_5] = \frac{(k_{36} + k_{56b})\,[CH_3][CH_3] + k_{58b}[C_2H_4][H]}{k_{56f}\,[H] + k_{57}\,[O_2] + k_{58f}}
\tag{8.12}
$$

which was obtained from the addition of the steady state relations of C_2H_5 and C_2H_6. The steady state relations for HO_2 and CH_2 are written as

$$
[HO_2] = \frac{(k_{5f}[H][M] + k_{23}[CHO] + k_{50}[C_2H_3] + k_{57}[C_2H_5] + k_{76}[i\text{-}C_3H_7])[O_2]}{k_{5b}\,[M] + (k_6 + k_7 + k_9)\,[H] + k_{10}\,[O] + k_8\,[OH]}
\tag{8.13}
$$

$$[CH_2] = \frac{(k_{45} + k_{46} Z_{43f})\,[C_2H_2][O]}{k_{25f}\,[H]\,(1 - Z_{25b}) + k_{26}\,[O] + (k_{27} + k_{28})\,[O_2] + k_{43b}\,[CO]\,(1 - Z_{43f})} \tag{8.14}$$

with

$$\begin{aligned} Z_{43f} &= \frac{k_{43f}\,[H]}{k_{43f}\,[H] + k_{44}\,[O]} \\ Z_{25b} &= \frac{k_{25b}\,[H_2]}{k_{25b}\,[H_2] + k_{19}\,[O_2] + k_{20}\,[CO_2]} \end{aligned} \tag{8.15}$$

The two factors represent the part of CH_2 coming from C_2H_2 and from the part of CH coming from the backward reaction 25b. Remaining explicit equation for hydrocarbon species are

$$[CH] = \frac{k_{25f}\,[CH_2][H]}{k_{19}\,[O_2] + k_{20}\,[CO_2] + k_{25b}\,[H_2]} \tag{8.16}$$

$$[C_3H_3] = \frac{k_{48}\,[C_2H_2][CH]}{k_{62}\,[O_2] + k_{63}\,[O] + k_{64b}\,[H]} \tag{8.17}$$

$$[CHCO] = \frac{k_{43b}\,[CH_2][CO] + k_{46}\,[C_2H_2][O]}{k_{43f}\,[H] + k_{44}\,[O]} \tag{8.18}$$

$$[CH_2O] = \frac{(k_{35}\,[O] + k_{37}\,[O_2])\,[CH_3]}{k_{29}\,[H] + k_{30}\,[O] + k_{31}\,[OH] + k_{32}\,[M]} \tag{8.19}$$

$$[CHO] = \frac{(k_{19}\,[O_2] + k_{20}\,[CO_2])\,[CH] + k_{24b}\,[CO][H][M] + K_{29}\,[CH_2O]}{k_{21}\,[H] + k_{22}\,[OH] + k_{23}\,[O_2] + k_{24f}[M]} \tag{8.20}$$

$$\text{where} \qquad K_{29} = (k_{29}\,[H] + k_{30}\,[O] + k_{31}\,[OH] + k_{32}\,M)$$

To calculate OH, we consider that reaction 3 is in equilibrium

$$[OH] = \frac{k_{3b}\,[H][H_2O]}{k_{3f}\,[H_2]} \tag{8.21}$$

For the calculation of O, it was found that the reactions where O and hydrocarbon react are very important in the first stage of the flame, and cannot be neglected. The most important are

35	$CH_3 + O \rightarrow CH_2O + H$
45	$C_2H_2 + O \rightarrow CH_2 + CO$
46	$C_2H_2 + O \rightarrow CHCO + H$

To take them into account in the calculation of O, it is necessary to have a simple expression of CH_3, which is

$$[CH_3]^* = \frac{(k_{77} + k_{78})\,[H]}{k_{35}\,[O]} \tag{8.22}$$

The steady state relation for O is then

$$w_1 - w_2 + w_4 - w_{35} - w_{45} - w_{46} = 0 \tag{8.23}$$

which leads to the expression

$$[O] = \frac{k_{1f}\,[O_2][H] + k_{2b}\,[OH][H] + k_{4f}\,[OH][OH] - (k_{77} + k_{78})\,[H][C_3H_8]}{k_{1b}\,[OH] + k_{2f}\,[H_2] + k_{4b}\,[H_2O] + (k_{45} + k_{46})\,[C_2H_2]} \tag{8.24}$$

The 6-step mechanism appears to be the optimal choice for propane flames, both, from the fundamental as from the numerical point of view. The steady state approximations used in deriving this mechanism are well justified. The same seems to be true for the truncations of steady state relations, which were introduced above. However, for some applications it is desirable to have even less global steps. Therefore the properties of a 4-step mechanism will be analysed as well.

8.4 The Four-step Reduced Mechanism

Starting from the 9-step mechanism, (8.1), a 4-step mechanism can be derived by assuming O, OH as well as C_2H_4, C_2H_2 and C_3H_6 in steady state. It requires that the production of C_2H_4 by reaction I is equal to the consumption by reaction VII and that the production of C_2H_2 by reaction VII is equal to the consumption by reaction VIII within locally the flame structure. The rates of these reactions are plotted in Fig. 8.12 for a stoichiometric propane flame at 1 atm. It is seen that the rate w_{I} peaks upstream of w_{VII} and that w_{VIII} is again shifted with respected to w_{VIII}. The magnitude of the rates are comparable. Assuming C_2H_4 and C_2H_2 in steady state therefore implies that one assumes that their formation and consumptions occurs at the same location withing the flame. Furthermore, Fig. 8.12 shows that the rate of the global step IX is fairly small in the entire flame structure indirating that the formation of C_3H_6 nearly equals the consumption. This again justifies the steady state assumption for C_3H_6.

Adding reactions I, three times V*b*, to VI*b*, VII and VIII of (8.1) leads to the first step in the 4-step mechanism. Reactions II and V*b* add to the second step, III remains the same and reaction IV, two times V*b* plus VI results in the fourth step. We thereby obtain

I	$C_3H_8 + O_2 + H_2O + 2H \rightarrow 3CO + 6H_2$
II	$CO + H_2O \rightarrow CO_2 + H_2$
III	$H + H + M' \rightarrow H_2 + M'$
IV	$O_2 + 3H_2 \rightarrow 2H_2O + 2H$

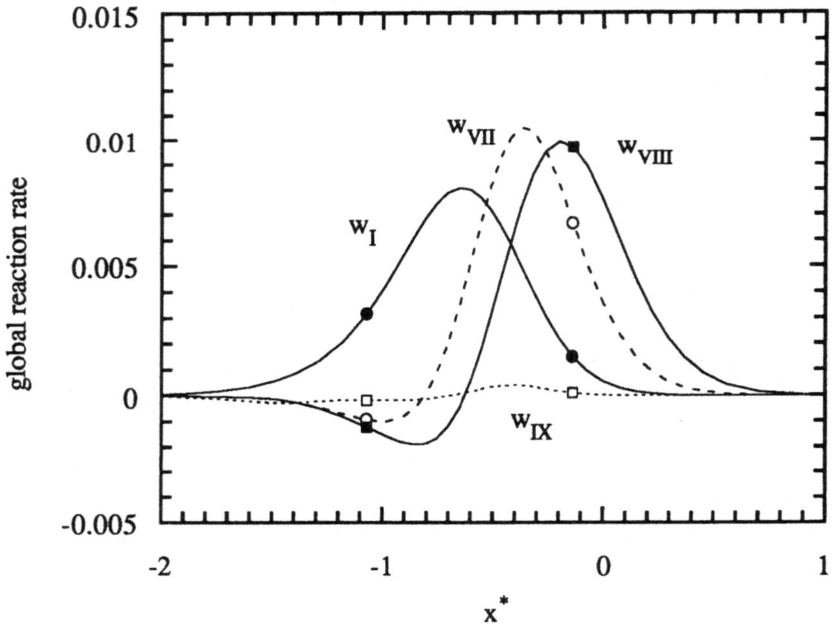

Fig. 8.12. Global reaction rates in the 9-step mechanism.

with the rates following also from the above addition of the rates of the 9-step mechanism

$$\begin{aligned}
w_{\mathrm{I}} &= w_{77} + w_{78} + w_{79} + w_{80} + w_{81} + w_{82} \\
w_{\mathrm{II}} &= w_{18} - w_{20} + w_{28} \\
w_{\mathrm{III}} &= w_5 + w_{15} + w_{16} + w_{21} + w_{22} + w_{23} + w_{34} + w_{48} + w_{49} \\
&\quad + w_{50} + w_{56} + w_{57} - w_{71} + w_{76} \\
w_{\mathrm{IV}} &= w_1 + w_6 + w_9 - w_{20} - w_{26} + w_{36} - w_{46} - w_{48} - w_{56}
\end{aligned} \tag{8.26}$$

In order to derive explicit algebraic expressions for the steady state species C_2H_2 the steady state expressions of C_2H_2, C_2H_3, C_2H_4, n-C_3H_7, i-C_3H_7, C_3H_6, C_2H_5 and C_2H_6 are added. One thereby obtains

$$[C_2H_2] = \frac{[C_3H_8]K_{77} + k_{36}\,[CH_3]^2}{(k_{45} + k_{46})\,[O]} \tag{8.27}$$

where $\quad K_{77} = (k_{77} + k_{78})\,[H] + (k_{79} + k_{80})\,[O] + (k_{81} + k_{82})\,[OH]$

Similarly, adding the steady state expressions for C_2H_3, C_2H_4, n-C_3H_7, i-C_3H_7, C_3H_6, C_2H_5 and C_2H_6 one obtains an explicit expression for C_2H_3

$$[C_2H_3] = \frac{[C_3H_8]K_{77} + k_{36}\,[CH_3]^2 + k_{51b}\,[C_2H_2][H]}{k_{49}\,[H] + k_{50}\,[O_2] + k_{51f}} \tag{8.28}$$

The concentrations of n-C_3H_7, i-C_3H_7 and C_3H_6 are calculated using (8.8)–(8.10). Then, from the addition of the steady state relations of C_2H_4, C_2H_5 and C_2H_6, C_2H_4 can be calculated as

$$[C_2H_4] = \frac{k_{73}\,[n\text{-}C_3H_7] + k_{75}\,[i\text{-}C_3H_7] + k_{36}\,[CH_3]^2 + K_{52b}\,[C_2H_3]}{k_{52f}\,[H] + k_{54f}\,[OH]} \tag{8.29}$$

$$\text{where} \qquad K_{52b} = k_{52b}\,[H_2] + k_{54b}\,[H_2O]$$

A different truncation than for the 6-step mechanism is used to obtain an explicit expression for the O-radical. Since C_2H_2 is not well predicted by the 4-step mechanism, its profile being too high in the first stage of the flame, an important error in the calculation of O would result by retaining reactions 45 and 46 in the expression for O. A similar argument holds for CH_3, which is in too high in the first stage of the flame due to the steady-state assumption of C_2H_4 and C_2H_2. To avoid negative concentrations of O, the following truncated form for CH_3 is used in calculating the concentration of the O-radical

$$[CH_3]^* = \frac{(k_{77} + k_{78})\,[H]}{k_{35}\,[O] + k_{37}\,[O_2]} \tag{8.30}$$

The truncated steady state relation for O involves then the following rates

$$w_1 - w_2 + w_4 - w_{35} = 0 \tag{8.31}$$

which leads with (8.30) to the quadratic form

$$(b_O \cdot k_{35})\,[O]^2 + (b_O \cdot k_{37}\,[O_2] + c_O \cdot k_{35} - a_O\,k_{35})\,[O] + a_O\,[O_2]\,k_{37} = 0 \tag{8.32}$$

with

$$\begin{aligned} a_O &= k_{1f}\,[O_2][H] + k_{2b}\,[OH][H] + k_{4f}\,[OH][OH] \\ b_O &= k_{1b}\,[OH] + k_{2f}\,[H_2] + k_{4b}\,[H_2O] \\ c_O &= (k_{77} + k_{78})\,[H][C_3H_8] \end{aligned} \tag{8.33}$$

All other steady state concentrations, as far as they are necessary, are calculated in the same way as in the 6-step mechanism.

8.5 Comparison of burning velocities

Burning velocities calculated with the full mechanism, the 6-step and the 4-step mechanism are shown for atmospheric flames as a function of equivalence ratio ϕ in Fig. 8.13. A comparison between the full mechanism, the 9-step mechanism and experimental data taken from Law [8.4] are shown in Fig. 8.14. It is seen that the three different reduced mechanisms differ little from the full mechanism, while the data show somewhat lower burning velocities with a maximum at 45 cm/sec for slightly rich flames. Even lower maximum burning velocities, namely around 40 cm/sec were found by previous experimental investigations [8.5]–[8.9] which were reproduced in [8.9]. The

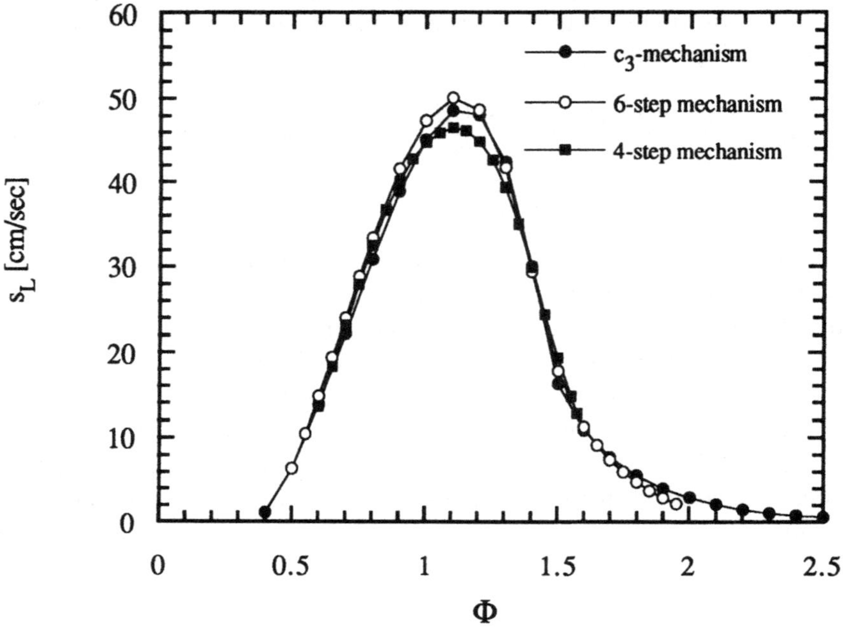

Fig. 8.13. Comparison of burning velocities calculated from the full kinetic mechanism, the 6-step and the 4-step mechanism as a function of equivalence ratio.

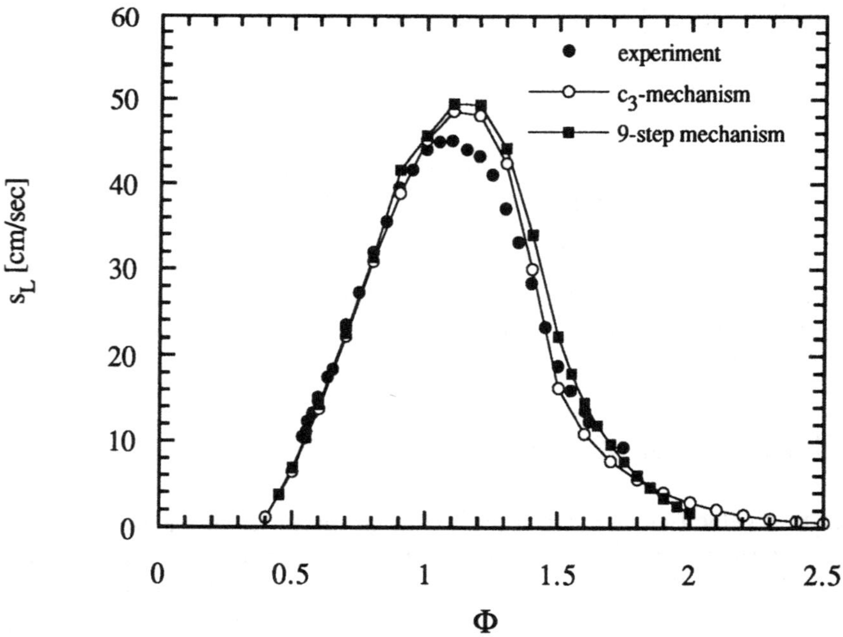

Fig. 8.14. Comparison of burning velocities calculated from the full kinetic mechanism, the 9-step mechanism and experimental data as a function of equivalence ratio.

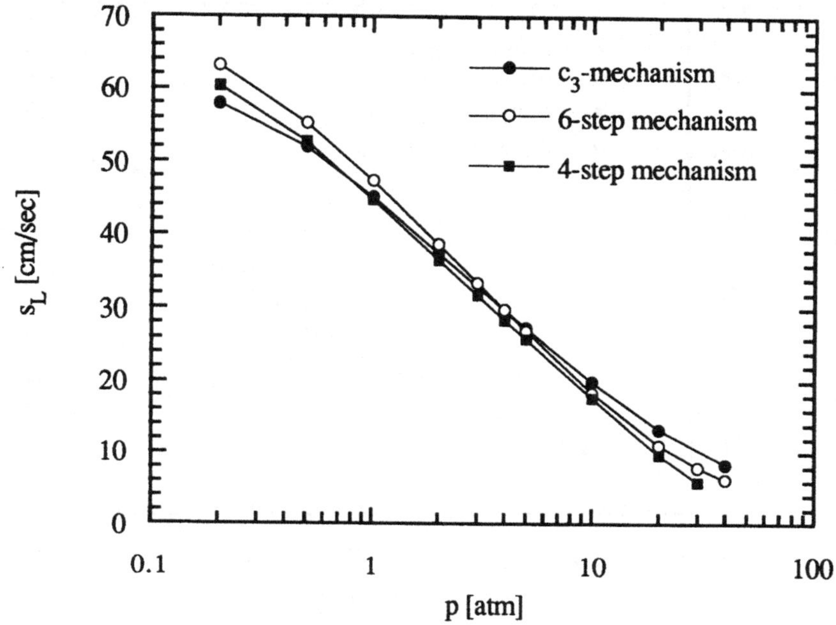

Fig. 8.15. Comparison of burning velocities calculated from the full kinetic mechanism, the 6-step mechanism and the 4-step mechanism as a function of pressure.

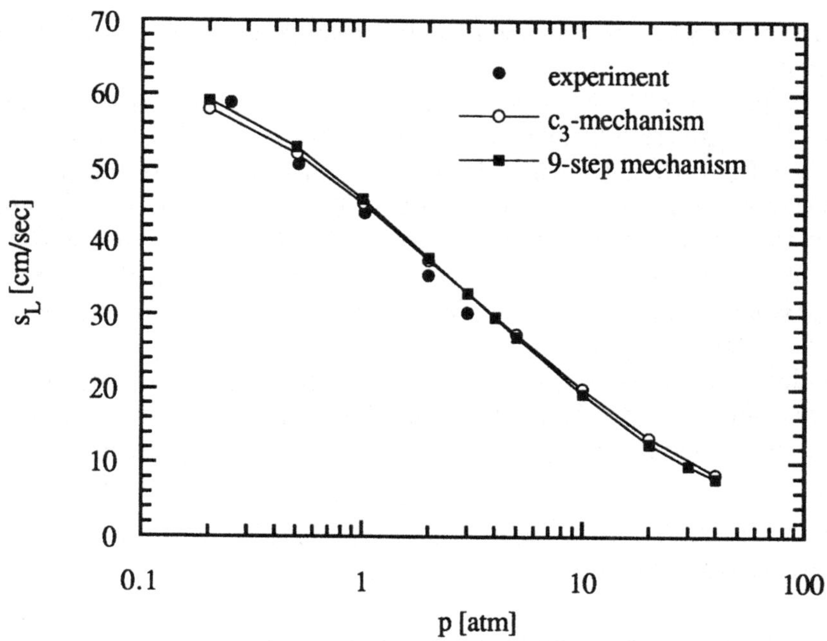

Fig. 8.16. Comparison of burning velocities calculated from the full kinetic mechanism and the 9-step mechanism with experimental data as a function of pressure.

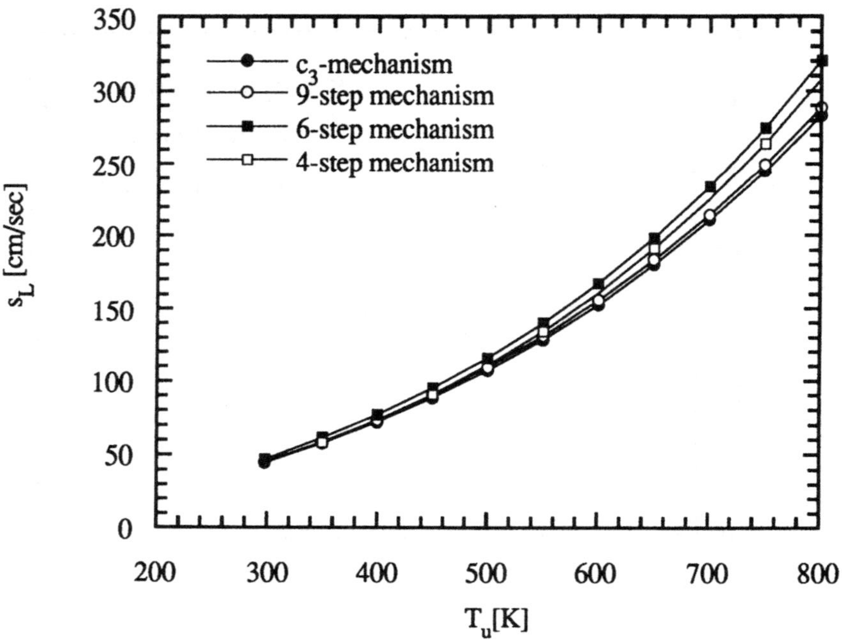

Fig. 8.17. Comparison of burning velocities calculated from the full kinetic mechanism, the 9-step mechanism, the 6-step mechanism and the 4-step mechanism as a function of preheat temperature.

discrepancy between the data and the full mechanism used here is not fully understood.

A comparison of burning velocities for stoichiometric flames at different pressures between the full kinetic mechanism, the 6-step and 4-step mechanism is shown in Fig. 8.15 and a comparison between the full mechanism, the 9-step mechanism and experimental data from Law [8.4] in Fig. 8.16 . The agreement between the 4-step, 6-step and the full mechanism is good for pressures above 1 atm until around 10 atm and becomes worse at lower and higher pressures. The agreement with experimental data which are available for pressures up to 3 atm is also quite good.

Finally, a comparison of burning velocities obtained for the different reduced mechanism for stoichiometric atmospheric flames with different preheat temperatures is shown in Fig. 8.17 showing a reasonable agreement.

8.6 Conclusions

Three different reduced kinetic mechanism were derived for propane flames. While the 9-step mechanism involves numerical iterations, the 6-step mechanism, which uses some truncations of steady state species seems to present on optimal choice, also in view of further extensions such as the prediction of benzene and soot formation.

The 4-step mechanism reproduces overall properties such as burning velocities quite well even for rich flames, but it is less accurate as far as minor species are concerned and it may prove to be numerically less stable.

References

[8.1] Warnatz, J.: Twentieth Symposium (International) on Combustion, The Combustion Institute, pp. 845–856, 1985.

[8.2] Paczko, G., Lefdal, P. M., Peters, N.: Twenty-First Symposium (International) on Combustion, The Combustion Institute, pp. 739–748, 1985.

[8.3] Djavdan, E., Darabiha, N., Giovangigli, V., Candel, S. M., Combust. Sci. Tech. 76, pp. 287–309, 1991.

[8.4] Law, C. K.: A Compilation of Recent Experimental Data of Premixed Laminar Flames, Chap. 2 of this book.

[8.5] Metghalchi, M., Keck, J. C.: Combustion Flame, **38**, p. 143, 1980.

[8.6] Smith, D. B., Taylor, S. C., Williams, A.: Burning Velocities in Stretched Flames, Poster P220 presented at the Twenty-Second Symposium (International) on Combustion, The Combustion Institute, 1988.

[8.7] Scholte, T. G., Vaags, P. B.: Combustion Flame, **3**, p. 495, 1980.

[8.8] Yamaoka, I., Tsuji, H.: Twentieth Symposium (International) on Combustion, The Combustion Institute, p. 1883, 1985.

[8.9] Kennel, C., Göttgens, J., Peters, N.: Twenty-Third Symposium (International) on Combustion, The Combustion Institute, p. 479–485, 1990.

9. Reduced Kinetic Mechanisms for Premixed Methanol Flames

C.M. Müller and N. Peters
Institut für Technische Mechanik, RWTH Aachen,
W-5100 Aachen, Germany

9.1 Introduction

Methanol is recently being discussed as a potentially most interesting alternative fuel for future use in spark ignition engines. It has the advantage over traditional hydrocarbon fuels which are derived from mineral oil, that it can be produced from biological sources and therefore represents a regenerative energy source. It also can be synthesized from coal for which fossil resources will continue to be available for many centuries to come. Many automotive tests have shown that it is indeed a suitable liquid fuel. It has important advantages over other alternative fuels like hydrogen which raise storage, safety and distribution problems. Compared to traditional fuels its main disadvantage is its volume pro unit energy which is about twice as large, but it also is somewhat more corrosive.

Concerning its kinetic properties it may safely be regarded as the most simple fuel other than H_2 or CO. Elementary mechanisms for methanol combustion have been proposed in [9.1]–[9.3] and a reduced mechanism has based on these has already been derived in [9.4]. The kinetics are essentially similar to those of the C_1-chain for methane combustion except that CH_2OH radical replaces the CH_3 as the first intermediate in the chain. Two CH_3's can recombine to produce C_2-hydrocarbon, but CH_2OH does not open a reaction path to higher hydrocarbon species. This is the reason why practically no soot is formed by the combustion of methanol.

9.2 Chemical Kinetic Mechanism

The numerical calculations of this chapter were performed using the C_1 mechanism shown in Table 1 of Chapter 1 except reactions involving CH_4, which were found to be unimportant for the problem adressed here. However, the kinetics to be used for methanol premixed flames are different from those used for methanol diffusion flames (Chap. 16). Premixed flames are more sensitive to chemical kinetics and it was found here that the mechanism which is sufficient to describe diffusion flames was not appropriate for premixed flames, especially for fuel rich flames. Additional reactions involving CH, CH_2 and CH_3 are introduced and newer kinetic data from [9.5] (reaction 83–90 of Table 9.1) are used. In [9.6] methanol oxidation is investigated using different kinetc data than used in this calculations and reactions including CH_3O are provided. It was found here that this is no appropriate way if calculating with Warnatz kinetic data. Thus the starting mechanism used includes the reactions 1–33, 35, 37, 83–90.

Table 9.1. Additional kinetic data for methanol oxidation from [9.5] (reaction 86 is taken from [9.13].).

No	Reaction	mole,cm^3,sec	n	E [kJ/mole]
83	$CH_2OH + H = CH_2O + H_2$	3.00E+13	0.0	0.0
84	$CH_2OH + O_2 = CH_2O + HO_2$	1.00E+13	0.0	30.0
85	$CH_2OH + M = CH_2O + H$	5.00E+13	0.0	105.0
86	$CH_3OH + M = CH_3 + OH + M$	3.16E+18	0.0	336.0
87	$CH_3OH + H = CH_2OH + H_2$	4.00E+13	0.0	25.5
88	$CH_3OH + O = CH_2OH + OH$	1.00E+13	0.0	19.6
89	$CH_3OH + OH = CH_2OH + H_2O$	1.00E+13	0.0	7.1
90f	$CH_3OH + HO_2 \rightarrow CH_2OH + H_2O_2$	0.62E+13	0.0	81.1
90b	$CH_2OH + H_2O_2 \rightarrow CH_3OH + HO_2$	0.10E+08	1.7	47.9

The resulting burning velocities calculated with the standard and the extended starting mechanism at 1 bar pressure and initial temperature 298 K are compared with several experiments [9.7]–[9.11] in Fig. 9.1. All experiments were summarized in [9.11]. There it is pointed out that the burning velocities do not only depend on the measurement technique used, but there are also differences between measurements executed with the same technique. Figure 9.1 shows that the calculations with the extended starting mechanism are in the lower range of the experiments at the lean side and in good agreement at the rich side. The standard starting mechanism yields too high burning velocities on the rich side.

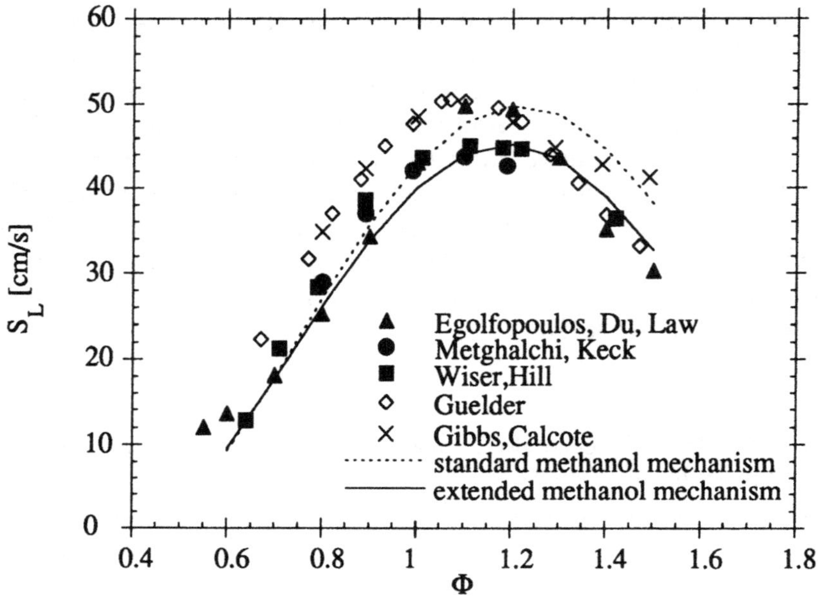

Fig. 9.1. Comparison between the measured [9.7]–[9.11] and numerically calculated burning velocities at p = 1 bar and T_u = 298 K.

9.3 Reduced Kinetic Mechanism

In order to reduce the kinetic steps to the smallest possible number which are necessary to describe the flame structure reasonably well, one has to identify those intermediate species for which steady state approximations hold. The terms representing production, consumption, diffusion and convection are shown in Figs. 9.2–9.5 for the chemical species H, OH, CH_2O, CH_2OH using the extended starting mechanism (which is called starting mechanism from here on) at 1 bar and initial temperature 298 K. The profiles are plotted as a function of the non-dimensional coordinate

$$x^* = (\rho_u s_L) \int_0^x \frac{c_p}{\lambda} \, dx. \tag{9.1}$$

Obviously the production and consumption terms dominate in the balance for the OH radical in Fig. 9.3 as compared to the diffusion and convection terms everywhere in the flame. Therefore it is reasonable to introduce the steady state approximation for OH. This is not as evident for H and CH_2OH shown in Figs. 9.2 and 9.5. Introducing steady state assumptions for these species leads to deviations from the starting mechanism depending on how important the species is for the entire mechanism. Here the steady state approximation is introduced for CH_2OH and not for the H radical, because the H radical takes part in the important chain branching reaction $H + O_2 \rightarrow OH + O$. Figure 9.4

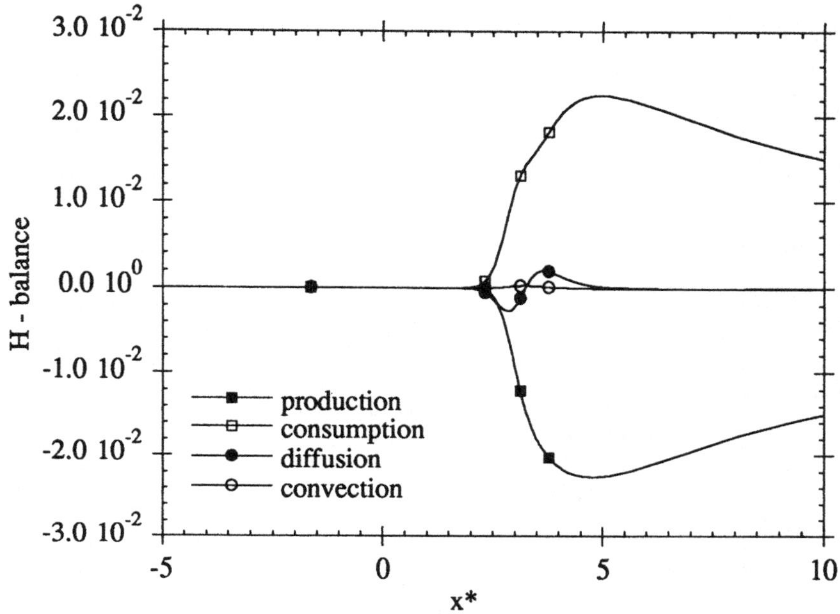

Fig. 9.2. Balance of production, consumption, diffusion and convection of H radical at p = 1 bar , $T_u = 298\,\mathrm{K}$ and $\Phi = 1$.

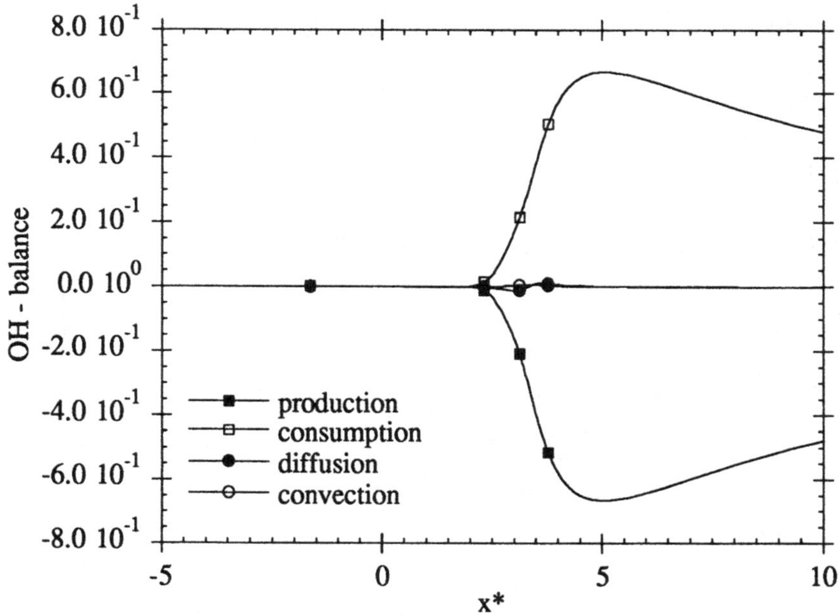

Fig. 9.3. Balance of production, consumption, diffusion and convection of OH radical at p = 1 bar , $T_u = 298\,\mathrm{K}$ and $\Phi = 1$.

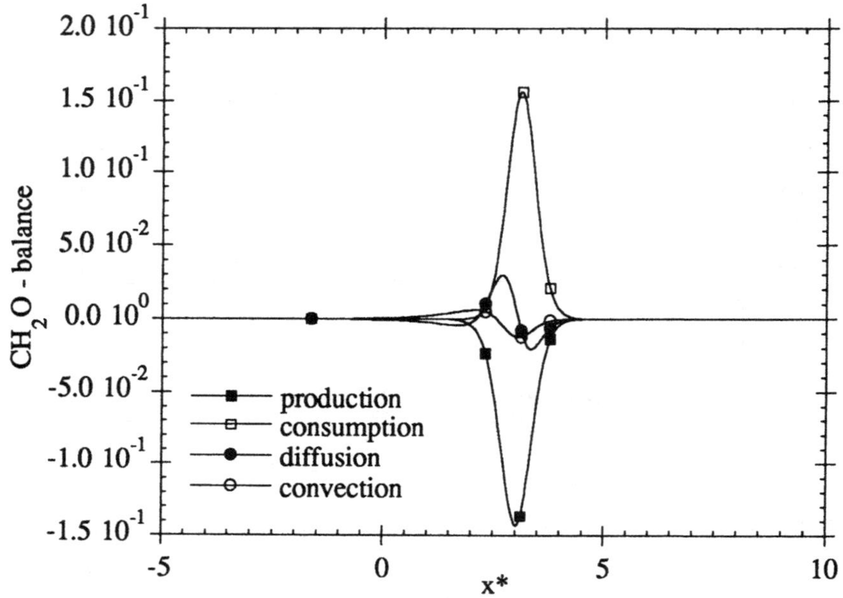

Fig. 9.4. Balance of production, consumption, diffusion and convection of CH_2O at $p = 1$ bar , $T_u = 298$ K and $\Phi = 1$.

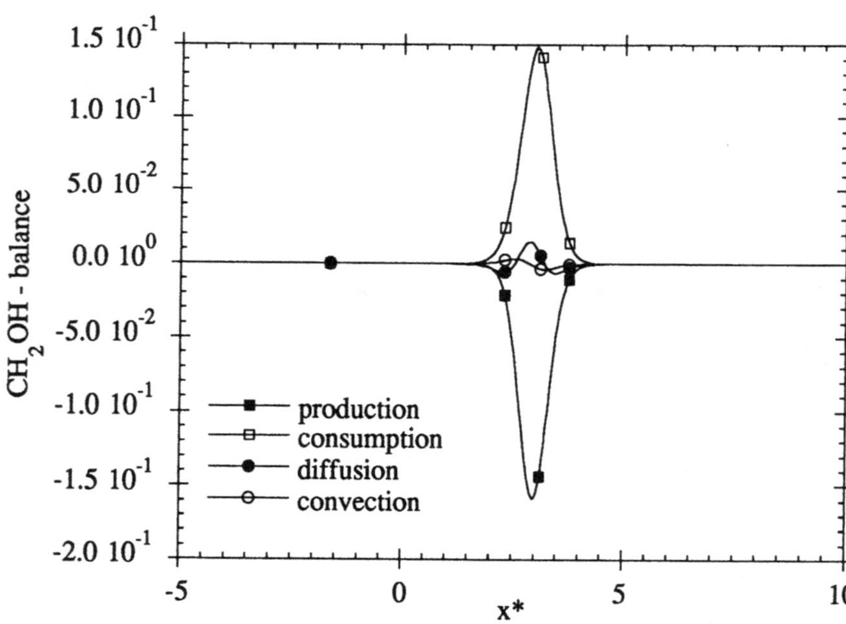

Fig. 9.5. Balance of production, consumption, diffusion and convection of CH_2OH at $p = 1$ bar , $T_u = 298$ K and $\Phi = 1$.

shows that CH_2O cannot be assumed in steady state and should be retained. This is different from [9.4] and will lead to a 5-step mechanism here rather then to a 4-step mechanism. Similar plots for the species O, HO_2, H_2O_2, CH, CH_2, CH_3, HCO (not shown here) suggest that all of them are in steady state.

By introducing the above steady state approximations and eliminating the reactions 2, 3, 8, 12, 19, 25, 35, 24 and 83 using the steady state relations for O, OH, HO_2, H_2O_2, CH, CH_2, CH_3, HCO, CH_2OH, respectively, a 5-step mechanism can be deduced. One obtains the following global steps

I	$CH_3OH + 2H$	$= 2H_2 + CH_2O$
II	CH_2O	$= CO + H_2$
III	$CO + H_2O$	$= CO_2 + H_2$
IV	$H + H + M$	$= H_2 + M$
V	$3H_2 + O_2$	$= 2H + 2H_2O$

The rates of the overall steps I – V can be written as

$$\begin{aligned} \omega_{I} &= \omega_1 + \omega_6 + \omega_9 + \omega_{11} - \omega_{14} - \omega_{17} \\ &\quad - \omega_{20} - \omega_{26} + \omega_{33} + \omega_{37} + \omega_{90} \\ \omega_{II} &= \omega_5 - \omega_{11} + \omega_{13} + \omega_{14} + \omega_{15} + \omega_{16} + \omega_{17} + \omega_{21} \\ &\quad + \omega_{22} + \omega_{23} - \omega_{32} - \omega_{85} - \omega_{86} - \omega_{90} \\ \omega_{III} &= \omega_{18} - \omega_{20} + \omega_{28} \\ \omega_{IV} &= \omega_{31} + \omega_{29} + \omega_{30} + \omega_{32} + \omega_{33} \\ \omega_{V} &= \omega_{87} + \omega_{86} + \omega_{88} + \omega_{89} + \omega_{90} \end{aligned} \tag{9.2}$$

A further reduction to a 4-step mechanism by introducing the additionally steady state assumption for CH_2O leads in contrast to the methanol diffusion flames not to satisfactory results for the large range of pressures, equivalence ratios and preheat temperatures considered here.

9.4 Steady State Relations

Introducing the steady state approximations into the starting mechanism leads to algebraic expressions for every steady state species. In general these expressions are coupled and can be solved iteratively. In order to obtain explicit expressions and to facilitate subsequent asymptotic analysis [9.12] some of the steady state relations need to be truncated. By neglecting the influence of reactions 6, 8, 10, 12, 13, 14, 22, 27, 37, 86b, 88b and 89b in the steady state relation for OH as well as using the steady state assumption of [O] (defined below), the steady state relation of [OH] can be written as

$$[OH] = \frac{-b + \sqrt{b^2 + 4ac}}{2a} \tag{9.3}$$

where

$$
\begin{aligned}
a = {} & 3k_{1b}k_{4f}[\mathrm{OH}] + k_{4f}(k_{2f}[\mathrm{H_2}] + k_{30}[\mathrm{CH_2O}] + k_{88f}[\mathrm{CH_3OH}]) \\
& + k_{1b}(2k_{2b}[\mathrm{H}] + k_{3f}[\mathrm{H_2}] + k_{16}[\mathrm{H}][\mathrm{M}] \\
& \qquad + k_{18f}[\mathrm{CO}] + k_{31}[\mathrm{CH_2O}] + k_{89f}[\mathrm{CH_3OH}]) \\
b = {} & (k_{3f}[\mathrm{H_2}] + k_{16}[\mathrm{H}][\mathrm{M}] + k_{18f}[\mathrm{CO}] + k_{31}[\mathrm{CH_2O}] + k_{89f}[\mathrm{CH_3OH}]) \\
& \qquad (k_{2f}[\mathrm{H_2}] + k_{4b}[\mathrm{H_2O}] + k_{30}[\mathrm{CH_2O}] + k_{88f}[\mathrm{CH_3OH}]) \\
& - k_{2b}k_{4b}[\mathrm{H_2O}][\mathrm{H}] \\
& - k_{1b}(k_{3b}[\mathrm{H_2O}][\mathrm{H}] + k_{18b}[\mathrm{CO_2}][\mathrm{H}] + k_{86f}[\mathrm{CH_3OH}][\mathrm{M}]) \\
c = {} & (2k_{1f}[\mathrm{O_2}][\mathrm{H}] + k_{3b}[\mathrm{H_2O}][\mathrm{H}] + k_{18b}[\mathrm{CO_2}][\mathrm{H}] + k_{86f}[\mathrm{CH_3OH}][\mathrm{M}]) \\
& \qquad (k_{2f}[\mathrm{H_2}] + k_{4b}[\mathrm{H_2O}] + k_{30}[\mathrm{CH_2O}] + k_{88f}[\mathrm{CH_3OH}]) \\
& + k_{1f}k_{4b}[\mathrm{H_2O}][\mathrm{O_2}][\mathrm{H}]
\end{aligned}
\tag{9.4}
$$

Here [M] represents the concentration of the third body. Since the quantity 'a' contains the [OH] the expression (9.3) is, in principle, a cubic equation. For algebraic simplicity the [OH] appearing in 'a' is replaced by $[\mathrm{OH}]^*$ similar to the procedure for the diffusion flames (Chap. 16). $[\mathrm{OH}]^*$ is the concentration of OH calculated from partial equilibrium of reaction 3

$$[\mathrm{OH}]^* = \frac{[\mathrm{H_2O}][\mathrm{H}]}{K_3[\mathrm{H_2}]} \tag{9.5}$$

where K_3 is the equilibrium constant of reaction 3.

The steady state relation for [O] can then be derived by neglecting the influence of reactions 9, 10, 17, 19, 26, 35, 88b.

$$[\mathrm{O}] = \frac{(k_{1f}[\mathrm{O_2}] + k_{2b}[\mathrm{OH}])[\mathrm{H}] + k_{4f}[\mathrm{OH}]^2}{k_{1b}[\mathrm{OH}] + k_{2f}[\mathrm{H_2}] + k_{4b}[\mathrm{H_2O}] + k_{30}[\mathrm{CH_2O}] + k_{88f}[\mathrm{CH_3OH}]} \tag{9.6}$$

The steady state relations for the other species can be written as

$$[\mathrm{CH_3}] = \frac{k_{86f}[\mathrm{CH_3OH}][\mathrm{M}]}{k_{33f}[\mathrm{H}] + k_{35}[\mathrm{O}] + k_{86f}[\mathrm{OH}][\mathrm{M}]} \tag{9.7}$$

$$[\mathrm{CH_2}] = \frac{k_{33f}[\mathrm{CH_3}][\mathrm{H}]}{k_{25f}[\mathrm{H}](1 - Z_{25b}) + k_{26}[\mathrm{O}] + (k_{27} + k_{28})[\mathrm{O_2}] + k_{33b}[\mathrm{H_2}]} \tag{9.8}$$

with

$$Z_{25b} = \frac{k_{25b}[\mathrm{H}]}{k_{19}[\mathrm{O_2}] + k_{20}[\mathrm{CO_2}] + k_{25b}[\mathrm{H_2}]} \tag{9.9}$$

$$[\mathrm{CH}] = \frac{k_{25f}[\mathrm{CH_2}][\mathrm{H}]}{k_{19}[\mathrm{O_2}] + k_{20}[\mathrm{CO_2}] + k_{25b}[\mathrm{H_2}]} \tag{9.10}$$

$$[\mathrm{CH_2OH}] = \frac{[\mathrm{CH_3OH}](k_{87f}[\mathrm{H}] + k_{88f}[\mathrm{O}] + k_{89f}[\mathrm{OH}] + k_{90f}[\mathrm{HO_2}])}{k_{83}[\mathrm{H}] + k_{84}[\mathrm{O_2}] + k_{85}[\mathrm{M}] + k_{87b}[\mathrm{H_2}] + k_{88b}[\mathrm{OH}] + k_{89b}[\mathrm{H_2O}]} \tag{9.11}$$

$$[HCO] = \frac{k_{19}[CH][O_2] + k_{20}[CO_2][CH] + k_{24b}[CO][H][M]}{k_{21}[H] + k_{22}[OH] + k_{23}[O_2] + k_{24f}[M]} + \frac{[CH_2O](k_{29}[H] + k_{30}[O] + k_{31}[OH] + k_{32}[M])}{k_{21}[H] + k_{22}[OH] + k_{23}[O_2] + k_{24f}[M]} \tag{9.12}$$

$$[H_2O_2] = \frac{k_{11}[HO_2][HO_2] + k_{12f}[OH][OH][M]}{k_{12b}[M] + k_{13}[H] + k_{14f}[OH] + k_{90b}[CH_2OH]} + \frac{k_{14b}[H_2O][HO_2] + k_{90f}[CH_3OH][HO_2]}{k_{12b}[M] + k_{13}[H] + k_{14f}[OH] + k_{90b}[CH_2OH]} \tag{9.13}$$

The $[HO_2]$ included here can be derived from the quadratic equation

$$[HO_2] = \frac{-b + \sqrt{b^2 + 4ac}}{2a} \tag{9.14}$$

where

$$\begin{aligned} a &= k_{11}(2k_{12b}k_{13}[H][M] + k_{14f}k_{90b}[OH][CH_2OH]) \\ b &= (k_{5b}[M] + (k_6 + k_7 + k_8)[H] + k_8[OH] + k_{10}[O] + k_{14b}[H_2O] \\ &\quad + k_{90f}[CH_3OH])(k_{12b}k_{13}[H][M] + k_{14f}k_{90b}[OH][CH_2OH]) \\ &\quad - (k_{14f}k_{90b}[OH][CH_2OH])(k_{14b}k_{90f}[H_2O][CH_3OH]) \\ c &= (k_{5f}[O_2][H][M] + k_{23}[HCO][O_2] + k_{84}[CH_2OH][O_2]) \\ &\qquad (k_{12b}k_{13}[H][M] + k_{14f}k_{90b}[OH][CH_2OH]) \\ &\quad + (k_{14f}k_{90b}[OH][CH_2OH])k_{12f}[OH][OH][M] \end{aligned} \tag{9.15}$$

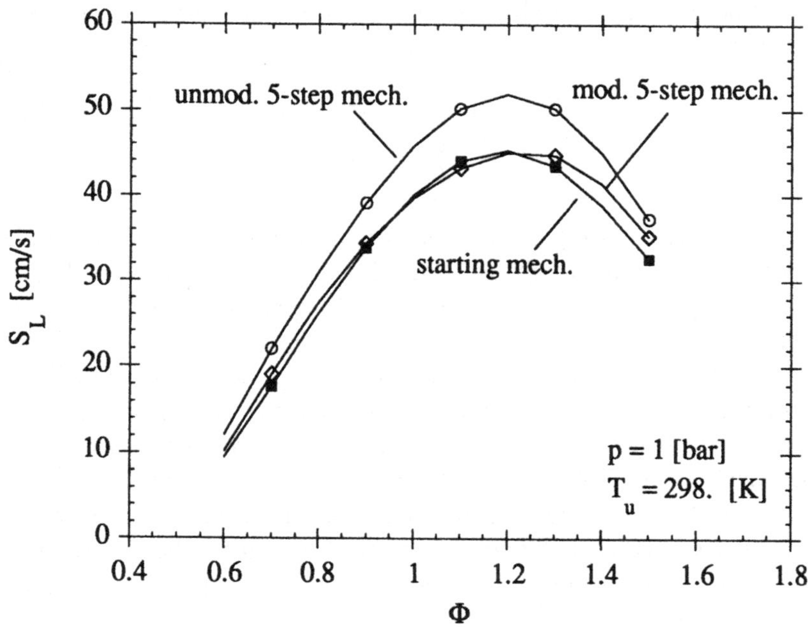

Fig. 9.6. Burning velocities calculated with the starting mechanism, the unmodified and the modified 5-step mechanism at p = 1 bar and $T_u = 298\,\mathrm{K}$.

9.5 Comparison between Starting and Reduced Mechanism

Numerical calculations were performed using the starting mechanism and the 5-step mechanism at different values of the equivalence ratio, pressure and inital temperature. Figure 9.6 shows the burning velocity at 1 bar and $T_u =$ 298K as a function of the equivalence ratio. The burning velocity calculated with the 5-step mechanism is higher than that of the starting mechanism. This is essentially due to the fact that the mass of the steady state species is not taken into account in the 5-step mechanism. This was first demonstrated for H_2 premixed flames (Chap. 3). A modified 5-step mechanism is therefore proposed, whose results are also in Fig. 9.6. The mass fractions which are used to determine the reaction rates are corrected by subtracting the atom mass defects calculated using the steady state species from the non-steady state species.

$$Y_i = Y_{i,\mathrm{O}} - \sum_{j=1}^{ne} dis_{i,j} \Delta_j, \quad (ne = 1,3) \tag{9.16}$$

Here $Y_{i,o}$ is the mass fraction of species i without correction, ne is the number of elements (C,O,H) used in the mechanism, Δ_j is the mass defect of atom j

$$\Delta_j = M_j \sum_{k=1}^{nss} a_{k,j} \frac{Y_k}{M_K} \tag{9.17}$$

and $dis_{i,j}$ describes the distribution of Δ_j among the different non steady state species i.

$$dis_{i,j} = \frac{a_{i,j} \frac{Y_{i,\mathrm{O}}}{M_i}}{\sum_{l=1}^{nis} a_{l,j} \frac{Y_{i,\mathrm{O}}}{M_l}} \tag{9.18}$$

In these two expressions $a_{i,j}$ is the number of atoms j in species i and nss stands for the number of steady state species and nis accounts for the number of non steady state species, which have to corrected. As Fig. 9.7 shows the concentrations of CH_3OH, CO_2 and H_2O are well represented by the uncorrected 5-step mechanism. For that reason they are not modified.

The corrected 5-step mechanism is in better agreement with the starting mechanism at all values of the equivalence ratio. Nevertheless for rich mixtures the burning velocity is still higher than calculated with the starting mechanism (Fig. 9.6). Figure 9.9 shows the mole fractions of H, O and OH calculated with the corrected and uncorrected 5-step mechanism compared with the starting mechanism at 1 bar and $T_u = 298$ K. The most important reason for the better behaviour of the corrected 5-step mechanism is the more accurate representation of the radical level. Figures 9.8, 9.10 and 9.11 show a good agreement of the modified 5-step mechanism compared with the starting mechanism at 1 bar and initial temperature of 298 K for the temperature

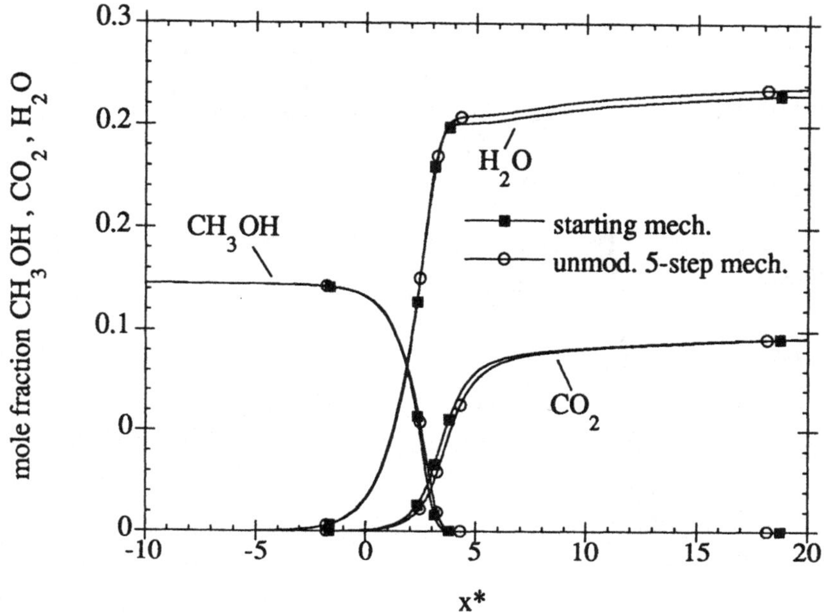

Fig. 9.7. Mole fractions of CH_3OH, CO_2 and H_2O calculated with the starting mechanism and the unmodified 5-step mechanism at p = 1 bar, T_u = 298 K and $\Phi = 1$.

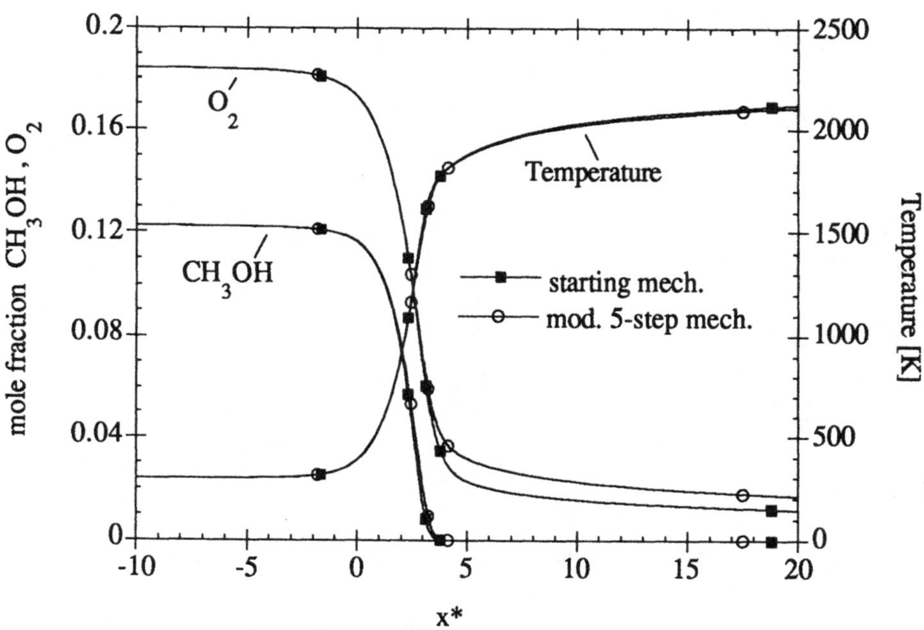

Fig. 9.8. Temperature profile and mole fractions of CH_3OH and O_2 calculated with the starting mechanism and the modified 5-step mechanism at p = 1 bar, T_u = 298 K and $\Phi = 1$.

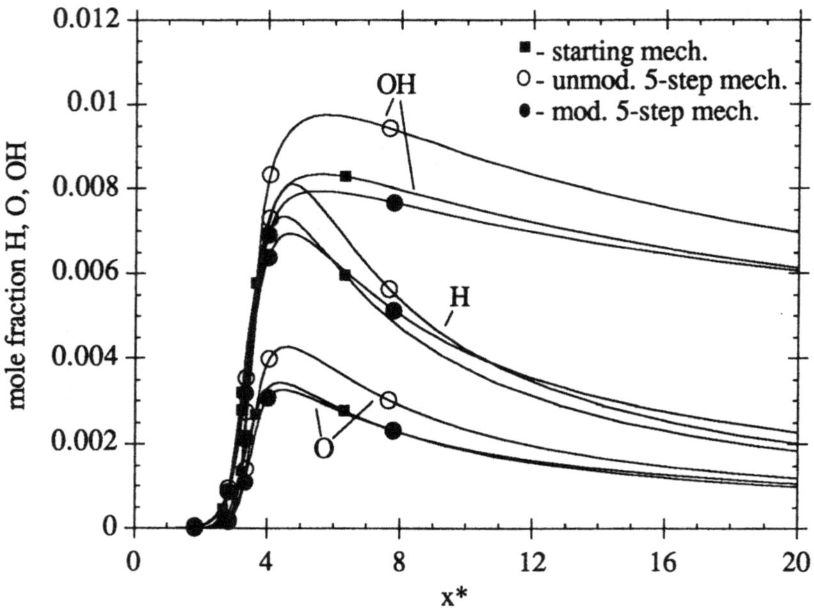

Fig. 9.9. Mole fractions of H, O and OH calculated with the starting mechanism, the unmodified and the modified 5-step mechanism at p = 1 bar, T_u = 298 K and $\Phi = 1$.

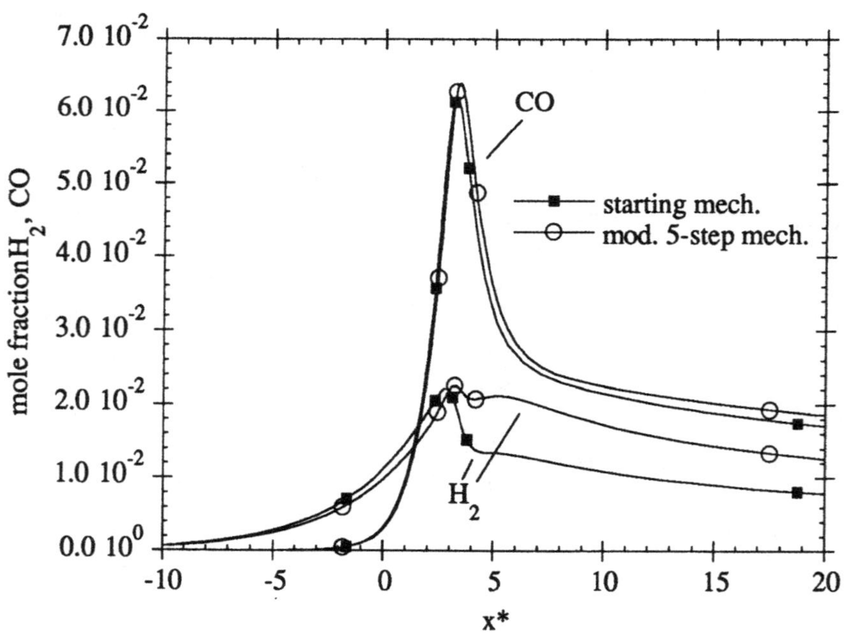

Fig. 9.10. Mole fractions of CO and H_2 calculated with the starting mechanism and the modified 5-step mechanism at p = 1 bar, T_u = 298 K and $\Phi = 1$.

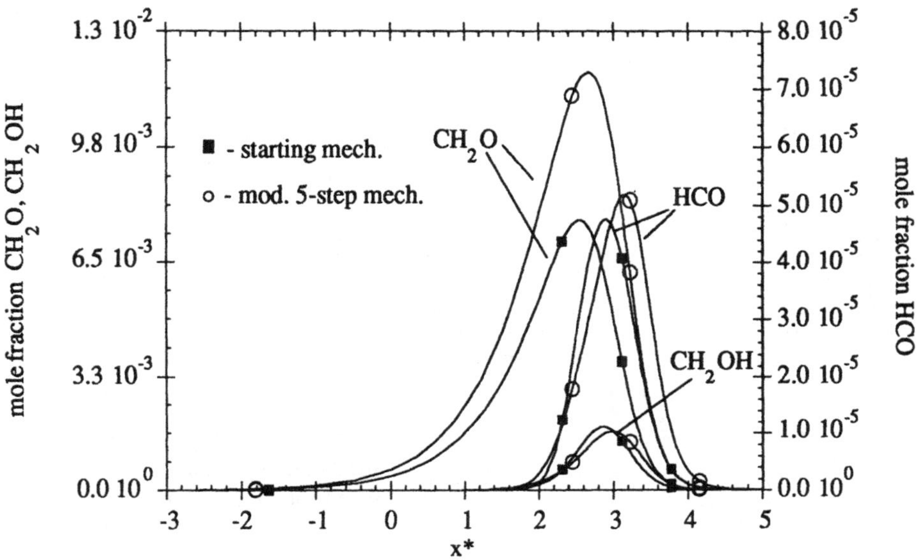

Fig. 9.11. Mole fractions of HCO, CH_2O, CH_2OH calculated with the starting mechanism and the modified 5-step mechanism at p = 1 bar, T_u = 298 K and $\Phi = 1$.

profile and the species CH_3OH and CO. However, the concentration of CH_2O is still too high, whereas O_2, H_2, HCO and CH_2OH are predicted fairly well.

Figure 9.12 shows the burning velocity at different values of the inital temperature T_u at $\Phi = 1$ and p = 1 bar. Some measurements [9.11] in the region between 298 K and 370 K are included. Again the corrected 5-step mechanism yields good results compared with the starting mechanism and reasonable agreement with the experiments. The uncorrected 5-step mechanism leeds ,as expected, to higher burning velocity at equivalence ratio $\Phi = 1$.

In Fig. 9.13 the burning velocity is plotted as a function of the pressure at $\Phi = 1$ and T_u = 298 K. At all pressures the corrected 5-step mechanism is better than the uncorrected mechanism, but at higher pressures both reduced mechanisms lead to nearly the same deviation from the burning velocity calculated with the starting mechanism. This can be explained by the fact that the truncations of the steady state relations are not valid with the same degree of accuracy as in the low pressure case.

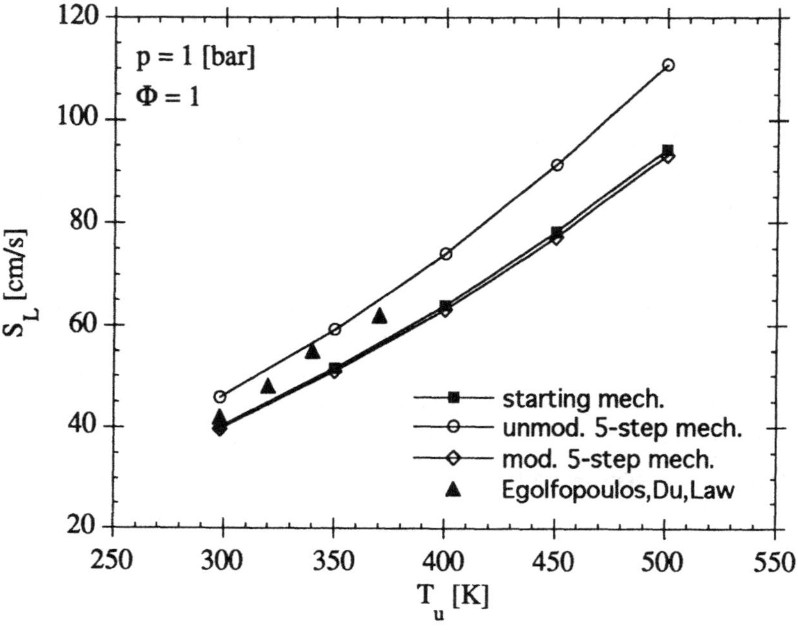

Fig. 9.12. Burning velocities as function of preheat temperature T_u calculated with the starting mechanism, the unmodified and the modified 5-step mechanism compared with experiments at p = 1 bar and $\Phi = 1$.

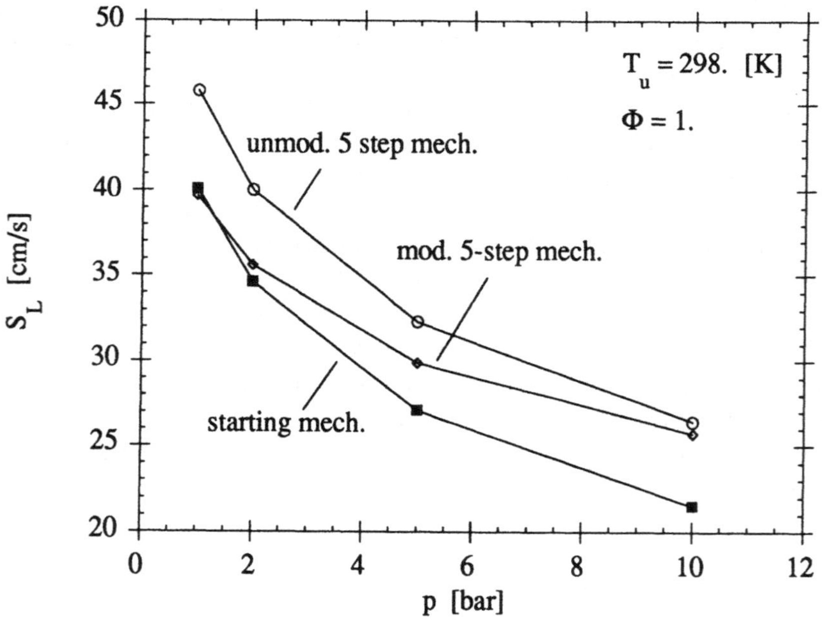

Fig. 9.13. Burning velocities as function of pressure p using the starting mechanism, the unmodified and the modified 5-step mechanism at $T_u = 298$ K and $\Phi = 1$.

9.6 Conclusions

A 5-step mechanism for methanol premixed flames has been derived and calculations of the burning velocity and the flame structure have been performed. The reduced mechanism was deduced from a starting mechanism, which is essentially the C_1-chain of the mechanism shown in Table 1 of Chap. 1 with the methanol reactions listed in Table 9.1. The starting mechanism was tested by comparing its results with experiments from different authors and was found to agree reasonably well.

It has been shown that the mass contribution of the steady state species can not be neglected in the reduced mechanism calculations and that a modified 5-step mechanism leads to reasonable results over a wide range of equivalence ratios (0.6–1.5) and initial temperatures (298 K–500 K). The calculation of the burning velocities for different pressures (1–10 bar) has shown, that the agreement of the corrected 5-step mechanism with the starting mechanism at higher pressure decreases. This is due to the truncations of the steady state relations. It has been found, that the flame structure obtained with the reduced mechanism is in good agreement with the starting mechanism in the case of the major species, the radicals and most of the intermediates. However, the CH_2O is predicted as too high. Nevertheless, it can be concluded that the starting mechanism and the corrected 5-step mechanism are capable of predicting the main features of methanol premixed flames.

References

[9.1] Westbrook, C. K., Dryer, F. L., Comb. Flame 37, 171, 1980.
[9.2] Dove, J. E., Warnatz, J., Bunsenges. Phys. Chem. 87, 1040, 1983.
[9.3] Norton, T. S., Dryer, F. L., Int. J. Chem. Kinet. 22, 219,1990.
[9.4] Paczko, G., Lefdal, P. M., Peters, N., 21th Symp. on Comb.,739, 1986.
[9.5] Warnatz, J., private communication.
[9.6] Bradley, D., Dixon-Lewis, G., El-Dim Habik, S., Kwa, L. K., El-Sherif, S., Comb. Flame 85, 105, 1991.
[9.7] Wiser, W. H., Hill ,G. R., Proc. 5th Symp. on Comb.,p 553, 1955.
[9.8] Gibbs, C. J., Calcote, H. C., J. Chem. Eng. Data 4, 226, 1959.
[9.9] Metghalchi, M., Keck, J. C., Comb. Flame 48, 191, 1982.
[9.10] Gülder, Ö. L., Proc. 19th Symp. on Comb., 275, 1982.
[9.11] Egolfopoulos, F. N., Du, D. X. and Law, C. K., A Comprehensive Study of Methanol Kinetics in Freely Propagating and Burner-Stabilized Flames, Flow and Statics Reactors, and Shock Tubes, Comb. Sci. Tech., to appear 1992.
[9.12] Yang, B., Seshadri, K., Peters N., The Asymptotic Structure of Premixed Methanol-Air Flames , submitted to Combustion and Flame.

Part III

Counterflow Diffusion Flames

10. Structure and Extinction of Non–Diluted Hydrogen–Air Diffusion Flames

C. Treviño
Facultad de Ciencias, UNAM, 04510 Mexico D.F., Mexico

F. Mauss
Institut für Technische Mechanik, RWTH Aachen, 5100 Aachen, Germany

10.1 Abstract

In this paper we study the structure of non-diluted hydrogen-air diffusion flames using both a starting kinetic mechanism and a reduced two-step mechanism with atomic hydrogen as an intermediate non-steady state species. The extinction conditions were calculated for pressures from 0.5 up to 40 atm. It is shown that a two-step mechanism is capable of reproducing the structure and extinction in the pressure range mentioned. Four different sets of approximations were used for the steady-state concentration of OH and HO_2. The errors introduced in these approximations are presented and the best choice is indicated.

10.2 Introduction

There is an increasing interest in the study of both premixed and diffusion flames of hydrogen, due to the fact that the kinetics is very well understood and because its key role in hydrocarbon combustion. A detailed numerical calculation of the governing equations in the planar counterflow configuration has been done by Dixon-Lewis and Missaghi [10.1]. They obtained the critical strain rate for extinction for atmospheric flames to be 13,000 s^{-1}. Tangirala et al. [10.2] performed numerical calculations of the hydrogen diffusion flame at low strain rates and developed an analysis of the structure of the flame valid for low strain rates (high Damköhler numbers). The results agree well with the numerical calculations in the main reaction zone. It was found that for low strain rates the hydrogen atom diffuses into the fuel side because of the absence of oxygen. The three body reaction $H + H + M \rightarrow H_2 + M$ is not strong enough to control diffusion at the fuel side. On the other side, the hydrogen atom concentration disappears at a finite value of the mixture fraction coordinate due to the very strong three body reaction $H + O_2 + M \rightarrow HO_2 + M$ at low strain rates. In the high Damköhler number limit, radicals disappear at the crossover temperature between the latter reaction and the chain branching reaction $H + O_2 \rightarrow OH + O$ at the lean side. Recently, Gutheil and Williams [10.3] analyzed the structure of a hydrogen/air diffusion flame in an axis-symmetrical counterflow configuration for different pressures and temperatures using numerical integration with a complete kinetic mechanism and by an asymptotic analysis based on a reduced one-step kinetic mechanism. They obtained reasonable agreement of the species concentration at low strain rates.

The purpose of the present work is to apply different two-step reduced kinetic mechanisms with different approximations for the steady-state of radicals and to compare with results of the starting mechanism involving only reaction 1–10 and 15–17 in Table 1 of Chap. 1.

10.3 Formulation

The flow studied in this work corresponds to the classical planar counterflow configuration. The governing equations are very well known and can be found elsewhere [10.4]. The formulation, transport description and numerical procedure have been described in the literature [10.4]–[10.8]. The calculations were made for the case of reactants at room temperature and pure hydrogen as the fuel. Pressure were varied from 0.5 atm to 40 atm. All the results are presented in the mixture fraction space, Z, computed from the equation given in Chap. 1.

10.4 Reaction Kinetics

HO_2 Chain

For the reactants at room temperature and not very large pressures (up to 40 atm), the H_2O_2 concentration is very low and does not play an important role in the structure of the flame. The species HO_2 can be assumed to be always in steady-state. HO_2 is produced only by the three-body reaction 5 and is consumed rapidly by reactions with hydrogen atom, 6, 7, 9, and with reaction 8 with OH. Reaction 10 does not have any influence on the structure of hydrogen/oxygen flames and can be neglected. The most important HO_2 consumption reaction is the chain branching reaction 6. However, it competes with the chain breaking reaction 8 in low temperature regions in the air side of the flame. Assuming all HO_2 is consumed by reaction 6, together with the very rapid chain propagating reaction 3, would give a global step

$$2H_2 + O_2 \rightarrow 2H_2O \qquad (10.1)$$

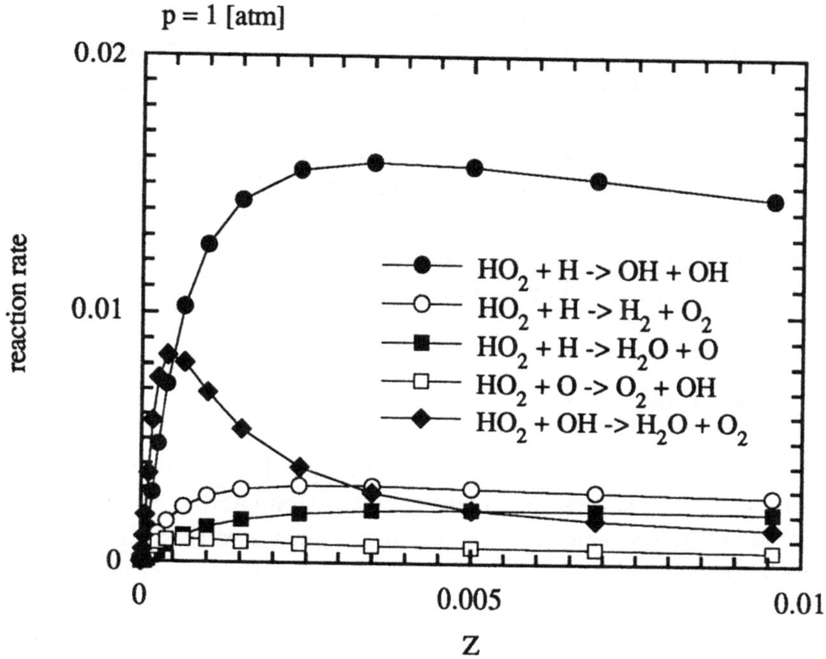

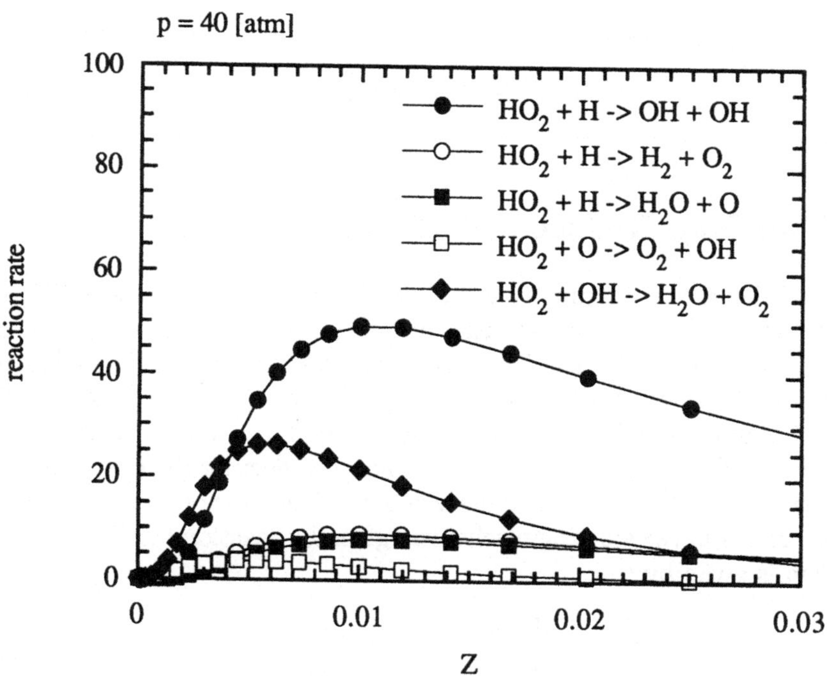

Fig. 10.1. HO_2 consumption rates of important reactions close to extinction conditions for two working pressures: 1 atm ($a_q = 14,260\ s^{-1}$) and 40 atm ($a_q = 570,000\ s^{-1}$)

which produces the most part of the heat release in the flame. However all other HO_2 consumption reactions are important and have to be retained in order to study the structure of the flame. Figure 10.1 shows the consumption rates of all important HO_2 reactions from the numerical calculations with the starting mechanism close to extinction for two different pressures, 1 and 40 atm. Due to the relative low activation energy of all the HO_2 consumption reactions, the rates of reactions 6–10 are almost proportionally related. Reaction 8, however, depends on the OH concentration and is very important only in the low temperature air side of the flame.

10.5 O Reactions

Figure 10.2 shows the global consumption and production rates for atomic oxygen, for two different pressures close to extinction conditions. The steady-state behavior of this species is fully justified. This steady-state is dictated mainly by the reactions 1 and 2. Close to extinction only the forward reactions are important because the low radical concentrations which make radical-radical reactions to be very slow. Therefore, the steady-state assumption for O gives

$$w_1 = w_2 \tag{10.2}$$

resulting the global step given by

$$H_2 + O_2 = OH + OH \quad . \tag{10.3}$$

10.6 Two-Step Reduced Mechanism

Figure 10.3 shows the consumption and production rates of radical OH, for two extreme pressures (1 and 40 atm) in the range adopted for this work, close to extinction conditions. The very strong forward reaction 3 makes the OH to be always in steady-state, except in the low temperature regions of the air side of the flame, where the concentration of H_2 is very low. Assuming further the steady-state for OH, HO_2 and O and neglecting the H_2O_2 chain, a global two-step mechanism can be derived with H as intermediate non-steady state species. From the starting mechanism given by reaction 1 in Table 1, this mechanism is given by

$$\text{(I)} \qquad 3H_2 + O_2 = 2H_2O + 2H \tag{10.4}$$

$$\text{(II)} \qquad H + H + M' \rightarrow H_2 + M' \tag{10.5}$$

with the global rates given by

$$w_I = w_1 + w_5 \left\{ \frac{k_6 + k_9}{k_6 + k_7 + k_9 + k_8\,[\mathrm{OH}]/[\mathrm{H}]} \right\} \tag{10.6}$$

$$w_{II} = w_5 + w_{15} \tag{10.7}$$

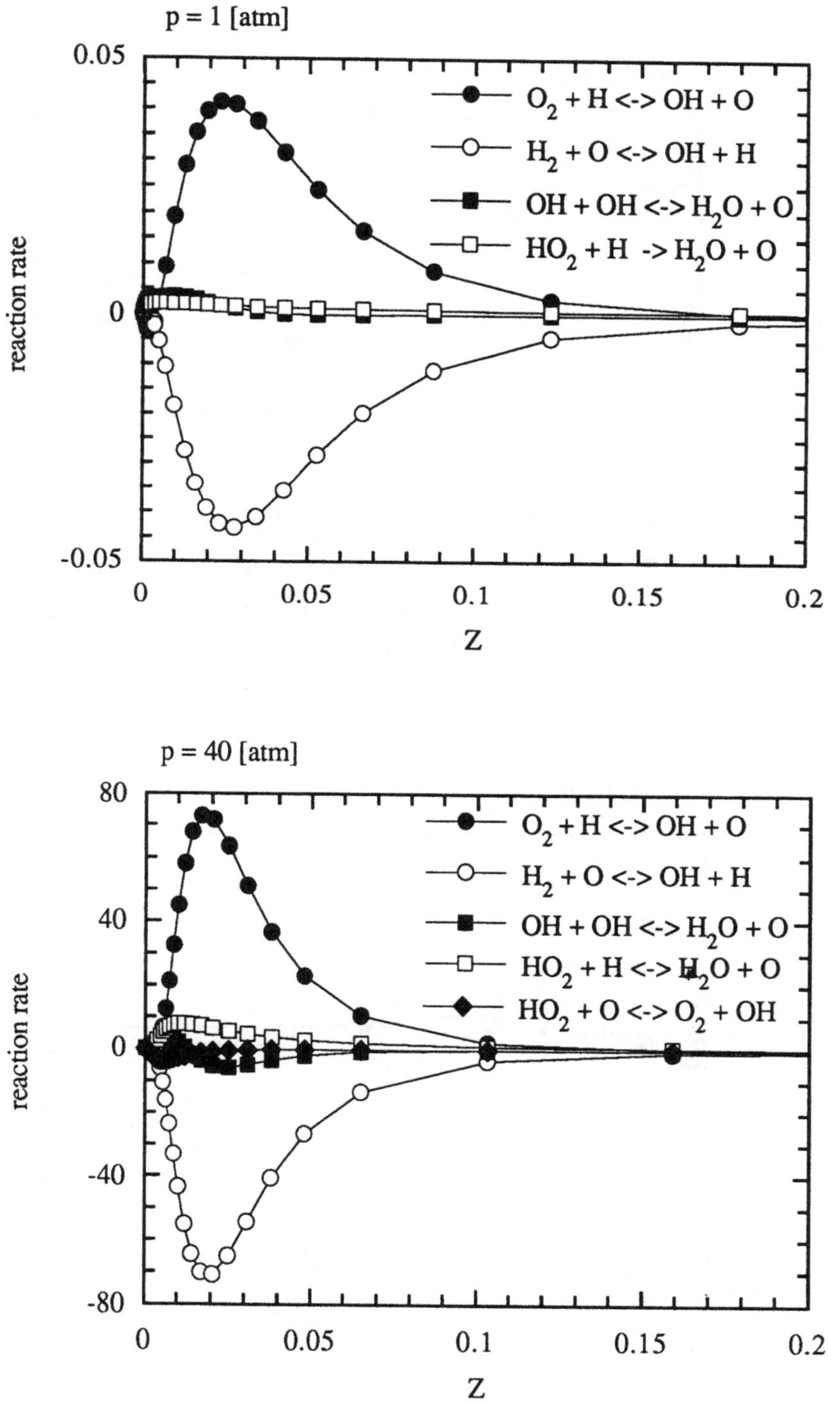

Fig. 10.2. O-atom Consumption-production rates for two different pressures:1 atm ($a_q = 14,260\ \mathrm{s}^{-1}$) and 40 atm ($a_q = 570,000\ \mathrm{s}^{-1}$)

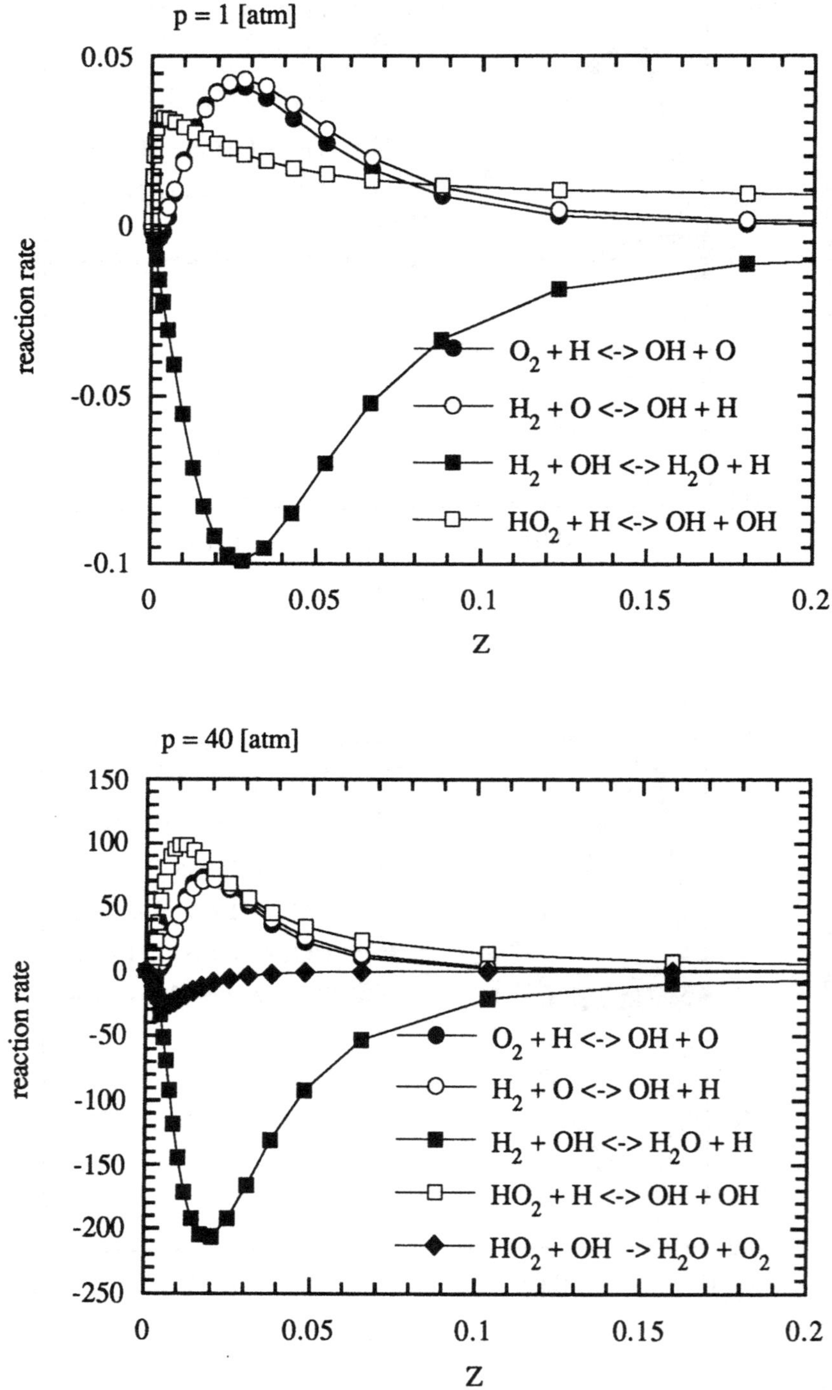

Fig. 10.3. OH Consumption-production rates for two different pressures: 1 atm ($a_q = 14,260\ s^{-1}$) and 40 atm ($a_q = 570,000\ s^{-1}$)

The OH concentration is related to the H concentration through its steady-state assumption as:

$$[\mathrm{OH}]/[\mathrm{H}] = \frac{2k_{1f}\,[\mathrm{O_2}]}{k_{3f}\,[\mathrm{H_2}]} + \frac{k_{3b}\,[\mathrm{H_2O}]}{k_{3f}\,[\mathrm{H_2}]} + \frac{k_5\,[\mathrm{M}]\,[\mathrm{O_2}]}{k_{3f}\,[\mathrm{H_2}]}\left[\frac{2\,k_6 + k_9 - k_8\,[\mathrm{OH}]/[\mathrm{H}]}{k_6 + k_7 + k_9 + k_8\,[\mathrm{OH}]/[\mathrm{H}]}\right] \tag{10.8}$$

which represents a quadratic algebraic equation for [OH]. Here we made use of the steady state assumption for O in terms $w_1 + w_9 = w_2 + w_{10}$. The first two terms on the right hand side are very important in the high temperature region of the flame. Therefore, in the low temperature air side of the flame the last term is important. However, in this zone the steady-state assumption for OH is questionable due to very low concentration of reactant H_2. From the steady-state assumption for HO_2, we have that the concentration of HO_2 is given by

$$[\mathrm{HO_2}] = \frac{k_{5f}\,[\mathrm{O_2}]\,[\mathrm{M}]}{[k_6 + k_7 + k_9 + k_8\,[\mathrm{OH}]/[\mathrm{H}]]} \quad . \tag{10.9}$$

In the low temperature region, the important term in the denominator is the last one inside the brackets. A small error introduced in this term through the steady-state assumption for OH, given in (10.18), causes a big error in the concentration of HO_2. This feature is the most important one related to the present two-step reduced mechanism. The steady-state for O gives

$$[\mathrm{O}]/[\mathrm{H}] = \frac{k_{1f}\,[\mathrm{O_2}] + k_{2b}\,[\mathrm{OH}]}{k_{2f}\,[\mathrm{H_2}] + k_{1b}\,[\mathrm{OH}]} \quad . \tag{10.10}$$

The steady-state calculation for the radical O is not a problem because the most important terms are the first ones in numerator and denominator in the whole region of the flame. Four different approximations were employed in the numerical calculations with this two-step mechanism, in order to maintain (10.8) and (10.9) in the same general form:

	Eq. (10.8)	Eq. (10.9)
Set 1	same as above	same as above
Set 2	$k_8 = 0.0$	same as above
Set 3	$k_9 = k_{13} = 0.0$	$k_9 = 0.0$
Set 4	$k_9 = 0.0$	$k_9 = 0.0$

10.7 Results

In this section the relevant results are shown for nondiluted H_2/Air diffusion flames with reactants at room temperature. The numerical results were obtained with the full starting mechanism on one side and the two-step reduced

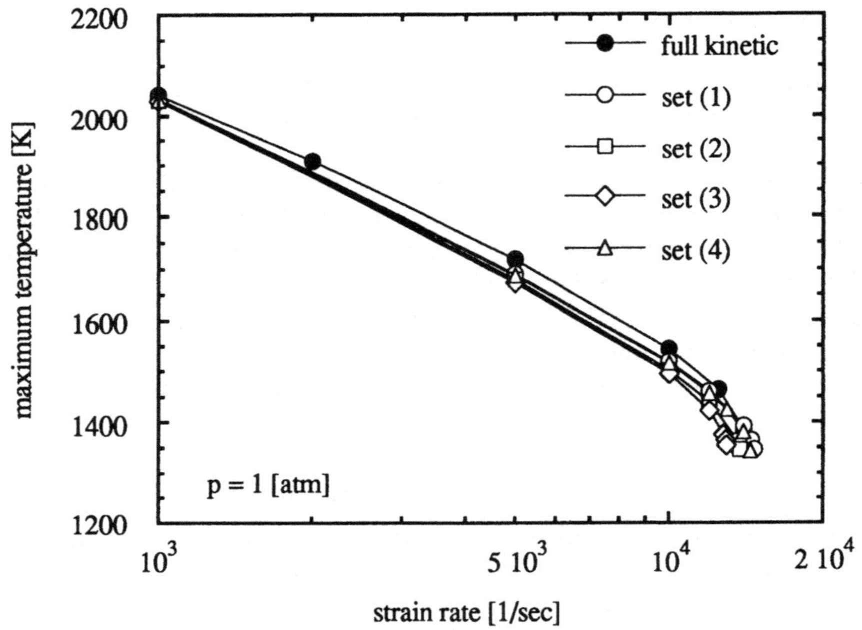

Fig. 10.4. Maximum flame temperature as a function of the strain rate, a, for $p =$ 1 atm for the starting mechanism and the four sets of approximations.

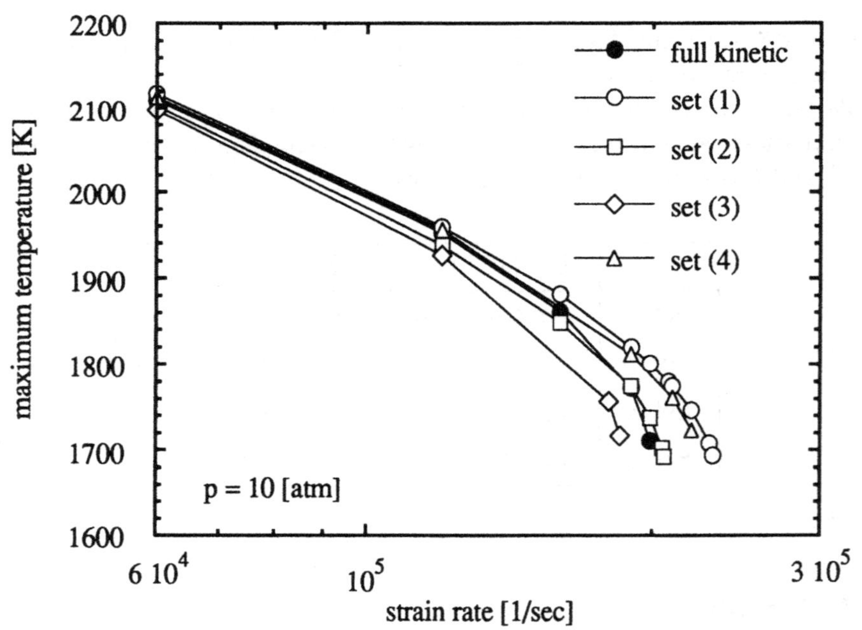

Fig. 10.5. Maximum flame temperature as a function of the strain rate, a, for $p =$ 10 atm for the starting mechanism and the four sets of approximations.

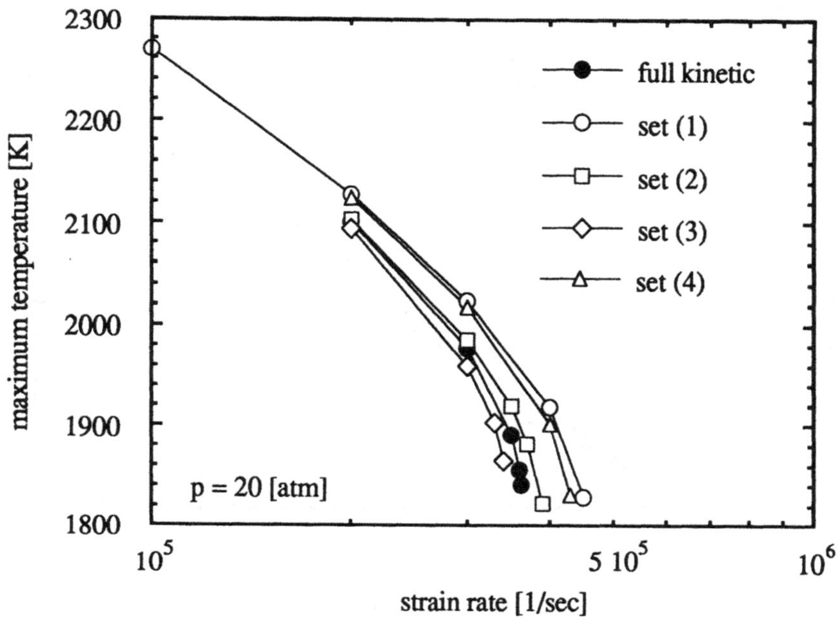

Fig. 10.6. Maximum flame temperature as a function of the strain rate, a, for $p =$ 20 atm for the starting mechanism and the four sets of approximations.

kinetic mechanism with four different sets of values used for the calculations of the concentrations of $[HO_2]$ and $[OH]$, on the other side. Figures 10.4–10.6 show the maximum temperature in the flame as a function of the strain rate a, for three different pressures 1, 10 and 20 atm, respectively. All four different sets mentioned above were employed in the calculations. For $p = 1$ atm (Fig. 10.4), the differences in all four sets were not very strong and compared well with the full starting mechanism. For very low strain rates there is no any difference in all the sets. However, close to extinction ($a_q = 14,260\,s^{-1}$) is the condition where the flame is sensitive to the steady-state formulation of radicals OH and HO_2. For $p = 10$ atm (Fig. 10.5), the differences are bigger increasing in an important way close to extinction ($a_q = 199,000\,s^{-1}$). For 20 atm (Fig. 10.6) the same trend was found increasing the differences bewteen all sets close to the extinction conditions ($a_q = 362,000\,s^{-1}$). In all the cases, for low values of the strain rate, the flame structure is insensitive to all four steady-state approximations.

Figure 10.7 shows the extinction strain rate as a function of the pressure for the case with full starting mechanism and the four different sets adopted. The best sets were (2) and (3). Set (3) gives always lower values for the strain rate of extinction. Set (2) gives very good results for all pressures up to 20 atm. In this case the chain branching reaction 9 is taken into account everywhere but the effect of reaction 8 in the calculation of the steady-state

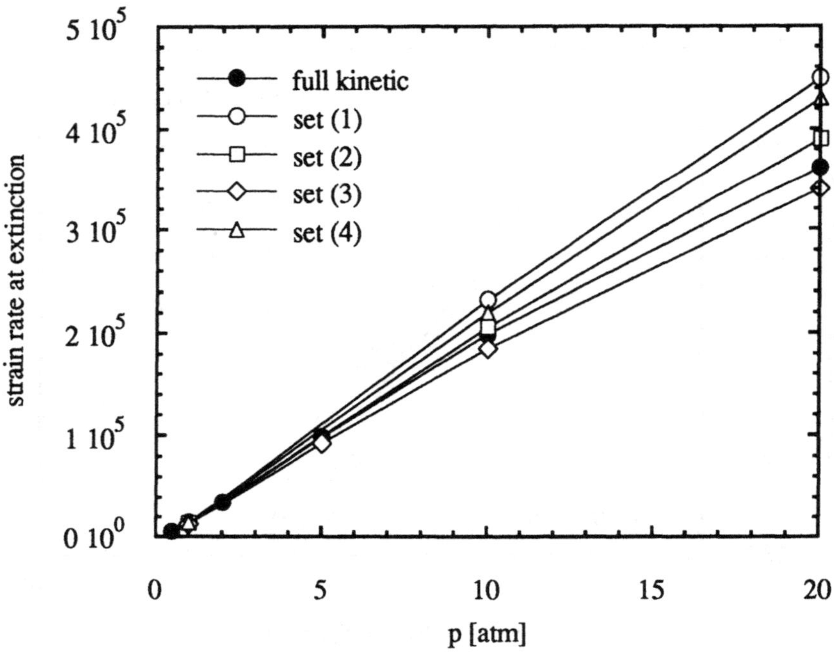

Fig. 10.7. Strain rate at extinction as a function of the pressure for the starting mechanism and the four sets of approximations.

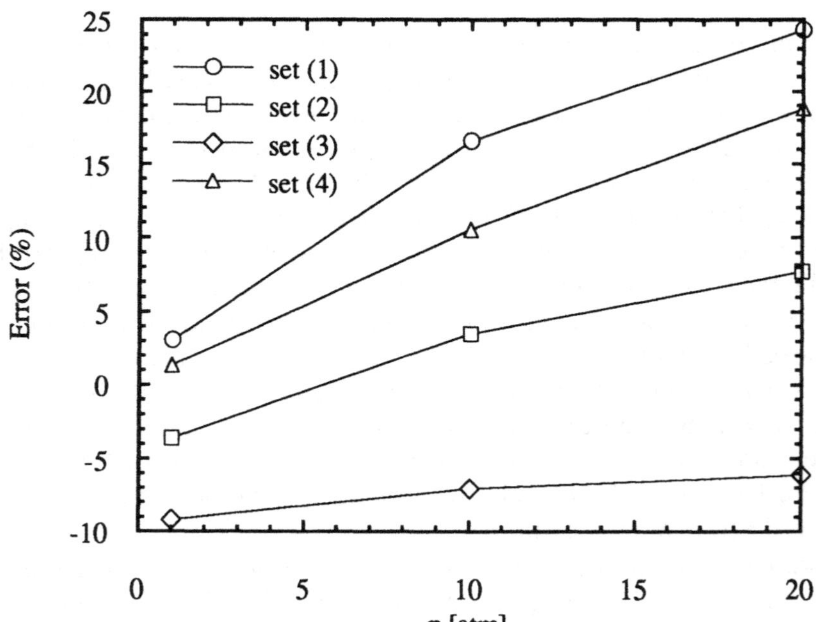

Fig. 10.8. Errors introduced at the extinction strain rate when using the four different sets of approximations.

value for OH is absent. Figure 10.8 shows the maximal relative error introduced (close to extinction conditions) using these approximations as a function of the pressure.

The structure of the flame can be explained as follows. For low strain rates, radicals are produced in the high temperature region, with the three shuffle reactions equilibrated. Radicals diffuse to both sides of the flame. On the air side they are consumed by the three body recombination reaction 5 together with reactions 7 and 8. In the limit of infinite Damköhler number (zero strain rate) hydrogen atom and oxygen molecule cannot coexist for temperatures below the cross-over temperature. The radicals then disappear at low temperatures on the air side. However on the fuel side, because of the absence of molecular oxygen, reactions 1 and 5 have not the importance as on the air side. Reaction 15 then takes over, reaching partial equilibrium for very low strain rates. Practically, hydrogen atom diffuses to the fuel side. As the strain rate increases because of its high diffusivity, hydrogen atom also diffuses to the air side. Partial equilibrium of the shuffle reactions breaks down in the radical production zone. Very close to extinction, the radical concentrations decrease in such a way that radical-radical reactions are very slow and can be neglected to first approximation.

There are three important reaction zones for the hydrogen diffusion flame close to extinction. A thin radical production zone is located around the maximum temperature, surrounded by two thick radical consumption-heat production layers. Molecular oxygen is fully consumed in the radical production zone. The leading order structure of the flame does not depend on the H_2 concentration as long the steady-state for O and OH is justified. In this flame, molecular hydrogen is one of the main products of radical recombimation, so reduction to a global one-step cannot be justified close to extinction.

Figure 10.9 shows the mole fractions of the three active radicals H, OH and O close to extinction for a pressure of 1 and 40 atm. It is seen that for atmospheric pressure, the hydrogen atom concentration is very large compared with the other active radicals OH and O, except in the low temperature region in the air side of the flame. Close to extinction, H is one of the main products of the diffusion flame, together with water vapor. Figure 10.10 shows the phase-space trajectories of molar concentration of H as a function of that of H_2O. This plot shows for a low pressure flame (Fig. 10.10a) a very good correlation with a straight behavior of slope close to unity if the effect of the Lewis number were included (X_H/L_H). This means that hydrogen atom as well water vapor are produced almost at the same rate in the main reaction zone as indicated by the global reaction I in the two-step mechanisms. Thus, hydrogen atom production and consumption has to be retained in the reaction mechanism for near extinction conditions. A reduction to one-step introduces a qualitative change in the flame structure. However, this is not the behaviour for large pressures where the three body reaction 5 becomes important in the high temperature region. The low temperature radical consumption reactions seems to be not very strong close to extinction to stop H diffusion to both fuel and air regions.

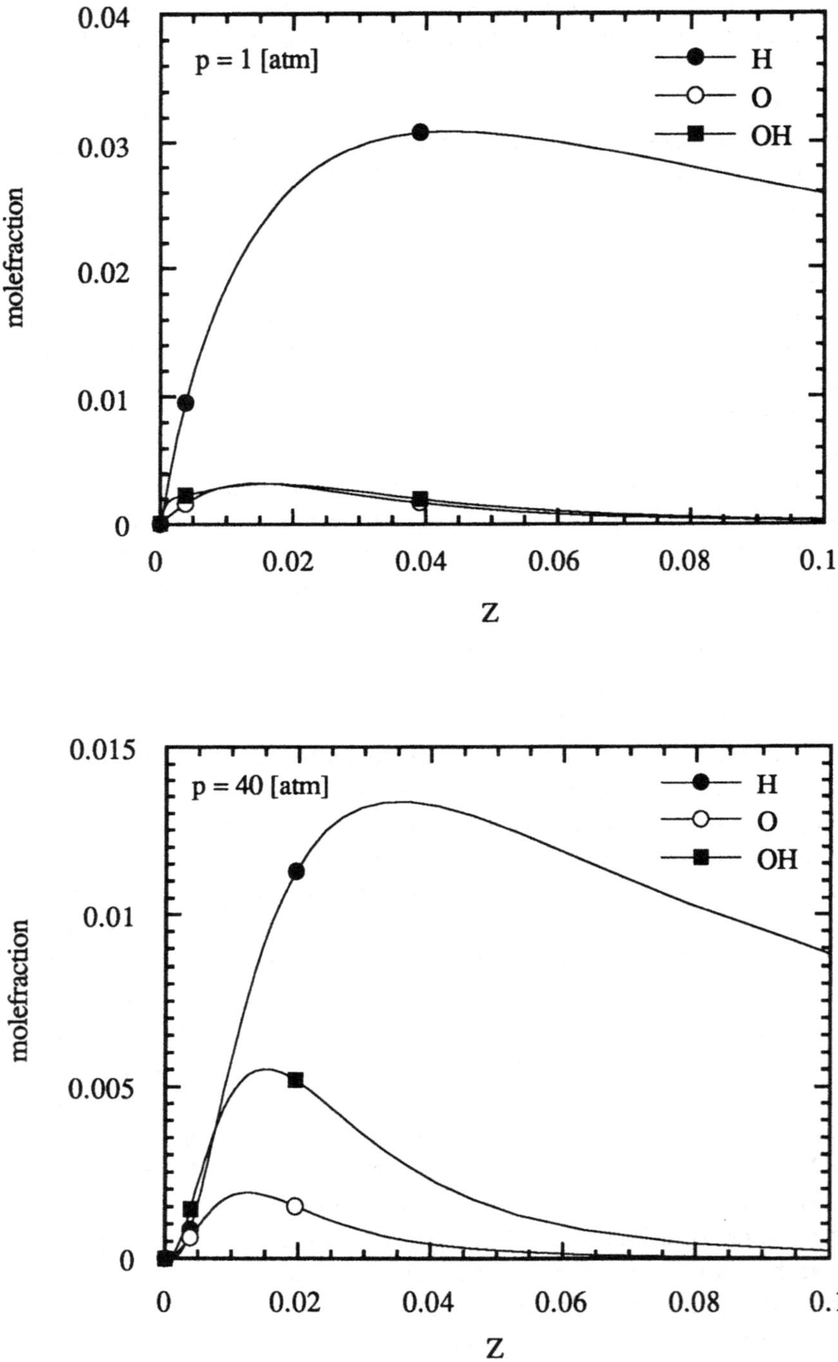

Fig. 10.9. Radical concentration profiles as a function of the mixture fraction coordinate, Z, for two different pressures: a) 1 atm, b) 40 atm.

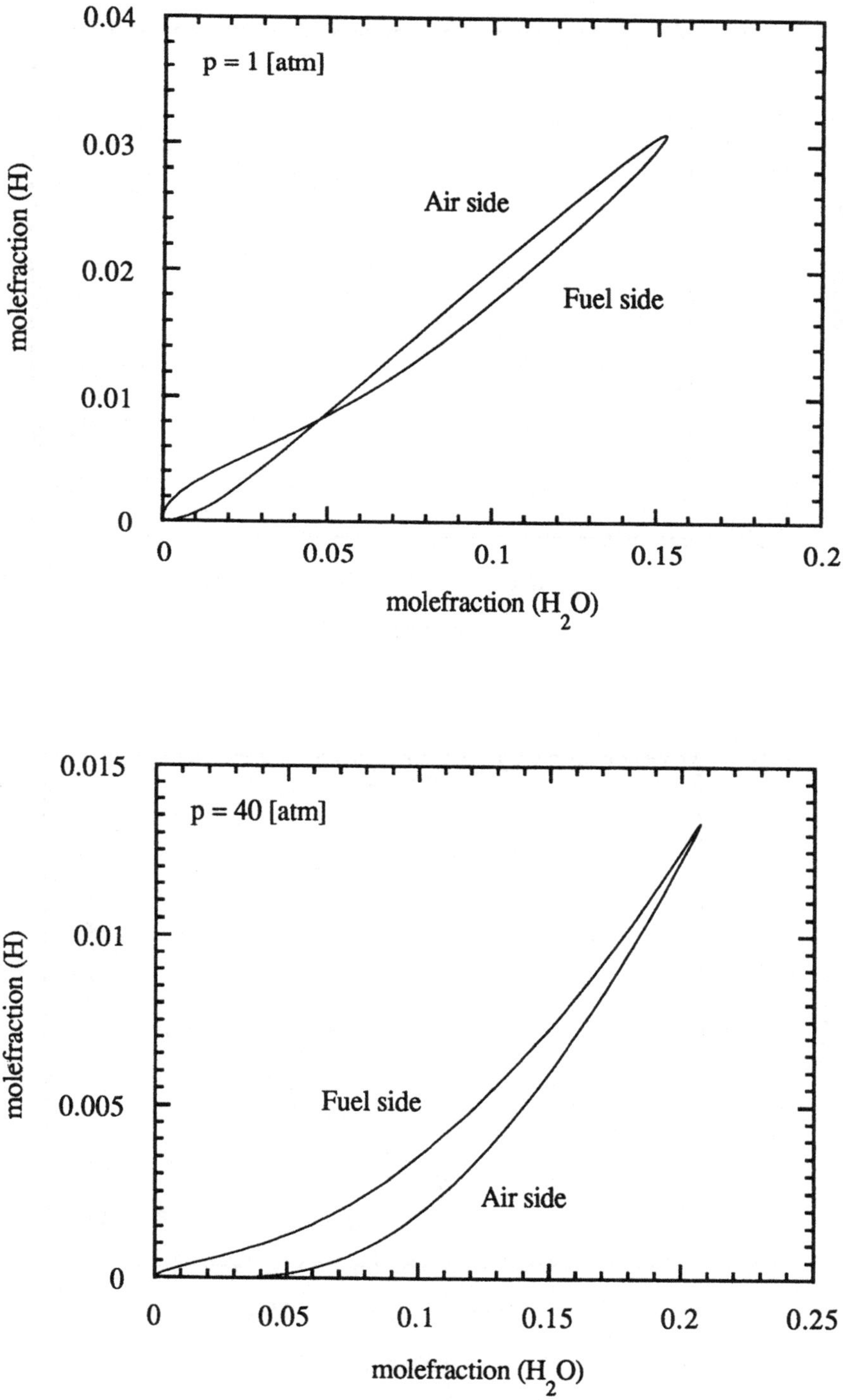

Fig. 10.10. Concentration of hydrogen atom as a function of the concentration of water vapor for two different pressures a) 1 atm, b) 40 atm.

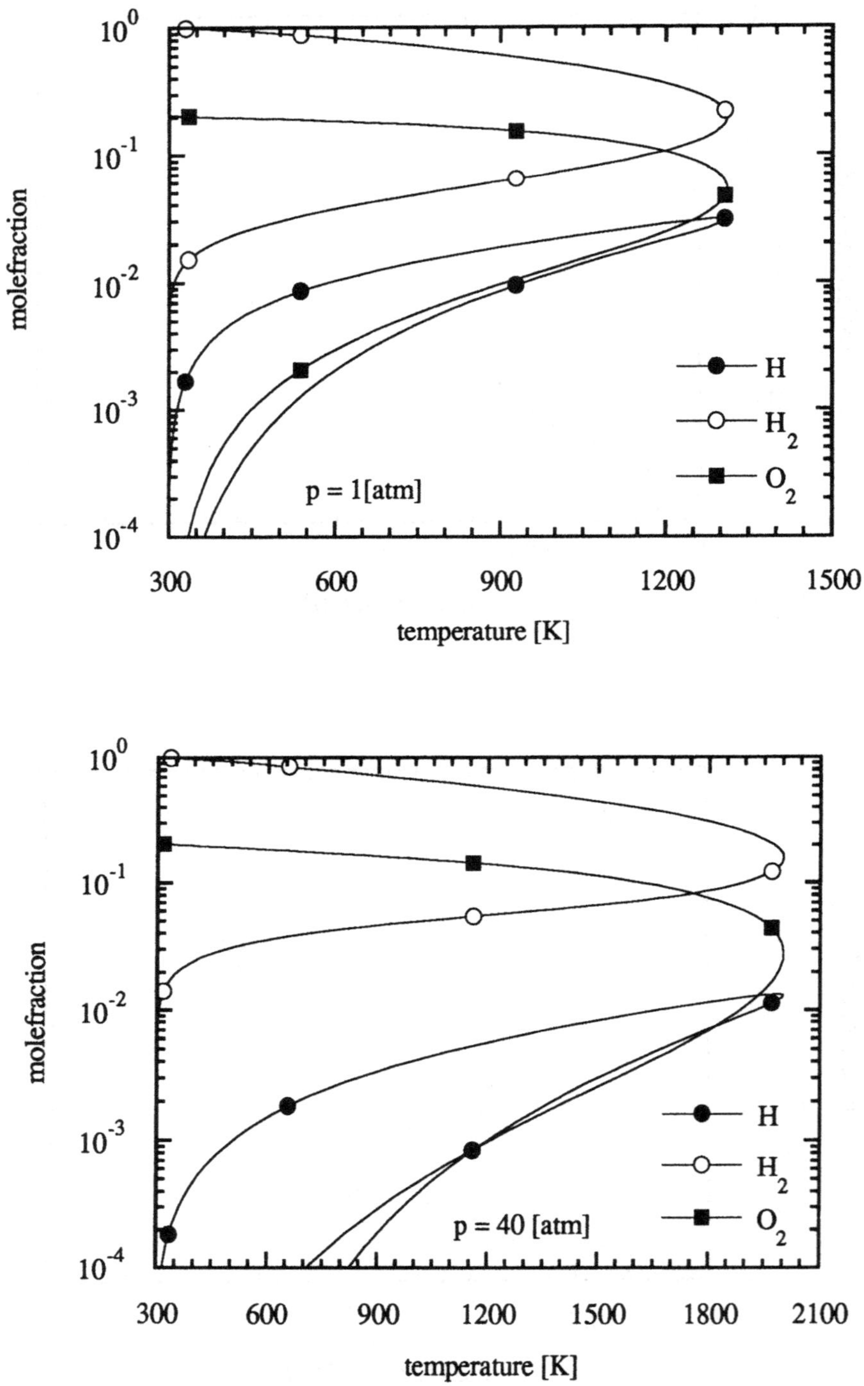

Fig. 10.11. Reactant concentration profiles as well atomic hydrogen profiles as a function of the temperature for two different pressures a) 1 atm, b) 40 atm.

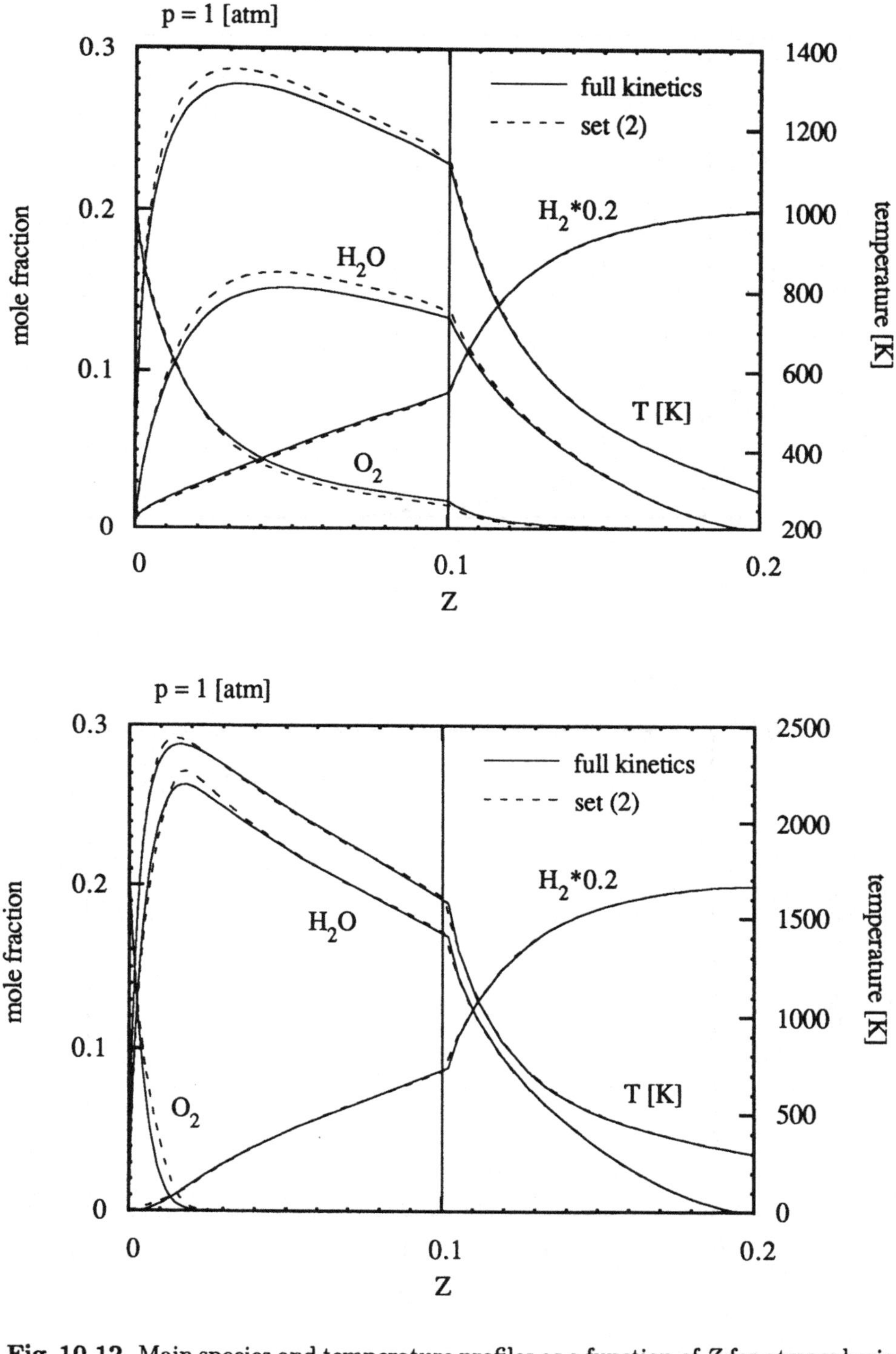

Fig. 10.12. Main species and temperature profiles as a function of Z for atmospheric flames with the full starting mechanism and set (2) for two different strain rates a) close to extinction, b) $a = 100\,\mathrm{s}^{-1}$.

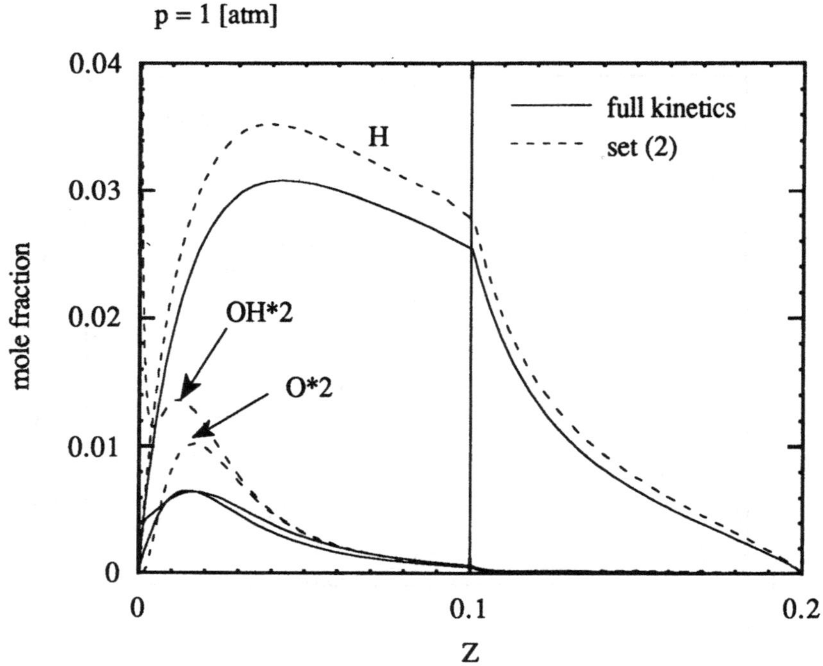

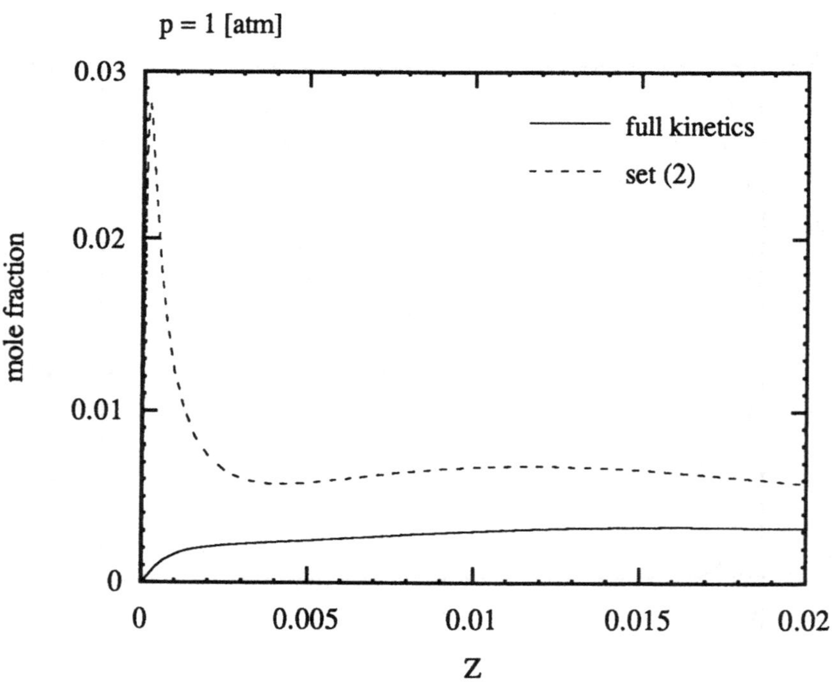

Fig. 10.13. Radical molar concentration profiles for atmospheric flames with the full starting mechanism and set (2).

Figure 10.11 shows the reactant and atomic hydrogen molar concentrations trajectories as a function of the temperature. In the high temperature region as well as on the fuel side of the flame, the O_2 concentration is very low. No leakage of O_2 is observed because of the low diffusivity of this species and since it is always attacked by the H atoms present everywhere. Including the effect of the Lewis numbers (X_i/L_i), the concentrations of species H_2, H_2O and H are much larger than that of O_2 in the main reaction zone. This figure confirms the following structure of the flame close to the extinction condition. There are two main products, H_2O and H. The flame structure is like Liñán's premixed flame with O_2 being totally consumed with leakage of H_2 and catalyzed by a main product, H. Hydrogen atom diffuses to both sides. The radical consumption in both air and fuel sides is important and has to be considered because of the heat release. The minimum reduced kinetic mechanism valid for this flame is the two-step, which can reproduce the flame structure and extinction conditions.

Figure 10.12 shows the profile as a function of the mixture fraction of the main species and temperature for $p = 1\,\text{atm}$ and two different strain rates: close to extinction (Fig. 10.12a) and low strain rates (Fig. 10.12b). Only set (2) was considered in the reduced mechanism. A very good agreement is found in both extremes of the strain rates. The agreement is better for low strain rates. There is only a small discrepancy on the high temperature region of the flame where equilibrium conditions are reached but not included in the model. Finally Fig. 10.13 shows the active radical concentration close to extinction as a function of the mixture fraction Z for the starting kinetics mechanism and for the set (2). The agreement is acceptable for the atomic hydrogen and the atomic oxygen. The discrepancy in larger for the hydroxyl radical (OH) in the low temperature air side of the flame. Figure 10.13b shows a stretched region close to the air side. Clearly, it is seen that the steady-state approximation for the OH breakes down in this region as explained before. The H_2 concentration is very low and reaction 3 is unable to consume the OH radicals produced by the consumption chain of HO_2. Fortunately, this has not a big influence on the flame structure.

10.8 Conclusions

In this paper we studied the structure and extinction of a nondiluted hydrogen/air diffusion flames with reactants at room temperature, using both a starting and a reduced two-step mechanisms with four different approximations for the calculations of the steady-state concentrations of radicals OH and HO_2. A three step mechanism, with OH as main product, was not employed because the two-step mechanism was capable to reproduce the flame structure from low strain rates up to extinction conditions. The steady state for O is fully justified and for OH is in general a good approximation, except in the low temperature region on the air side of the flame where the chain branching reaction 6 competes with the chain braking reaction 5 and with a weak chain propagating reaction 3, because the low concentration of molecular hydrogen. A small error introduced in the calculation of OH gives a large error in the concentration of HO_2. It was found that the best choice for the concentration of the steady-state of OH was (10.8) with $k_8 = 0$. The maximal errors introduced in the extreme conditions ($p = 20$ atm close to extinction) were 7% in the extinction strain rates. The differences are reduced for lower pressures.

10.9 Acknowledgements

C. Treviño acknowledges the support of the Alexander von Humboldt Stiftung of Germany. The authors appreciate interesting discussions about this work with N. Peters.

References

[10.1] Dixon Lewis, G. and Missaghi, M., Structure and extinction limits of counterflow diffusion flames of hydrogen-nitrogen mixtures in air, XXII Symposium (international) on Combustion, The Combustion Institute, pp. 1461–1470, 1988.

[10.2] Tangirala, V., Seshadri, K., Treviño, C. and Smooke, M. D., Progress in Astronautics and Aeronautics, 131, pp. 89–110,1991.

[10.3] Gutheil, E., and Williams, F. A., XXIII Symposium (international) on Combustion, p. 513, The Combustion Institute, 1991.

[10.4] Smooke, M. D., Puri, I. K. and Seshadri, K., XXII Symposium (International in Combustion), p. 1783, The Combustion Institute, 1986.

[10.5] Keyes, D. E. and Smooke, M. D., J. Comp. Physics, **73**, p. 267, 1987.

[10.6] Kee, J., Warnatz, J. and Miller, J. A., Sandia Report SAND83-8209, UC-32, Livermore, 1983.

[10.7] Smooke, M. D. (Ed.), Reduced Kinetic Mechanisms and Asymptotic Approximations for Methane-Air Flames, Springer Verlag, 1991.

[10.8] Rogg, B., RUN-1DL: A Computer Program for the Simulation of One-Dimensional Chemically Reacting Flows, Technical Report CUED/A-THERMO/TR39, Cambridge University Engineering Department, April 1991.

11. Structure and Extinction of Hydrogen–Air Diffusion Flames

E. Gutheil
Institut für Technische Verbrennung, Universität Stuttgart,
W-7000 Stuttgart 80, Federal Republic of Germany

G. Balakrishnan and F.A. Williams
Department of Applied Mechanics and Engineering Sciences,
University of California, San Diego, La Jolla, CA 92093, U.S.A.

11.1 Introduction

Hydrogen is a fuel of practical interest in high-speed air-breathing combustion, where effects of finite-rate chemical kinetics are expected to play a significant role as a consequence of the short characteristic flow times. Associated combustion problems may be investigated by considering laminar, counterflow configurations, where high strain rates encountered in high-speed combustion cause effects of chemical nonequilibrium to become important. The present chapter concerns the structures and extinction characteristics of these flames. It expands on material of the preceding chapter, in which background references are reviewed.

Numerical computations of flame structures with detailed 21-step chemistry are performed for a wide range of pressures and inlet temperatures. Strain rates from $60\,s^{-1}$ to extinction are included, and attention is paid to influences of dilution of the hydrogen with nitrogen. A reduced four-step mechanism involving steady-state approximations for HO_2 and H_2O_2 has been derived; assuming a steady state for O atoms then leads from this four-step mechanism to a three-step reduced scheme, and introducing a steady state for OH then results in a two-step approximation. A one-step approximation, obtained by introducing an H-atom steady state into this two-step scheme, has been investigated previously [11.1] and was shown to provide poor extinction predictions. Therefore attention here is focussed mainly on the two-step mechanism. Numerical calculations are reported that exhibit extinction strain rates close to those obtained with the detailed scheme.

The effect of nitrogen chemistry is investigated by considering a detailed mechanism for NO formation. A sixteen-step mechanism involving the species

N, NO, N_2O and HNO, which was derived from an initial mechanism employing 38 elementary reactions and the species NO_2 and NH in addition to the previous ones, is considered. Numerical calculations are performed with as much as 10^4 ppm NO added to the air stream. In all cases the nitrogen chemistry is found to exert only small influences on the flame structure and extinction.

After the numerical formulation and the full kinetic scheme are presented, the reduced kinetic mechanisms are given. The results of numerical integrations are then shown and discussed. The effects of nitrogen chemistry are investigated next. Finally, the computational results are compared with previous integrations and with experimental data.

11.2 Formulation

The numerical calculations are performed employing a computer code of Smooke [11.2], [11.3]. A steady counterflow problem is considered, with oxidizer approaching from $y = \infty$ and fuel from $y = -\infty$, where y is the axial coordinate. Although most of the computations concerned axisymmetric counterflows to facilitate comparisons with experiments, some attention was paid to planar, two-dimensional counterflows for comparison with some previous numerical results as those of the preceding chapter. The system of ordinary differential equations for mass, energy, and chemical species is the same as given in Chap. 1. The external stagnation flow is irrotational in both fuel and oxidizer streams. The transport description [11.4] and numerical procedure [11.5] have been described in the literature. Thermal diffusion in hydrogen-air flames is included for species H and H_2; its effect on flame structure is negligible at low strain and delays extinction only slightly for the highest boundary temperatures [11.1].

The specific reaction-rate constants for the hydrogen-oxygen system employed in the present chapter are given in Table 11.1, where the rate of reaction k is written as given in Chap. 1. Table 11.1 gives the constants for the forward rates; backward rates for reversible reactions were calculated through equilibrium constants. Some comparisons will be made with results that use rate constants employed elsewhere in this volume. The mechanism of Table 11.1 comprises eight reactive species: H_2, O_2, H_2O, O, H, OH, HO_2 and H_2O_2. Tests showed that steps 14 to 21 of the mechanism were negligible to a great extent [11.1], so that a 13-step mechanism was obtained. Nevertheless, the comparisons here employed the 21-step mechanism. The results differ negligibly from those for the 13-step mechanism; the largest difference is about 20% in the peak H_2O_2 concentration, and differences are observable only for HO_2 and H_2O_2 profiles. Since the 21-step mechanism differs from the 17 hydrogen-oxygen steps of Chap. 1 only through the presence of the unimportant steps 16, 20 and 21 in Table 11.1, as well as step 19 (which influences HO_2 and H_2O_2 only at high strain rates [11.1]), discrepancies attributable to different selection of steps are not anticipated.

Table 11.1. Specific Reaction-Rate Constants for the Reaction System.

No.[a]	Reactions	A[b]	n[b]	E[b]
1	$H+O_2 \rightleftharpoons OH+O$	2.00E14	0.0	70.30
2	$H_2+O \rightleftharpoons OH+H$	1.80E10	1.0	36.93
3	$H_2O+O \rightleftharpoons OH+OH$	5.90E09	1.3	71.25
4	$H_2+OH \rightleftharpoons H_2O+H$	1.17E09	1.3	15.17
5	$H+O_2+M^c \rightarrow HO_2+M^c$	2.30E18	-0.8	0.00
6	$H+HO_2 \rightarrow OH+OH$	1.50E14	0.0	4.20
7	$H+HO_2 \rightarrow H_2+O_2$	2.50E13	0.0	2.93
8	$OH+HO_2 \rightarrow H_2O+O_2$	2.00E13	0.0	4.18
9	$H+H+M^c \rightarrow H_2+M^c$	1.80E18	-1.0	0.00
10	$H+OH+M^c \rightarrow H_2O+M^c$	2.20E22	-2.0	0.00
11	$HO_2+HO_2 \rightarrow H_2O_2+O_2$	2.00E12	0.0	0.00
12	$H_2O_2+M \rightleftharpoons OH+OH+M$	1.30E17	0.0	190.38
13	$H_2O_2+OH \rightleftharpoons H_2O+HO_2$	1.00E13	0.0	7.53
14[d]	$O+HO_2 \rightleftharpoons OH+O_2$	2.00E13	0.0	0.00
15[d]	$H+HO_2 \rightleftharpoons O+H_2O$	5.00E12	0.0	5.90
16[d,e]	$H+O+M \rightleftharpoons OH+M$	6.20E16	-0.6	0.00
17[f]	$O+O+M \rightleftharpoons O_2+M$	6.17E15	-0.5	0.00
18[f]	$H_2O_2+H \rightleftharpoons H_2O+OH$	1.00E13	0.0	15.02
19[f]	$H_2O_2+H \rightleftharpoons HO_2+H_2$	4.79E13	0.0	33.26
20[g]	$O+OH+M \rightleftharpoons HO_2+M$	1.00E16	0.0	0.00
21[h]	$H_2+O_2 \rightleftharpoons OH+OH$	1.70E13	0.0	200.00
	Nitrogen Chemistry			
22	$O+N_2 \rightleftharpoons N+NO$	1.82E14	0.0	319.02
23	$O+NO \rightleftharpoons N+O_2$	3.80E09	1.0	173.11
24	$H+NO \rightleftharpoons N+OH$	2.63E14	0.0	210.94
25	$NO+M \rightleftharpoons N+O+M$	3.98E20	-1.5	627.65
26	$N_2+M \rightleftharpoons N+N+M$	3.72E21	-1.6	941.19
27	$N_2O+O \rightleftharpoons NO+NO$	6.92E13	0.0	111.41
28	$N_2O+O \rightleftharpoons N_2+O_2$	1.00E14	0.0	117.23
29	$N_2O+N \rightleftharpoons N_2+NO$	1.00E13	0.0	83.14
30	$N+HO_2 \rightleftharpoons NO+OH$	1.00E13	0.0	8.31
31	$N_2O+H \rightleftharpoons N_2+OH$	7.60E13	0.0	63.19
32	$HNO+O \rightleftharpoons NO+OH$	5.01E11	0.5	8.31
33	$HNO+OH \rightleftharpoons NO+H_2O$	1.26E12	0.5	8.31
34	$NO+HO_2 \rightleftharpoons HNO+O_2$	2.00E11	0.0	8.31
35	$HNO+HO_2 \rightleftharpoons NO+H_2O_2$	3.16E11	0.5	8.31
36	$HNO+H \rightleftharpoons NO+H_2$	1.26E13	0.0	16.63
37	$HNO+M \rightleftharpoons H+NO+M$	1.78E16	0.0	203.70

[a] Forward rate constants for steps 1-13 from Ref. [11.6], for steps 22-37 from Ref. [11.7].

[b] Units: mol/cm^3, s^{-1}, K, kJ/mol; rates for reverse steps obtained from JANAF thermochemical equilibrium data.

[c] Chaperon efficiencies H_2:1, H_2O:6.5, O_2:0.4, N_2:0.4.

[d] Rate constants from Ref. [11.8].

[e] Chaperon efficiency H_2O:5, others 1.

[f] Rate constants from Ref. [11.9].

[g] Rate constants from Ref. [11.10].

[h] Rate constants from Ref. [11.11].

11.3 Reduced Mechanisms

Although calculations with both full and reduced mechanisms to be reported here will include all 21 steps, the analytical development of the reduced mechanisms will be based on the 13-step description, an excellent approximation to the 21-step mechanism. Results of numerical calculations show the steady-state approximations for HO_2 and H_2O_2 to be very good. Since the contributions of steps 11 and 13 to the HO_2 concentration are shown by numerical integration to be negligible, these steady states enable the mole fractions X_i of these minor species to be expressed as

$$X_{HO_2} = k_5(\rho/\overline{W})X_H X_{O_2}/[(k_6 + k_7)X_H + k_8 X_{OH}] \tag{11.1}$$

and

$$X_{H_2O_2} = \frac{k_{11}X_{HO_2}^2 + k_{12b}(\rho/\overline{W})X_{OH}^2 + k_{13b}X_{H_2O}X_{HO_2}}{k_{12f} + k_{13f}X_{OH}}. \tag{11.2}$$

Here k_{if} and k_{ib} denote the rate of step i in forward or backward direction, respectively, ρ denotes density and $\overline{W}$ the average molecular weight.

The HO_2 and H_2O_2 steady states lead to a four-step mechanism, namely

(I) $O + H_2 \rightleftharpoons H + OH,$

(II) $OH + H_2 \rightleftharpoons H + H_2O,$

(III) $H + H + M \rightarrow H_2 + M,$

(IV) $H + O_2 \rightleftharpoons OH + O,$

where the reaction rates are

$$\begin{aligned} w_I &= w_2 + w_3 + w_6 + w_{12}, \\ w_{II} &= -w_3 + w_4 + w_8 + w_{10} + w_{11} - w_{12}, \\ w_{III} &= w_5 + w_9 + w_{10} - w_{12}, \\ w_{IV} &= w_1 + w_6 + w_{12}. \end{aligned} \tag{11.3}$$

Here w_i denotes the reaction rate of elementary reaction i given in Table 11.1. The concentrations of HO_2 and H_2O_2 appearing in the reaction rates need to be replaced by their steady-state values calculated from (11.1) and (11.2).

The numerical calculations show the steady-state assumption for O atoms to be acceptable above temperatures of about 1000 K. This leads to the scheme

(I′) $H_2 + O_2 \rightleftharpoons OH + OH,$

(II′) $OH + H_2 \rightleftharpoons H + H_2O,$

(III′) $H + H + M \rightarrow H_2 + M,$

where the rates are given by the first three expressions in (11.3). The O-atom concentration is obtained from the steady-state condition

$$X_{\mathrm{O}} = \frac{k_{1f}X_{\mathrm{H}}X_{\mathrm{O}_2} + k_{2b}X_{\mathrm{OH}}X_{\mathrm{H}} + k_{3b}X_{\mathrm{OH}}^2}{k_{1b}X_{\mathrm{OH}} + k_{2f}X_{\mathrm{H}_2} + k_{3f}X_{\mathrm{H}_2\mathrm{O}}}. \tag{11.4}$$

The steady-state assumption for OH radicals is somewhat better than that for O atoms in the main reaction zone as well as closer to the fuel side of the flame. Additional introduction of the steady-state approximation for OH leads to the two-step mechanism

(I″) $\quad 3H_2 + O_2 \rightleftharpoons 2H_2O + 2H,$

(II″) $\quad H + H + M \rightarrow H_2 + M,$

with rates $w_{\mathrm{I}''} = w_{\mathrm{I}'} = w_{\mathrm{I}}$ and $w_{\mathrm{II}''} = w_{\mathrm{III}'} = w_{\mathrm{III}}$. The formula for the evaluation of the OH concentration is lengthy and will not be given here. In the computations X_{OH} was obtained by solving algebraic equations through iteration.

The one-step mechanism

(I‴) $\quad 2H_2 + O_2 \rightleftharpoons 2H_2O$

can be obtained from the two-step mechanism by introducing a steady-state approximation for H atoms. Since this is useful only for estimating concentrations in the hot reaction zone at low rates of strain [11.1], it is not tested further here.

11.4 Results of Numerical Integrations

11.4.1 Full Mechanisms for Undiluted Flames

Extensive results for undiluted flames have been given previously [11.1] and in the preceding chapter. A few additional points and comparisons will be made here. As the notation throughout the chapter, p denotes pressure, T_∞ and $T_{-\infty}$ are the temperatures in the external oxidizer and fuel streams, respectively, and the strain rate a is defined as the derivative of the component of velocity parallel to the flame sheet with respect to the coordinate parallel to the flame sheet, evaluated in the external oxidizer stream. When results are shown as functions of mixture fraction, the element H is employed to define the mixture fraction Z as

$$Z = \sum_{j=1}^{N} \mu_j (Y_j - Y_{j\infty}) \Big/ \sum_{j=1}^{N} \mu_j (Y_{j-\infty} - Y_{j\infty}), \tag{11.5}$$

where μ_j is the mass of H atoms in molecule j, and Y_j is the mass fraction of species j. This selection provides a Z that is monotonic in y [11.1] and avoids integration of the additional Z equation in Chap. 1. The rate of scalar dissipation associated with this mixture fraction is defined as

$$\chi = \left[\frac{2\lambda}{\rho c_P}\right]\left(\frac{dZ}{dx}\right)^2 . \tag{11.6}$$

It is instructive to employ the full numerical results to test the accuracy with which the popular highly simplified approximation of complete chemical equilibrium can be applied. One measure of this is the equilibrium constant for step I‴. This equilibrium constant can be calculated in two ways from the numerical results, namely from the temperature profile and from concentration profiles. Figure 11.1 is a graph of the equilibrium constant for partial pressures for this step, calculated in both ways, for a low strain rate, under which condition an equilibrium approximation is at its best. There are two branches of the curve from concentrations, one for concentrations on the fuel side and the other for concentrations on the air side. From Fig. 11.1 it is seen that, even at the highest temperature, departures from equilibrium exceed 20%, and the differences exceed a factor of two for temperatures below about 1500 K. The difference at low temperature is expected because of radical removal, but the extend of the difference at higher temperatures may be surprising at these low strain rates. The poor agreement with equilibrium predictions under all conditions of experimental interest is a consequence of the negligible rates of the high-activation-energy dissociation processes, which must occur if equilibrium is to be achieved. Even the one-step approximation, with its associated complete steady states, provides a substantial improvement over the equilibrium approximation [11.1]. Moreover, as will be seen further below, the two-step approximation provides a substantial improvement over the one-step approximation.

Figure 11.2 compares the dependence of the peak temperature on the strain rate a, obtained in the previous chapter, with that calculated according to the procedure of the present chapter, with full chemistry. Both computations were performed for the planar problem rather than for axisymmetric flow; this is the only figure in the present chapter that pertains to planar, two-dimensional flow. The agreement is seen to be excellent; although temperatures are slightly lower with the present rate constants, especially at the lower strain rates, the temperature difference never reaches 50 K. The small differences observed are attributable to the use of different rate constants - in some cases differing by as much as a factor of three, although the rate data for the key steps, 1 and 5, are identical. In this context, it seems worth reporting here that some computations also were made at atmospheric pressure with the new compilation of rate data [11.12], in some cases employing functional forms different from the Arrhenius forms of the present volume, to improve accuracy, and the results differed negligibly. It therefore seems safe to conclude that for the hydrogen-oxygen system the rate constants are now known and parameterized well enough that rate uncertainties no longer introduce significant differences in predictions of diffusion-flame temperatures, concentrations of major species and extinction except possibly at high pressures.

The strain rate a for extinction in Fig. 11.2, identified from the present calculations, is 14,400 s^{-1}, very close to the value of about 14,300 s^{-1}, obtained

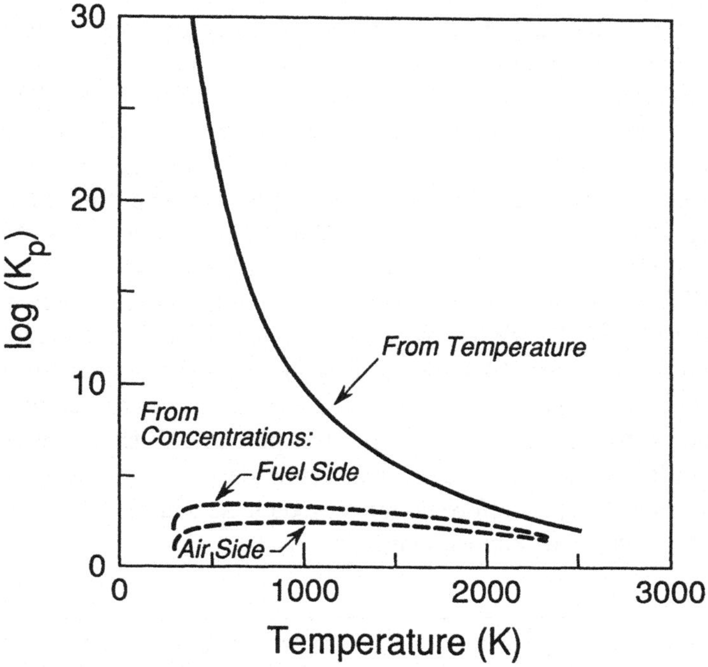

Fig. 11.1. The equilibrium constant for partial pressures, K_p, for $H_2 + 1/2\ O_2 \rightleftharpoons H_2O$, evaluated from the temperature and from the computed profiles of partial pressures with the full mechanism, for $a = 60\,s^{-1}$, $p = 1\,atm$ and $T_{-\infty} = T_{\infty} = 300\,K$.

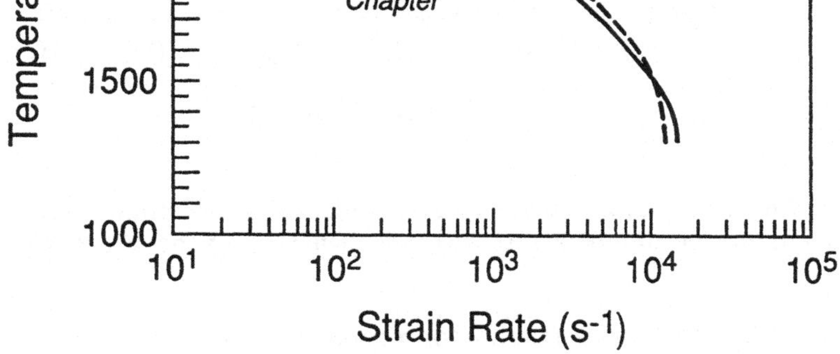

Fig. 11.2. The dependence of the maximum temperature on the strain rate for the planar stagnation flow with $T_{-\infty} = T_{\infty} = 300\,K$, $p = 1\,atm$, as obtained here with full chemistry and as reported in the previous chapter.

in the previous chapter. It is of interest to compare this number with the corresponding value of $8140\,s^{-1}$, given previously [11.1] for the axisymmetric problem. If density is constant, then there is an exact correspondence between the two problems, such that the planar strain rate is twice the axisymmetric value. Therefore $16{,}280\,s^{-1}$ would have been obtained, instead of $14{,}400\,s^{-1}$, if a constant-density approximation were applicable; the difference is slightly above 10 %. It was suggested [11.1] that the difference between $8140\,s^{-1}$ and a value of $13{,}000\,s^{-1}$, reported by Dixon-Lewis and Missaghi [11.8], may be due to use of different rate data. However, such a comparison is incorrect because the earlier study [11.8] addressed the planar problem. It is proper instead to compare $13{,}000\,s^{-1}$ with $14{,}400\,s^{-1}$, showing a difference near ten percent. It may be concluded that the extinction strain rates, computed independently in different laboratories, now agree within ten percent. With sufficient care and effort, the differences become less than a few percent. The large uncertainty cited previously [11.1] therefore no longer exists, as will be further emphasized later in comparing with experimental results.

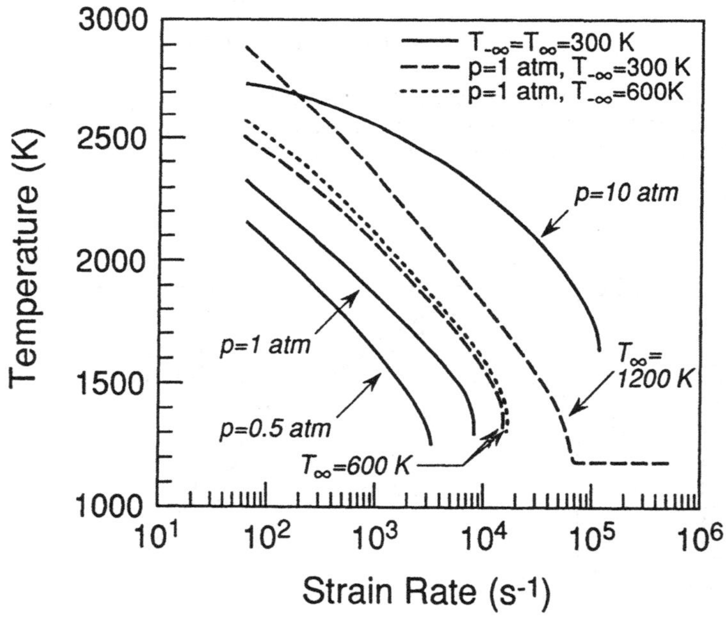

Fig. 11.3. The dependence of maximum temperature on the strain rate for various pressures and ambient temperatures with full chemistry for the axisymmetric flow.

Figure 11.3 summarizes results concerning the calculated dependence of the maximum flame temperature on the strain rate, employing full chemistry for axisymmetric flow. The curves for $p = 0.5\,\text{atm}$, $p = 1\,\text{atm}$ and $p = 10\,\text{atm}$ show that maximum temperatures and extinction strain rates are lower at lower pressures. This is consistent with the fact that the strain rate at extinction is inversely proportional to the diffusion coefficient, which again is proportional to p^{-1}. The results also are consistent with crossover temperature T_c (at which the rate of forward step 1 equals that of step 5) increasing with increasing pressure. For $p = 0.5$ atm extinction occurred at $a = 3320\,s^{-1}$ with a maximum temperature of 1260 K, for $p = 1$ atm it occurred at $a = 8140\,s^{-1}$ with 1332 K, and at $p = 10$ atm the flame extinguished at $a = 118,800\,s^{-1}$ with a maximum temperature of 1665 K. Curves also are shown in Fig. 11.3 at $p = 1$ atm for other boundary temperatures. The calculation for $T_{-\infty} = T_{\infty} = 600$ K gave $a = 16,510\,s^{-1}$ and a maximum temperature of 1330 K at extinction which is close to the temperature at extinction of the flame at $p = 1$ atm, $T_{-\infty} = T_{\infty} = 300$ K; the strain rate for extinction at elevated temperature is much higher, as anticipated. The extinction conditions for $T_{-\infty} = 300$ K, $T_{\infty} = 600$ K are nearly identical to those for $T_{-\infty} = T_{\infty} = 600$ K, as expected from the fact that the flame lies far on the oxidizer side of the mixing layer. The calculation for $T_{-\infty} = 300$ K, $T_{\infty} = 1200$ K gave marginal indication of abrupt extinction; abrupt extinction is not anticipated at this high air temperature since $T_{\infty} > T_c$ here at $p = 1$ atm; at $p = 10$ atm and $T_{\infty} = 1200$ K it is estimated that $T_{\infty} < T_c \approx 1300$ K, so that abrupt extinction is anticipated.

11.4.2 The Two-Step Mechanism for Undiluted Flames

In the previous chapter results for a two-step mechanism were compared with those for a full mechanism for undiluted flames in planar flows at both atmospheric and elevated pressures. Related comparisons for axisymmetric flows at atmospheric pressure are shown in Figs. 11.4–11.7.

Figure 11.4 shows the peak temperature as a function of the strain rate, according to predictions of full and two-step chemistry. It is seen that the two-step chemistry gives higher maximum temperatures, by as much as 200 K at the lower strain rates, yet the extinction strain rates are nearly identical. Corresponding curves for three-step and four-step mechanisms are not shown but generally lie between those plotted, giving a continual increase in temperature with reduction in mechanism. These results are in conflict with those of the previous chapter, which show slightly lowered temperatures and slightly lowered extinction strain rates for the two-step approximation. The discrepancy is associated mainly with the further approximations introduced in the preceding chapter to achieve formulas that facilitate asymptotic analysis; the present reduced-mechanism results were obtained by iterations that eventually include rate constants of all 21 steps of Table 11.1 in the steady-state relations.

Figure 11.5 gives representative temperature profiles as functions of the mixture fraction Z that was defined in (11.5). It is seen that the higher temperatures associated with the two-step approximation are restricted mainly to the region of maximum temperature. At the higher strain rate (near extinction) the temperature differences extend over a somewhat broader range of mixture fraction, but a narrower range of the physical coordinate y. Also

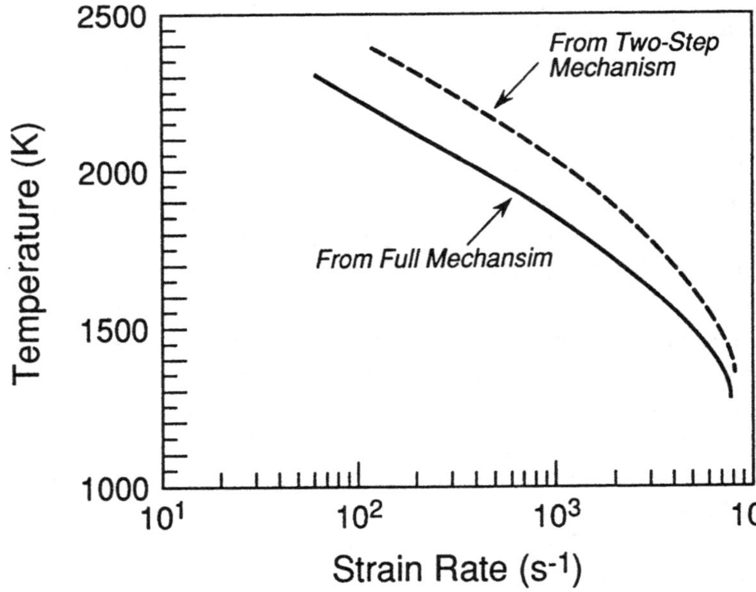

Fig. 11.4. The dependence of the maximum flame temperature on the strain rate for axisymmetric flow with $T_{-\infty} = T_\infty = 300\,\mathrm{K}$, $p = 1\,\mathrm{atm}$, as obtained with full chemistry and with the two-step mechanism.

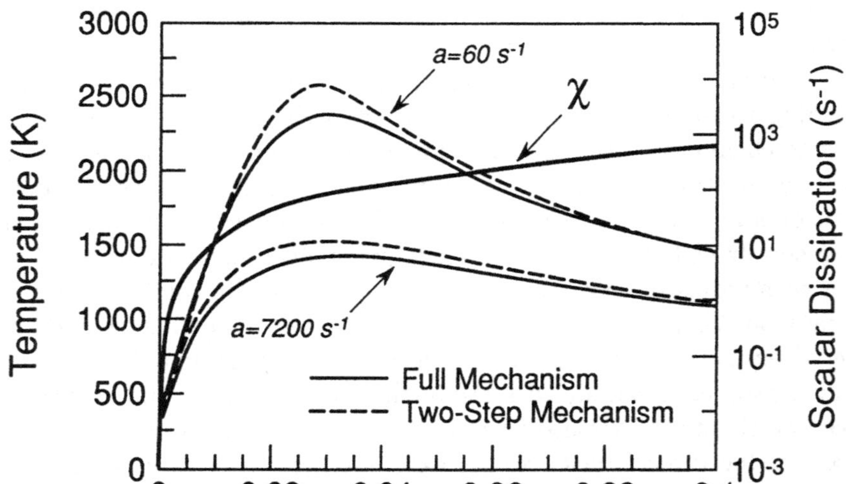

Fig. 11.5. The variation of the temperature with the mixture fraction for full and two-step chemistry with $T_{-\infty} = T_\infty = 300\,\mathrm{K}$, $p = 1\,\mathrm{atm}$, at two strain rates, and the variation of the scalar dissipation at the higher strain rate.

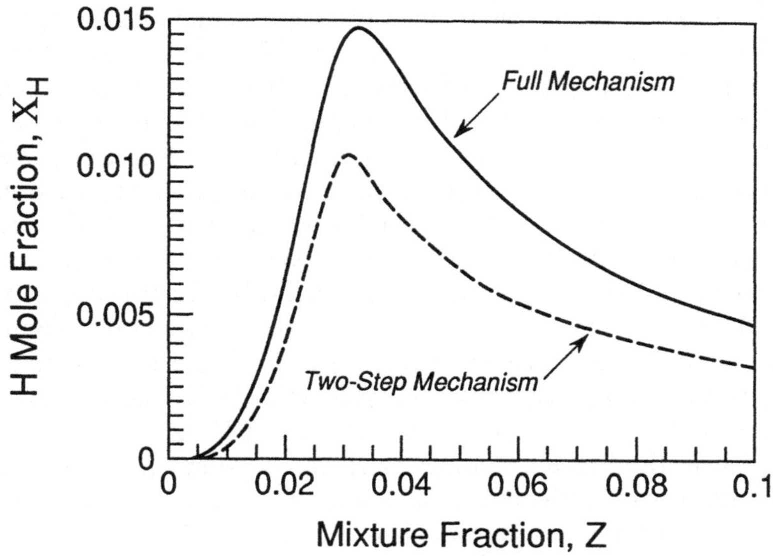

Fig. 11.6. The variation of the H-atom mole fraction with the mixture fraction for full and two-step chemistry with $T_{-\infty} = T_\infty = 300\,\mathrm{K}$, $p = 1\,\mathrm{atm}$ and $a = 60\,\mathrm{s}^{-1}$.

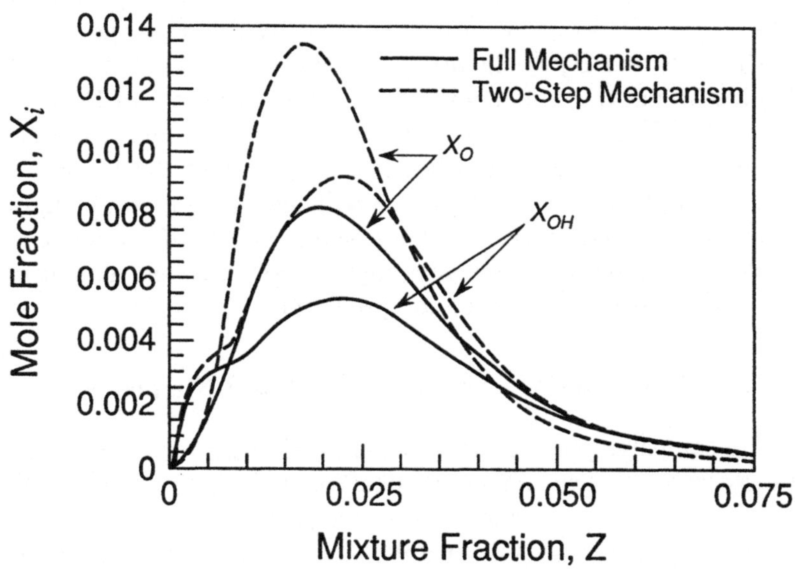

Fig. 11.7. The variations of the O-atom and OH-radical mole fractions with the mixture fraction for full and two-step chemistry with $T_{-\infty} = T_\infty = 300\,\mathrm{K}$, $p = 1\,\mathrm{atm}$ and $a = 7200\,\mathrm{s}^{-1}$.

shown in Fig. 11.5 for the higher strain rate are profiles of the scalar dissipation defined in (11.6); the differences in the predictions for full and two-step chemistry are less than the width of the line. The largest discrepancies between predictions of full chemistry and reduced mechanisms occur in the radical concentrations. Figure 11.6 compares the dependence of the H-atom mole fraction on Z predicted by the full chemistry with that given by the two-step mechanism. It is seen that the two-step approximation consistently predicts H concentrations about 50 % lower, consistent with the predictions of a higher temperature. Figure 11.6 corresponds to a low strain rate and is representative of the largest discrepancy found by the two-step mechanism. The difference maintains about the same ratio as extinction is approached. Figure 11.7 is representative of the discrepancies found near extinction. This figure pertains to the steady-state radicals, O and OH, and it shows that the magnitudes of their discrepancies reflect that of H, as do the discrepancies reflect that of H, although contrary to H, the two-step approximation underpredicts the values of these concentrations. The full and two-step chemistry both predict that the concentrations of HO_2 and H_2O_2 (not shown) continually increase with decreasing Z, at least until the temperature is below 600 K, and the maximum discrepancies for these minor species occur at a temperature of about 700 K, around $Z = 0.002$, and reach about a factor of five for HO_2 and a factor of ten for H_2O_2, reflecting inapplicability of the two-step mechanism in predicting concentrations of minor species at low temperatures. Figure 11.7 demonstrates a shoulder on the OH curves at very lean conditions; this has been identified previously [11.1] and results from the production of OH in the HO_2 and H_2O_2 chemistry, through steps like 6, 12 and 18, since the shoulder lies in the region of the largest concentrations of HO_2 and H_2O_2. Comparison of Figs. 11.6 and 11.7 shows that the locations of the O and OH peaks are nearly coincident and lie on the air side of the H peak; this has been indicated previously [11.1] and is predicted by asymptotic analysis with a one-step approximation [11.13]. Thus, some aspects of the flame structure are reproduced correctly in one-step approximations, even though two steps are essential for reasonable extinction predictions.

11.4.3 The Two-Step Mechanism for Diluted Flames

Figure 11.8 compares predictions of the extinction strain rate as a function of dilution, given by the two-step approximation, with those of the full chemistry. The excellent agreement evident here is much better than that of the previous chapter, probably through taking differences of additional elementary steps into account. A small deterioration of the agreement is seen at high dilutions, signaling onset of failure of the two-step mechanism at low temperatures. A two-step mechanism is not likely to be accurate for ignition problems, for example, although the four-step and possibly the three-step mechanism may work there. It would be of interest to investigate the performance of the two-step mechanism at elevated pressures with dilution. Its performance is likely to improve with increasing feed-stream temperatures, but its behavior with pressure is less clear, although results of the previous chapter suggest degradation in accuracy with increasing pressure.

11.5 Effect of Nitrogen Chemistry on Flame Structure

The flame temperatures in hydrogen-air flames at high pressures or elevated feed temperatures are high so that production of oxides of nitrogen may be expected in the flames investigated. Therefore a chemical reaction mechanism was identified to include nitrogen chemistry, and the influence of this mechanism on the flame structure was investigated.

At first, calculations with reactions involving species N, NO, NO_2, N_2O, NH, and HNO were included; all 38 elementary reactions and forward rates given by Hanson and Salimian [11.7] were considered, with backward reaction rates obtained through equilibrium constants. Later various steps were removed. Results of the computations are summarized in Table 11.2, where entries (iii), (iv) and (v) refer to removal of the species NO_2 and/or NH from the system, and entries (viii) and (ix) correspond to using steps 22–37 with addition of NO to the air stream. The greatest effect of nitrogen chemistry on flame structure occurs in flames at low strain rate and at elevated air temperature where the maximum flame temperature is highest. Therefore Table 11.2 corresponds to this condition. Including nitrogen chemistry reduces the peak temperature by less than 1% and changes maximum mole fractions of major species by less than 1%. The peak values of O, OH, and H change up to 3%. The mole fraction of NO_2 contributes less than 0.01% to NO_x production and may be neglected. An analysis of the mechanism identifies reactions 22 to 37 of Table 11.1 to be sufficient to describe NO formation; the mechanism involves the species N, NO, N_2O, and HNO. Neglecting the other species, as well as some of the reactions involving the species mentioned, causes only small changes in concentrations of all species except the minor component N_2O. Nitrogen chemistry barely affects flame structure at standard conditions, and at decreased pressures for all rates of strain, the peak flame temperature changes by less than 0.1% at the lowest strain rate. The flame at $p = 1\,\text{atm}$, $T_{-\infty} = T_{\infty} = 300\,\text{K}$ produces 181 ppm NO as the maximum concentration at $a = 60\,\text{s}^{-1}$ whereas this value decreases in the same flame burning at 0.5 atmospheres to 50 ppm NO. Calculations employing different rate data than the ones used here showed that the changes in peak values for NO concentration are below 10%. In particular, nitrogen chemistry is negligible at high strain rates for all flames investigated, and extinction is not affected.

The maximum NO concentration observed is 4620 ppm at $60\,\text{s}^{-1}$, atmospheric pressure and $T_{-\infty} = 300\,\text{K}$, $T_{\infty} = 1200\,\text{K}$; the peak NO value in the flame at $a = 60\,\text{s}^{-1}$, $p = 10\,\text{atm}$ and $T_{-\infty} = T_{\infty} = 300\,\text{K}$ is 1061 ppm. Increasing the strain rate to $6000\,\text{s}^{-1}$ in the latter case causes a tremendous decrease of NO formation to 23 ppm. Figure 11.9 displays the profiles of nitrogen-containing species for the flame at $a = 60\,\text{s}^{-1}$, $p = 1\,\text{atm}$ and $T_{-\infty} = T_{\infty} = 300\,\text{K}$. The NO and N_2O concentrations peak approximately

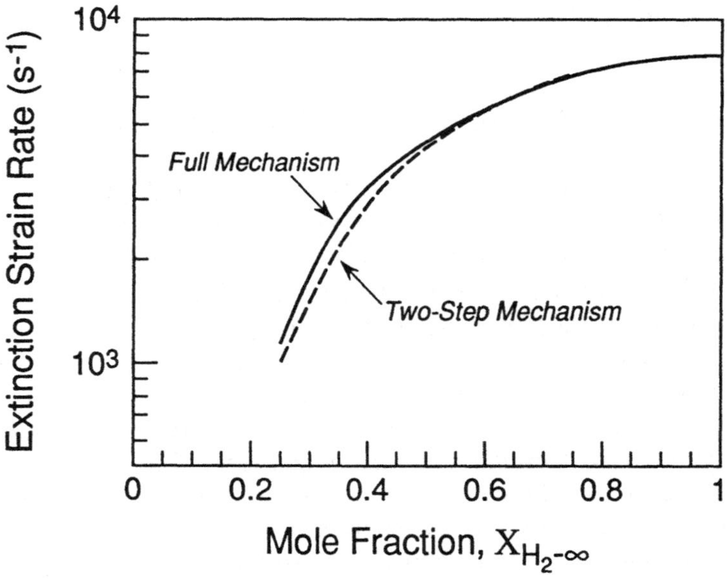

Fig. 11.8. Comparison of predictions of full chemistry and two-step chemistry for the dependence of the air-side extinction strain rate on the hydrogen mole fraction in the fuel stream for axisymmetric counterflow mixing of hydrogen-nitrogen mixtures with air at $T_{-\infty} = T_{\infty} = 300\,\mathrm{K}$, $p = 1\,\mathrm{atm}$.

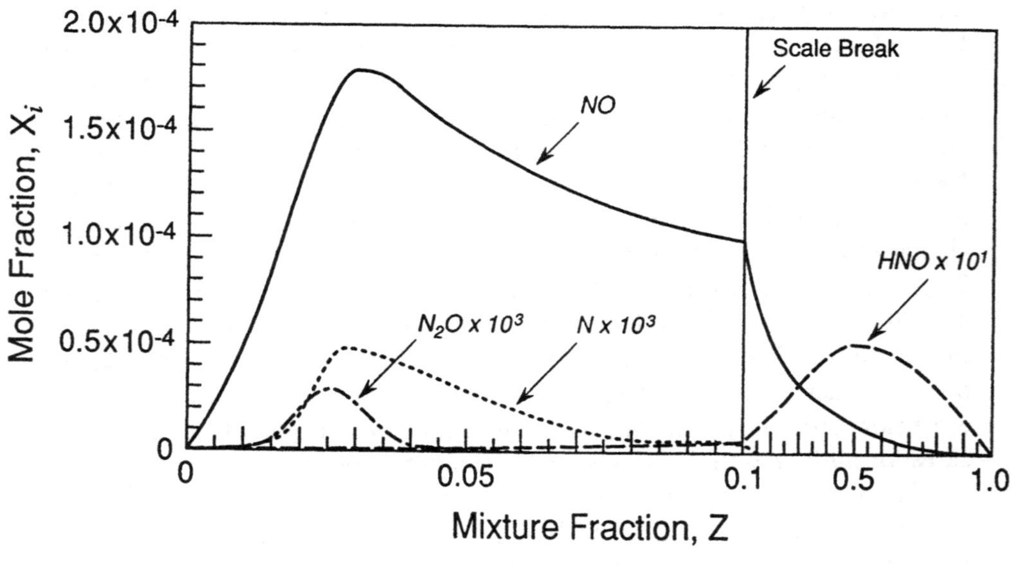

Fig. 11.9. Profiles of mole fractions of NO, HNO, N and N_2O for axisymmetric counterflow hydrogen-air flames at $T_{-\infty} = T_{\infty} = 300\,\mathrm{K}$, $p = 1\,\mathrm{atm}$ and $a = 60\,\mathrm{s}^{-1}$, according to the mechanism in Table 11.1.

at the stoichiometric value of the mixture fraction. Results of calculations employing all 38 elementary reactions [11.7] for nitrogen chemistry show a second peak for N_2O on the fuel side of the flame. This peak is considered unimportant for the present flames, since it occurs in the cold regions of the flame and since the species N_2O is of minor importance for NO formation; the order of magnitude of the second peak, however, increases at elevated pressures and might become more important in high-pressure situations. The figure also shows that the concentration of N peaks somewhat on the fuel side of the flame and HNO even closer to the fuel inlet. The profiles shown in Fig. 11.9 are characteristic for all flames at low strain rate.

Sometimes NH_3 is used to reduce NO by homogeneous reactions, the thermal $DeNO_x$ process. Calculations using a more complete mechanism [11.7] show that the peak value of NO is altered by less than 2%, including reactions and species that enable prediction of NH_3 production. The NH_3 profile peaks at about 1600 to 1700 K, a temperature range that is not relevant for NO production nor is useful for the thermal $DeNO_x$ process. Thus, the mechanism shown in Table 11.1 comprising steps 1 to 13 and 22 to 37, appears to be sufficient to predict NO production for practical purposes.

Table 11.2. Effects of Nitrogen Chemistry on Maximum Temperatures and Maximum Mole Fractions at $T_\infty = 1200$ K, $T_{-\infty} = 300$ K, $p = 1$ atm and $a = 60$ s^{-1}.

	Mechanism	Max.Temp.	X_H	X_O	X_{OH}
(i)	No N_2 Chemistry	2847 K	1.522×10^{-2}	7.806×10^{-3}	2.714×10^{-2}
(ii)	38 N_2 Steps	2839 K	1.494×10^{-2}	7.681×10^{-3}	2.644×10^{-2}
(iii)	No NO_2 Chemistry	2839 K	1.494×10^{-2}	7.681×10^{-3}	2.644×10^{-2}
(iv)	No NH Chemistry	2839 K	1.490×10^{-2}	7.682×10^{-3}	2.641×10^{-2}
(v)	No NO_2 or NH Chemistry	2839 K	1.490×10^{-2}	7.682×10^{-3}	2.641×10^{-2}
(vi)	Only Steps 22, 23, 24	2844 K	1.482×10^{-2}	7.702×10^{-3}	2.648×10^{-2}
(vii)	Steps 22–37	2839 K	1.493×10^{-2}	7.684×10^{-3}	2.642×10^{-2}
(viii)	10^3 ppm NO in Air	2822 K	1.477×10^{-2}	7.602×10^{-3}	2.605×10^{-2}
(ix)	10^4 ppm NO in Air	2816 K	1.519×10^{-2}	7.558×10^{-3}	2.610×10^{-2}

	Mechanism	X_{HO_2}	X_{H_2O}	$X_{H_2O_2}$
(i)	No N_2 Chemistry	1.730×10^{-5}	2.725×10^{-1}	1.048×10^{-6}
(ii)	38 N_2 Steps	1.720×10^{-5}	2.694×10^{-1}	1.080×10^{-6}
(iii)	No NO_2 Chemistry	1.747×10^{-5}	2.694×10^{-1}	1.080×10^{-6}
(iv)	No NH Chemistry	1.721×10^{-5}	2.695×10^{-1}	1.083×10^{-6}
(v)	No NO_2 or NH Chemistry	1.757×10^{-5}	2.695×10^{-1}	1.083×10^{-6}
(vi)	Only Steps 22, 23, 24	1.744×10^{-5}	2.698×10^{-1}	1.068×10^{-6}
(vii)	Steps 22–37	1.720×10^{-5}	2.690×10^{-1}	1.081×10^{-6}
(viii)	10^3 ppm NO in Air	1.784×10^{-5}	2.694×10^{-1}	1.116×10^{-6}
(ix)	10^4 ppm NO in Air	1.784×10^{-5}	2.678×10^{-1}	1.092×10^{-6}

	Mechanism	X_N	X_{NO}	X_{N_2O}
(i)	No N_2 Chemistry	-	-	-
(ii)	38 N_2 Steps	1.945×10^{-6}	4.62×10^{-3}	1.24×10^{-6}
(iii)	No NO_2 Chemistry	1.945×10^{-6}	4.62×10^{-3}	-
(iv)	No NH Chemistry	2.063×10^{-6}	4.58×10^{-3}	1.23×10^{-6}
(v)	No NO_2 or NH Chemistry	2.063×10^{-6}	4.58×10^{-3}	-
(vi)	Only Steps 22, 23, 24	2.094×10^{-6}	4.59×10^{-3}	-
(vii)	Steps 22–37	1.944×10^{-6}	4.62×10^{-3}	-
(viii)	10^3 ppm NO in Air	1.958×10^{-6}	4.80×10^{-3}	-
(ix)	10^4 ppm NO in Air	2.807×10^{-6}	1.00×10^{-2}	-

	Mechanism	X_{N_2O}	X_{NH}	X_{HNO}
(i)	No N_2 Chemistry	–	–	–
(ii)	38 N_2 Steps	5.83×10^{-7}	1.10×10^{-6}	2.27×10^{-5}
(iii)	No NO_2 Chemistry	5.83×10^{-7}	1.15×10^{-6}	2.27×10^{-5}
(iv)	No NH Chemistry	6.32×10^{-7}	–	5.60×10^{-5}
(v)	No NO_2 or NH Chemistry	5.88×10^{-7}	–	5.60×10^{-5}
(vi)	Only Steps 22, 23, 24	–	–	–
(vii)	Steps 22–37	5.83×10^{-7}	–	2.27×10^{-5}
(viii)	10^3 ppm NO in Air	5.87×10^{-7}	–	5.97×10^{-5}
(ix)	10^4 ppm NO in Air	6.27×10^{-7}	–	7.64×10^{-5}

11.6 Comparison with Experiment

The predicted dependence of the extinction strain rate on dilution in axisymmetric configurations can be compared with results of recent experiments [11.14]. These comparisons are shown in Fig. 11.10, where results of an independent numerical computation [11.15] with a different full kinetic scheme also are plotted. The experiments represent modifications of earlier experiments [11.16], carried out with the objective of improving the correspondence between the experimental and theoretical boundary conditions in evaluating the strain rate on the oxidizer side of the flame. The general agreement among all three sets of results shown in Fig. 11.10 is seen to be quite good. Over most of the range of dilution, the extinction strain rate of Ho and Isaac [11.15] is slightly low, which is consistent with the observation that the crossover temperature T_c for their rate data is about 50 K below that of the rates employed here.

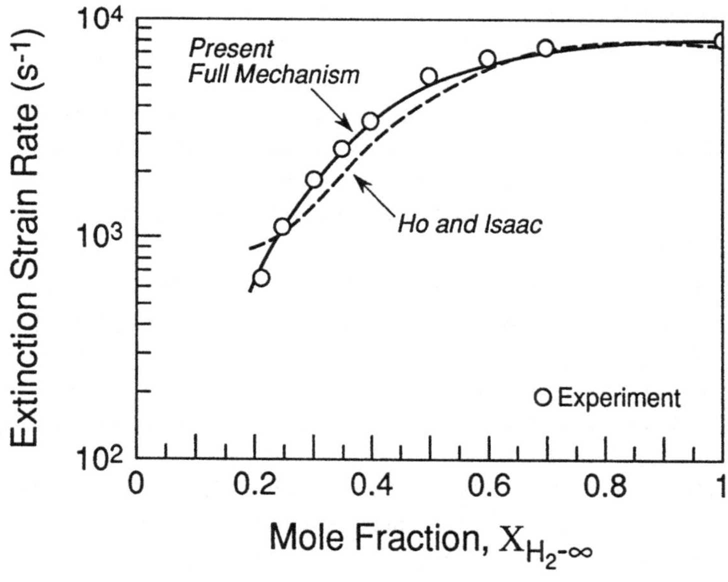

Fig. 11.10. Comparison between theory employing full chemistry and experiment for the dependence of air-side extinction strain rate on the hydrogen mole fraction in the fuel stream for axisymmetric counterflow mixing for hydrogen-nitrogen mixtures with air at $T_{-\infty} = T_{\infty} = 300\,\mathrm{K}$, $p = 1\,\mathrm{atm}$.

It is noteworthy from these results that uncertainties in extinction strain rates of these flames have decreased dramatically in recent years. When the extinction value of $a = 8140\,\mathrm{s}^{-1}$ at normal conditions was first published [11.1], disagreement with experiment amounted to about a factor of two. The revised experimental value of $a = 8250\,\mathrm{s}^{-1}$ is within 2% of the computed value. The numerical computations of the effect of dilution were performed after the new experimental results became available, and the close agreement over the entire range of dilution proved encouraging. Moreover, the excellent agreement in Fig. 11.8 between the full-chemistry and two-step predictions implies that the experimental results can be calculated very well from a two-step mechanism. Remaining to be done is an asymptotic analysis with the two-step approximation, to obtain formulas for predicting experimental results.

11.7 Conclusions

The present investigation identifies a thirteen-step kinetic mechanism for the description of structures and extinction of hydrogen-air diffusion flames over the full range of practical interest. An additional sixteen elementary kinetic steps have been added to model the production of oxides of nitrogen in pure-hydrogen and in nitrogen-diluted flames. Neglect of steps 14 to 21 of the reactions given in Table 11.1, as well as neglecting species NO_2 and NH, does not change the structure or extinction of the flames significantly, so that the mechanism including steps 1 to 13 and 23 to 37 is adequate for practical purposes.

Parametric dependences of extinction strain rates were obtained to show, for example, that the strain rate for extinction increases with increasing pressure. Abrupt extinction does not occur when the air temperature is above a critical value that increases with increasing pressure, for example above 1200 K for $p = 1\,\mathrm{atm}$.

Numerical integrations employing reduced mechanisms of four, three and two steps show good agreement with results obtained employing the detailed scheme concerning structures and extinction strain rates. Results of a one-step approximation also exhibit agreement in some respects, but not in regard to extinction strain rates. Full chemical equilibrium is always a poor approximation.

The effect of nitrogen chemistry and nitrogen dilution of the fuel feed has been investigated. Although there is a considerable production of NO in flames at high pressures and elevated feed temperatures at low strain, nitrogen chemistry does not influence extinction.

A comparison of extinction data obtained by numerical integrations shows excellent agreement with very recent experimental extinction data. Two-step mechanisms thus can perform well predicting experimental extinction results.

The reduced kinetic mechanisms investigated by numerical means provide a basis for developing an extinction theory for hydrogen/nitrogen-air flames by means of asymptotic methods. The characteristics of the structure according to the two-step mechanism, as summarized at the end of the preceding chapter, point to the direction of the asymptotic development that is needed.

Acknowledgements

This work was supported in part by the U.S. Air Force Office of Scientific Research through Grant No. AFSOR-89-0310. The authors are indebted to M. Smooke for his help with the computational procedures.

References

[11.1] Gutheil, E. and Williams F. A., Twenty-Third Symposium (International) on Combustion, The Combustion Institute, Pittsburgh, 513, 1991.

[11.2] Smooke, M. D., Puri, I. K. and Seshadri, K., Twenty-First Symposium (International) on Combustion, The Combustion Institute, Pittsburgh,1461, 1988).

[11.3] Keyes, D. E. and Smooke, M. D., Journal of Computational Physics **73**, 2, 267, 1987.

[11.4] Kee, R. J., Warnatz, J. and Miller, J. A., Sandia Report SAND83-8209, UC-32, Livermore, 1983.

[11.5] Smooke, M. D., Journal of Computational Physics, **48**, 1, 72, 1982.

[11.6] Smooke, M. D., Reduced Kinetic Mechanisms and Asymptotic Approximations for Methane-Air Flames, Springer 1991.

[11.7] Hanson, R. K. and Salimian, S. in: Combustion Chemistry (W. C. Gardiner, Ed.), Springer, 361, 1984.

[11.8] Dixon-Lewis, G. and Missaghi, M., Twenty-Second Symposium (International) on Combustion, The Combustion Institute, Pittsburgh, 1161, 1988.

[11.9] Yetter, R. A., Vadja, S., and Dryer, F. L., Princeton University Report No. MAE1853, Princeton, 1989.

[11.10] Yetter, R. A. and Dryer, F. L., Combust. Sci. and Tech., 1989, submitted.

[11.11] Puri, I. K., Seshadri, K. , Smooke, M. D., and Keyes, D. E., Combust. Sci. and Tech. **56**, 1, 1987.

[11.12] Yetter, R. A., Rabitz, H., and Hedges, R. M. : "A Combined Stability-Sensitivity Analysis of Weak and Strong Reactions of Hydrogen/Oxygen Mixtures", Department of Mechanical and Aerospace Engineering, Princeton University, Princeton, NJ, 1990.

[11.13] Chung, S. H. and Williams, F. A., Combust. Flame **82**, 3/4, 389, 1990.

[11.14] Pellett, G. L., Northam, G. B., and Wilson, L. G., AIAA Preprint No. 91-0370, 1991.

[11.15] Ho, Y. H. and Isaac, K. M., Technical Memorandum MAE-TM-25, Mechanical and Aerospace Engineering and Mechanics Department, University of Missouri-Rolla 1989.

[11.16] Pellett, G. L., Northam, G. B., Wilson, L. G., Jerrett Jr., O., Antcliff, R. R., Dancey, C. L. and Wang, J. A., AIAA Preprint No. 89-2522, 1989.

12. CO–H_2–N_2/Air Diffusion Flames: Thermal Radiation and Transient Effects

J.-Y. Chen[1]
Combustion Research Facility, Sandia National Laboratories,
Livermore, CA 94551-0969, U.S.A

Y. Liu
Department of Engineering, University of Cambridge, Trumpington Street,
Cambridge CB2 1PZ, England

B. Rogg
Department of Engineering, University of Cambridge, Trumpington Street,
Cambridge CB2 1PZ, England

12.1 Introduction

Flames of $CO/H_2/N_2$ fuels have been investigated for various reasons. For instance, laminar, premixed flames were investigated experimentally by van Tiggelen and coworkers to determine the rate constant of the water-gas shift reaction [12.1] and the inhibition induced by CF_3BR [12.2], [12.3] , and numerically by Rogg and Williams [12.4] who derived a reduced kinetic mechanism for wet CO flames. Laminar, non-premixed flames were studied, for instance, by Drake and Blint [12.5], who investigated experimentally and numerically the structure of counterflow diffusion flames in the Tsuji geometry paying special attention to NO_x formation. For the latter geometry, Chen et al [12.6] derived a reduced kinetic mechanism which they used in the numerical simulation of turbulent diffusion flames. For further information on work relevant to flames of $CO/H_2/N_2$ fuels the cited papers should be consulted.

In this chapter we investigate counterflow diffusion flames with a molar fuel composition of 40/30/30 percent $CO/H_2/N_2$ and with air as oxidizer. Flames in the Tsuji geometry and, alternatively, in the opposed-jet geometry are considered; schematics of these geometries are shown in Fig. 1. Attention is focussed on transient effects and on the effects of thermal radiation. Calculations have been performed with a detailed mechanism of elementary reactions

[1] Now with the University of California at Berkeley, Mechanical Engineering Department

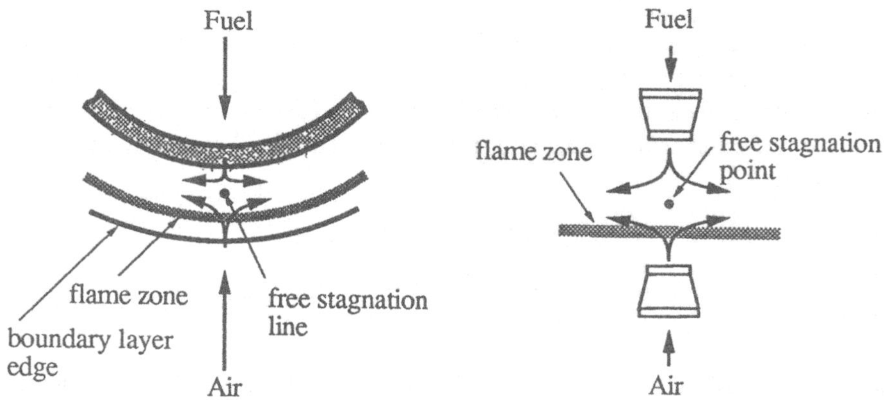

Fig. 12.1. Schematic of the Tsuji (left) and the opposed-jet (right) counterflow geometry.

and, alternatively, with a systematically reduced three-step mechanism. For the opposed-jet geometry, a flamelet library has been generated for pressures ranging from 1 to 10 bars.

The structure of the remainder of this chapter is as follows. In Sect. 12.2 we specify and derive, respectively, the detailed mechanism of elementary reactions, subsequently termed the "short mechanism", and the three-step mechanism used in the numerical calculations. In Sect. 12.3 we define the governing equations relevant for the flow geometries considered in the present work. In Sect. 12.4 the derivation of the model of thermal radiation is given. Section 12.5 gives a short overview over the numerical methods used to solve the governing equations, and in Sect. 12.6 results are presented and discussed. A summary and conclusions are given in Sect. 12.7.

12.2 Chemistry Models

12.2.1 The Short Mechanism

As a starting point we consider reactions 1 to 24 of the detailed mechanism of elementary reactions given in Chap. 1. Since in none of these reactions the CH radical is generated, we disregard the CH consuming steps 19 and 20. Thus, the detailed mechanism of elementary reactions used in the numerical calculations comprises 22 elementary reactions (30 if forward and backward steps are counted separately), the reacting species CO, H_2, O_2, CO_2, H_2O, HCO, H, O, OH, HO_2 and H_2O_2 and inert nitrogen. Subsequently this mechanism will be termed the "short mechanism"; in the literature various synonyms for short mechanism can be found such as "starting mechanism" and "skeletal mechanism". The three-step mechanism presented below is based on the short mechanism.

12.2.2 A Three-Step Mechanism

The chemical system under consideration comprises eleven reacting species and three elements. Therefore, with five species assumed in steady state, three-step kinetic mechanisms exist for this system. If HCO, O, OH, HO_2 and H_2O_2 are selected as the steady-state species, using standard methods the three-step mechanism

$$\text{(I)} \qquad CO + H_2O \rightleftharpoons CO_2 + H_2 \,,$$
$$\text{(II)} \qquad 2H + M \rightarrow H_2 + M \,,$$
$$\text{(III)} \qquad O_2 + 3H_2 \rightleftharpoons 2H_2O + 2H$$

can be derived. In this mechanism, step I is the overall CO consumption step which neither creates nor destroys reaction intermediaries. Step II represents an overall recombination step. Step III is an overall radical-production, oxygen-consumption step. Note that this three-step mechanism is formally identical to the mechanisms derived by Rogg and Williams [12.4] and Chen et al [12.6]. In terms of the rates of the elementary steps contained in the short mechanism, the global rates of the three-step mechanism can be written as

$$\omega_{I} = \omega_{18} \,, \tag{12.1}$$
$$\omega_{II} = \omega_5 + \omega_{12} + \omega_{15} + \omega_{16} + \omega_{17} - \omega_{24} \,, \tag{12.2}$$
$$\omega_{III} = \omega_1 + \omega_6 + \omega_9 - \omega_{12} + \omega_{13} - \omega_{17} \,. \tag{12.3}$$

In the elementary rates appearing on the right-hand-sides of (12.1)–(12.3), concentrations of the steady-state species must be expressed in terms of the concentrations of the species appearing in the three-step mechanism. Therefore, the global rates are algebraically more or less complex expressions containing rate data of many of the elementary steps of the short mechanism; the degree of complexity of the global rates depends, of course, on the specific assumptions introduced in their derivation.

In the following we assume that the concentrations of the species appearing explicitly in the three-step mechanism, i.e., [CO], [H_2], [O_2], [CO_2], [H_2O] and [H], are known by solving numerically the equations governing conservation of overall mass, species mass, momentum and energy for a specific problem in laminar non-premixed combustion. The concentrations of OH, O, HCO, HO_2 and H_2O_2, which do not appear explicitly in the reduced mechanism but are required to evaluate the global rates according to (12.1)–(12.3), are expressed in terms of the known species concentrations as follows.

The OH-radical concentration is obtained from the partial-equilibrium assumption for reaction 3, i.e.,

$$[OH] = \frac{k_{3b}}{k_{3f}} \frac{[H_2O][H]}{[H_2]} \,. \tag{12.4}$$

With [OH] known from (12.4), the concentrations of O and HCO are calculated by assuming steady-states for these species. These steady states result in

$$[O] = \frac{k_{1f}[H][O_2] + k_{2b}[H][OH] + k_{4f}[OH]}{k_{1b}[OH] + k_{2f}[H_2] + k_{4b}[H_2O]}, \tag{12.5}$$

$$[HCO] = \frac{k_{24b}[CO][H][M]}{k_{21}[H] + k_{22}[OH] + k_{23}[O_2] + k_{24f}[M]}, \tag{12.6}$$

where to derive [O] we have neglected the elementary recombination step 17. Note that if the latter step is taken into account a quadratic equation results for [O] whose solution too can be expressed explicitly in terms of known concentrations.[2]

Assuming steady states for HO_2 and H_2O_2 results in two nonlinear, coupled equations for the two unknowns $[HO_2]$ and $[H_2O_2]$. If these equations are solved for $[HO_2]$, the result

$$[HO_2] = \frac{\sqrt{B_{HO_2}^2 + 4A_{HO_2}C_{HO_2}} - B_{HO_2}}{2A_{HO_2}} \tag{12.7}$$

is obtained, where

$$A_{HO_2} = k_{11}\left(2 - \frac{k_{14f}[OH]}{D_{HO_2}}\right), \tag{12.8.a}$$

$$B_{HO_2} = (k_6 + k_7 + k_9)[H] + k_{5b}[M] + k_8[OH] + k_{10}[O] + k_{14b}[H_2O]\left(1 - \frac{k_{14f}[OH]}{D_{HO_2}}\right), \tag{12.8.b}$$

$$C_{HO_2} = k_{5f}[H][O_2][M] + k_{23}[HCO][O_2] + \frac{k_{12f}k_{14f}[M][OH]^3}{D_{HO_2}}, \tag{12.8.c}$$

$$D_{HO_2} = k_{12b}[M] + k_{13}[H] + k_{14f}[OH]. \tag{12.8.d}$$

The relationship resulting from the steady-state assumption for H_2O_2 can be written as

$$[H_2O_2] = \left(k_{11}[HO_2]^2 + k_{12f}[M][OH]^2 + k_{14b}[H_2O][HO_2]\right)/D_{HO_2}, \tag{12.9}$$

where D_{HO_2} is given by (12.8.d). Equations (12.4)–(12.7) and (12.9) complete the derivation of the three-step mechanism.

12.3 The Governing Equations

The governing equations presented in Chap. 1 describe the special case of steady flames subject to a constant, time-independent rate of strain, and do not take radiation heat loss into account. The work presented in this chapter requires a set of equations which allows the simulation of transient effects with

[2] If step 17 is taken into account in the reduced mechanism, then higher flame temperatures result than without this step.

a time-varying rate of strain and the simulation of radiation heat loss. This more general set of equations is given in Subsect. 12.3.1. A simplified set of governing equations is given in Subsect. 12.3.2.

12.3.1 General Equations

Boundary-layer equations are adopted to describe the variations of density, velocity, pressure, energy and species mass fractions in a laminar, chemically reacting, low-Mach-number stagnation-point diffusion flame. Specifically, the flow field is modelled in the Tsuji or, alternatively, the opposed-jet counterflow geometry. Schematics of these geometries are shown in Fig. 1. An ideal-gas mixture of N chemical species is considered, and Dufour effects, diffusion caused by pressure gradients, and external forces are neglected. In terms of the accumulative-convective operator L,

$$L(\phi) \equiv \frac{\partial(\rho\phi)}{\partial t} + \frac{\partial(\rho v\phi)}{\partial y} + (j+1)\,\rho G\,\phi\,, \tag{12.10}$$

the equations governing conservation of overall mass, momentum, energy and species mass can be written as

$$L(1) = 0\,, \tag{12.11}$$

$$L(G) = \frac{\partial}{\partial y}\left(\mu\frac{\partial G}{\partial y}\right) - \rho G^2 + P'(t)\,, \tag{12.12}$$

$$c_p L(T) = \frac{\partial}{\partial y}\left(\lambda\frac{\partial T}{\partial y}\right) - \frac{\partial T}{\partial y}\sum_{i=1}^{N} c_{pi} j_i - \sum_{i=1}^{N} h_i \dot{m}_i + \frac{\partial p}{\partial t} - \frac{\partial q_R}{\partial y}\,, \tag{12.13}$$

$$L(Y_i) = -\frac{\partial j_i}{\partial y} + \dot{m}_i\,, \qquad i = 1, ..., N\,. \tag{12.14}$$

Note that, in generalization of the definition given in Chap. 1, the term P' appearing in the momentum equation (12.12) represents a temporal forcing term defined by

$$P'(t) \equiv \rho_\infty \left(\frac{da}{dt} + a(t)^2\right)\,.$$

In the energy equation (12.13), the term q_R is the y component of the radiant-heat-flux vector. The other symbols in (12.11)–(12.14) have the meaning discussed in Chap. 1. For both the Tsuji and the opposed-jet geometry the boundary conditions at the oxidiser side are

$$G - a(t) = T - T_\infty = Y_i - Y_{i,\infty} = 0 \qquad \text{as} \qquad y \to \infty\,. \tag{12.15}$$

For the Tsuji geometry the boundary conditions at the fuel side, i.e., at the burner surface, are

$$G = T - T_w = Y_i - Y_{i,-\infty} + j_i/(\rho v)_w = 0 \qquad \text{at} \qquad y = 0\,. \tag{12.16}$$

In (12.16) and below in the context of the Tsuji geometry, the subscripts w and $-\infty$ are used to identify conditions or quantities at the burner surface and in the fuel supply upstream of the burner, respectively; the quantity $(\rho v)_w$ is the mass flux at the burner surface which is taken as a known constant. For the opposed-jet geometry two further sets of conditions must be imposed, viz., the boundary conditions at the fuel side,

$$G - a(t) \cdot \left(\frac{\rho_{+\infty}}{\rho_{-\infty}}\right)^{1/2} = T - T_{-\infty} = Y_i - Y_{i,-\infty} = 0 \qquad \text{as} \qquad y \to -\infty\,, \tag{12.17}$$

and the condition at the stagnation point,

$$v = 0 \qquad \text{at} \qquad y = 0\,. \tag{12.18}$$

In (12.17) and below in the context of the opposed-jet geometry, the subscript $-\infty$ is used to identify conditions or quantities taken at $y = -\infty$, i.e., in the fuel stream. In (12.18) we have assumed without loss of generality that the stagnation point is located at $y = 0$.

If thermal radiation is taken into account, boundaries values for the emmisivities must be specified. This is discussed in detail below.

12.3.2 Simplified Equations

The above set of governing equations has been solved for steady-state problems, together with the equation for the mixture fraction Z given in Chap. 1. Although (12.11)–(12.14) represent the most general set of similarity equations allowing for a time-varying strain rate, herein the numerical simulations for a time-varying strain rate have been carried out with a simplified set of equations used previously by Mauß et al [12.7], viz.,

$$\frac{\partial T}{\partial t} = \frac{\chi_{st}}{2}\frac{\partial^2 T}{\partial Z^2} - \sum_{i=1}^{N} \frac{h_i}{c_p}\frac{\dot{m}_i}{\rho}\,, \tag{12.19}$$

$$\frac{\partial Y_i}{\partial t} = \frac{\chi_{st}}{2}\frac{1}{Le_i}\frac{\partial^2 Y_i}{\partial Z^2} + \frac{\dot{m}_i}{\rho}\,, \tag{12.20}$$

$i = 1, \cdots, \mathrm{N}$. In (12.19) and (12.20), χ_{st} is the scalar dissipation rate taken at stoichiometric conditions. The scalar dissipation rate χ is defined as

$$\chi = 2\,D\left(\frac{\partial Z}{\partial y}\right)^2 = 2\frac{\lambda}{\rho c_p}\left(\frac{\partial Z}{\partial y}\right)^2\,, \tag{12.21}$$

where D denotes a representative, suitably defined diffusion coefficient. Note that the second equality in (12.21) holds only if a value of one is assumed for the Lewis number $Le = \lambda/(\rho c_p D)$. By writing (12.21) in terms of suitable similarity variables it can be shown that χ_{st} is proportional to the strain rate a.

In (12.19) and (12.20), χ_{st} is viewed as a known quantity to be specified. Specifically, in the present work χ_{st} is prescribed as a function varying sinusoidally with time. Comparisons of transient results obtained from the simplified set of (12.19) and (12.20) with results obtained from the more general set of (12.11)–(12.14) have not been performed; results of such comparisons will be presented in future work.

The numerical values used for the Lewis numbers appearing on the r.h.s. of (12.20) are 1.10, 0.30, 1.11, 1.39, 0.83, 1.27, 0.18, 0.70, 0.73, 1.10, 1.12 and 1.00 for CO, H_2, O_2, CO_2, H_2O, HCO, H, O, OH, HO_2 and H_2O_2 and N_2, respectively. The initial and boundary conditions, as well as other details, used in the numerical solution of (12.19) and (12.20) will be discussed in Sect. 12.6.

12.4 Radiation Model

In the present work both transient and steady-state combustion phenomena are studied. Thermal radiation, however, is considered only for steady-state flames and, therefore, time-dependencies are neglected in the derivation of the radiation model. Since details of the derivation were published elsewhere [12.8], herein only a short outline is given.

Radiating flames are considered in the Tsuji geometry only. Assuming the optically thin limit and adopting the grey-medium approximation, it can be shown that the derivative of the radiation heat-flux appearing in the energy equation is given by

$$\frac{dq_R}{dy} = 2k_P\left(2\sigma T^4 - B_w - B_\infty\right). \tag{12.22}$$

In (12.22), k_P denotes the mean Planck absorption coefficient, σ is the Stefan-Boltzmann constant, and B is given by $B = \epsilon\sigma T^4$, where ϵ denotes the mean emissivity. At the boundary-layer edge ϵ is that of the incoming air stream, at the burner surface ϵ represents the joint emissivity of the burner-surface and the gas mixture; herein the numerical values 0.2 and 0.1 have been used for ϵ_w and ϵ_∞, respectively.

The Planck mean absorption coefficient accounts for the absorption and emission from the gaseous species CO_2 and H_2O; it is expressed as

$$k_P = p[X_{CO_2}k_{P,CO_2}(T) + X_{H_2O}k_{P,H_2O}(T)]. \tag{12.23}$$

Here p is the pressure in units atm, and $k_{P,i}$ and X_i denote the mean absorption coefficient in units $1/(\mathrm{m}\cdot\mathrm{atm})$ and mole fraction, respectively, of species i. In writing (12.23) we have neglected the partial overlapping of the CO_2 and H_2O bands, and the contributions to thermal radiation of CO. Whilst it is generally recognized that the contributions to thermal radiation of CO are negligibly small in flames of hydrocarbon fuels, such contributions may be more important for the flames of $CO/H_2/N_2$ fuels considered herein. Nevertheless, in the present work CO has not been taken into account on the r.h.s.

of (12.23) because of the general lack of radiation data for this species. In the literature estimates are given for a positive term to be subtracted from the r.h.s. of (12.23) in order to account for the band overlapping of CO_2 and H_2O. Such negative corrections have not been employed in the present work because (i) they are small and (ii) are taken into account implicitly by neglecting in (12.23) the small positive contributions to thermal radiation of CO.

The mean Planck absorption coefficients of CO_2 and H_2O required to evaluate the r.h.s. of (12.23) are calculated using the polynomial fits [12.8]

$$\log_{10}\left(\frac{k_{p,i}}{k_{p,i,\mathrm{ref}}}\right) = \sum_{n=0}^{6} a_{i,n} \cdot \left(\frac{T}{300\,\mathrm{K}}\right)^n \quad , \quad i = \mathrm{CO_2}, \mathrm{H_2O}\,. \tag{12.24}$$

In (12.24), T is the temperature in kelvins and $k_{p,i,\mathrm{ref}} = 1/(\mathrm{m} \cdot \mathrm{atm})$; the polynomial coefficients $a_{i,n}$ are presented in Table 12.1.

Table 12.1. Polynomials coefficients for the evaluation of $k_{p,\mathrm{CO_2}}$ and $k_{p,\mathrm{H_2O}}$; see (12.24).

	Polynomial Coefficients	
	CO_2	H_2O
a_0	0.22317E + 01	0.38041E + 01
a_1	−0.15829E + 01	−0.27808E + 01
a_2	0.13296E + 01	0.11672E + 01
a_3	−0.50707E + 00	−0.28491E + 00
a_4	0.93334E − 01	0.38163E − 01
a_5	0.83108E − 02	−0.26292E − 02
a_6	0.28834E − 03	0.37774E − 04

12.5 Numerical Methods

The calculations performed in Cambridge have employed the computer code RUN-1DL by Rogg [12.9], [12.13] which has been developed for the numerical solution of premixed burner-stabilized and freely propagating flames, strained premixed, non-premixed and partially premixed flames subject to a variety of boundary conditions, and tubular flames. RUN-1DL employs fully self-adaptive gridding; it solves both transient and steady-state problems in physical space[3] and for diffusion flames, alternatively, in mixture fraction space;[4] models for thermodynamics and molecular transport are implemented that range from trivially simple to very sophisticated; also the model of thermal

[3] such as problems described by (12.11)–(12.14)

[4] such as problems described by (12.19) and (12.20)

radiation described above is implemented. Chemistry models that can be handled range from overall global one-step reactions over reduced kinetic mechanisms to detailed mechanisms comprising an arbitrary number of chemical species and elementary steps. Furthermore, RUN-1DL is able simulate two-phase combustion problems, such as droplet and spray combustion. A copy of RUN-1DS is available from the author upon request [12.9], [12.13].

The calculations performed at Sandia National Laboratories have employed the CHEMKIN II package for evaluation of the thermochemical data [12.10]. The steady version of (12.11)–(12.14) were integrated numerically with a computer program used previously by Peters and Kee [12.11]. The numerical integration of (12.19) and (12.20) was accomplished as follows. Using central differences the 2nd derivatives with respect to Z were discretized on a mesh consisting of 40 grid points with dense distribution around the stoichiometric value. The set of ordinary differential equations resulting from this discretization was solved with Hindmarsh's package LSODE [12.12].

12.6 Results and Discussion

In this section results obtained for flames in the Tsuji geometry and, alternatively, in the opposed-jet geometry are presented and discussed. The section is organized as follows. In Subsect. 12.6.1 we present and discuss results that have been obtained during validation of the 3-step mechanism. Steady-state results obtained with the model of thermal radiation are presented and discussed in Subsect. 12.6.2, results obtained from the time-dependent calculations in Subsect. 12.6.3.

All results were obtained with a molar fuel composition of 40/30/30 percent $CO/H_2/N_2$ and with air as oxidizer. The pressure was varied between 1 and 10 bars. Unless stated otherwise, in the calculations the temperatures of fuel and air were taken as 300 K.

12.6.1 Validation of the 3-Step Mechanism; Flamelet Library

The three-step mechanism has been validated in the Tsuji counterflow geometry by comparing computational results obtained with both the short mechanism and the reduced mechanism. Shown in Fig. 12.2 is the maximum temperature in the flame, $T_{\max}$, as a function of the reciprocal of the strain rate, a^{-1}, for a fuel temperature of 300 K and a pressure of 1 bar and 10 bars, respectively; results for intermediate values of the pressure are similar. It is seen that the reduced mechanism predicts a slightly higher maximum temperature than the short mechanism. At 1 bar, extinction occurs at $a_q = 5000\,s^{-1}$ for the short mechanism, and at $a_q = 5400\,s^{-1}$ for the reduced mechanism. For a pressure of 5 bars the extinction strain rates $a_q = 25 \cdot 10^3\,s^{-1}$ and $a_q = 29 \cdot 10^3\,s^{-1}$ were calculated for the short mechanism and the 3-step mechanism, respectively; the values obtained for 10 bars are $a_q = 42 \cdot 10^3\,s^{-1}$

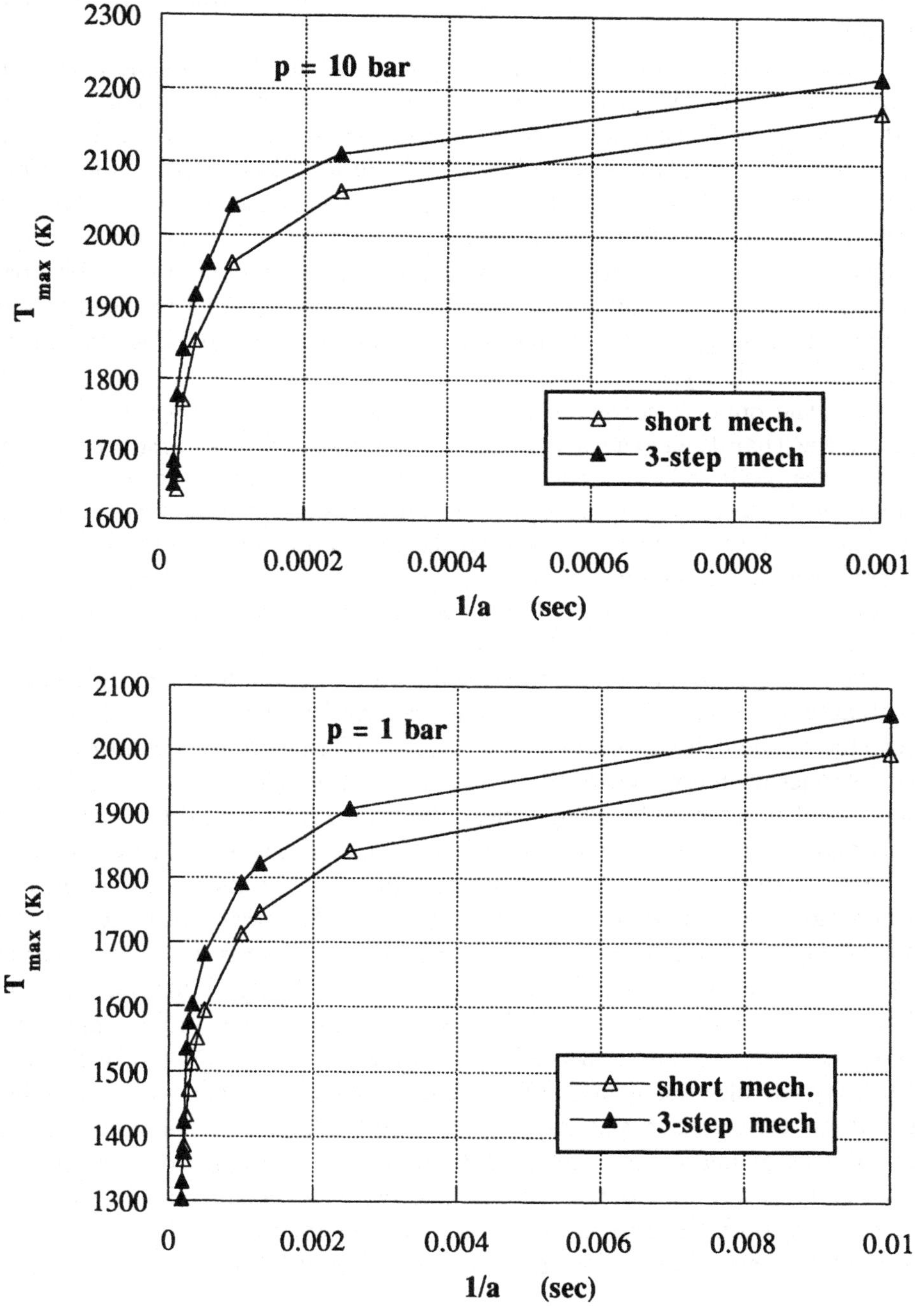

Fig. 12.2. Maximum temperature T_{max}, versus the reciprocal of the strain rate, a^{-1}, for the Tsuji geometry for $p = 1$ bar (left) and $p = 10$ bars (right). Void triangles: short mechanism; solid triangles: 3-step mechanism.

for the short mechanism and $a_q = 53 \cdot 10^3\,s^{-1}$ for the 3-step mechanism. Thus, for the Tsuji geometry flame structures obtained with the reduced mechanism and the short mechanism agree very well in the entire range of pressures and strain rates.

After validation of the 3-step mechanism in the Tsuji geometry, calculations were performed for the opposed-jet geometry for pressures of 1, 3, 6 and 10 bar using both the short mechanism and the 3-step mechanism. The results obtained with the short mechanism have been used to construct a flamelet library. For each pressure the flamelet library includes a number of flames for strain rates ranging from small values to values at which extinction occurs. The extinction strain rates, in units 1/s, are 5 890 for 1 bar, 18 150 for 3 bars, 33 450 for 5 bars and 49 300 for 10 bars. Shown in Fig. 12.3 is the maximum temperature T_{max} as a function of the strain rate a for the flames included in the flamelet library.

For the opposed-jet geometry, at all pressures and values of the strain rate lower than the extinction value, good agreement was found in the flame structures predicted using the short mechanism and the reduced mechanism. However, with the 3-step mechanism strain rates at extinction were predicted in excess of 50 to 100 % compared to the values predicted on the basis of the short mechanism. The latter, unexpected finding can be explained using the following physical arguments. For a specified value of the pressure and specified thermochemical states at the fuel and oxidiser boundaries, calculations in the Tsuji geometry require the specification of two additional independent parameters, namely the strain rate at the oxidizer side, a, and the blowing velocity at the cylinder surface, v_w. This is different from calculations in the opposed-jet geometry where the strain rate at the oxidiser side, a, is the only additional parameter to be specified; the strain rate at the fuel side is proportional to a. In the present work the calculations in the Tsuji geometry were performed keeping the blowing velocity v_w at a constant value; only the strain rate was varied, i.e., the velocity of the fuel was not increased with increasing velocity of the air. In the opposed-jet geometry the increase of the fuel velocity is proportional to that of the air velocity. Thus, although the Tsuji geometry and the opposed-jet geometry are similar, in general they do differ with respect to the flow-field and, hence, the flame. Only if in the Tsuji geometry the blowing velocity of the fuel is chosen sufficiently high, flow field and flame structure in this geometry evolve towards those predicted by the opposed-jet geometry. In conclusion, to eliminate the influence of the blowing velocity as a second independent parameter, it is recommended that in future studies both the derivation of reduced mechanisms and the generation of laminar-flamelet libraries be based on the opposed-jet geometry rather than on the Tsuji geometry.

Figures 12.4 to 12.5 are based on computations performed for the opposed-jet geometry. Shown in Fig. 12.4 is the maximum temperature in the flame, T_{max}, as a function of the pressure, p, for $a = 200\,s^{-1}$ and $a = 8\,000\,s^{-1}$, respectively, as obtained from the short mechanism (void triangles) and the 3-step mechanism (solid triangles). It is seen that over the entire pressure range

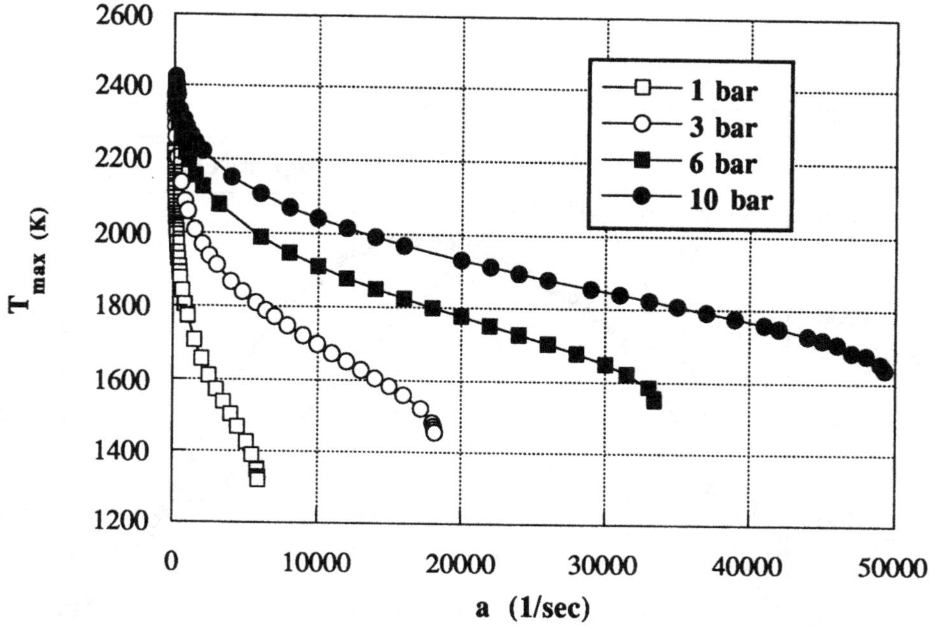

Fig. 12.3. Maximum temperature, $T_{\max}$, versus strain rate, a, for the opposed-jet geometry for pressures of 1, 3, 6 and 10 bars. The data are based on the short mechanism. The corresponding flames are contained in the flamelet library.

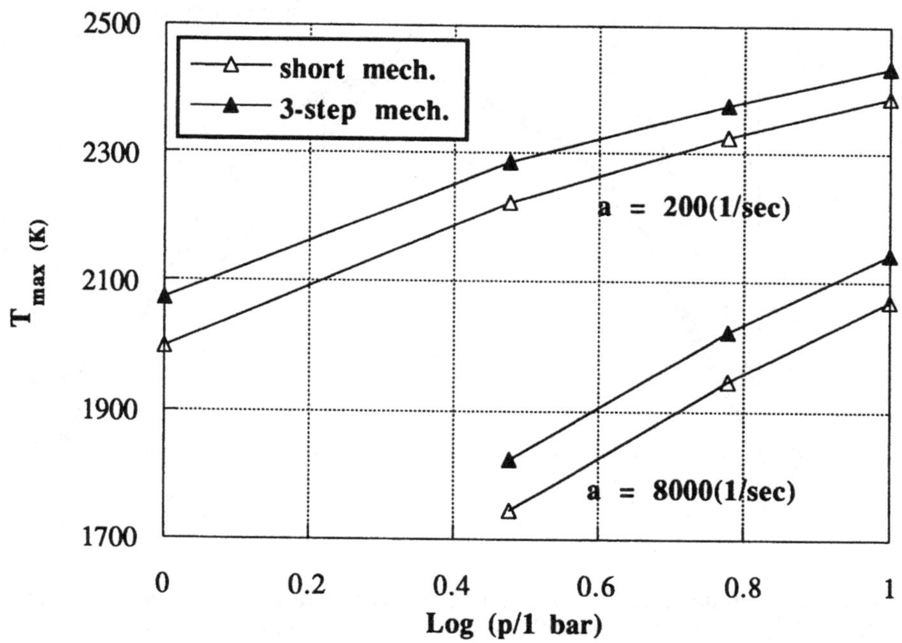

Fig. 12.4. Maximum temperature, $T_{\max}$, versus pressure, p, for the opposed-jet geometry for $a = 200\,s^{-1}$ bar (left) and $a = 8000\,s^{-1}$ (right). Void triangles: short mechanism; solid triangles: 3-step mechanism.

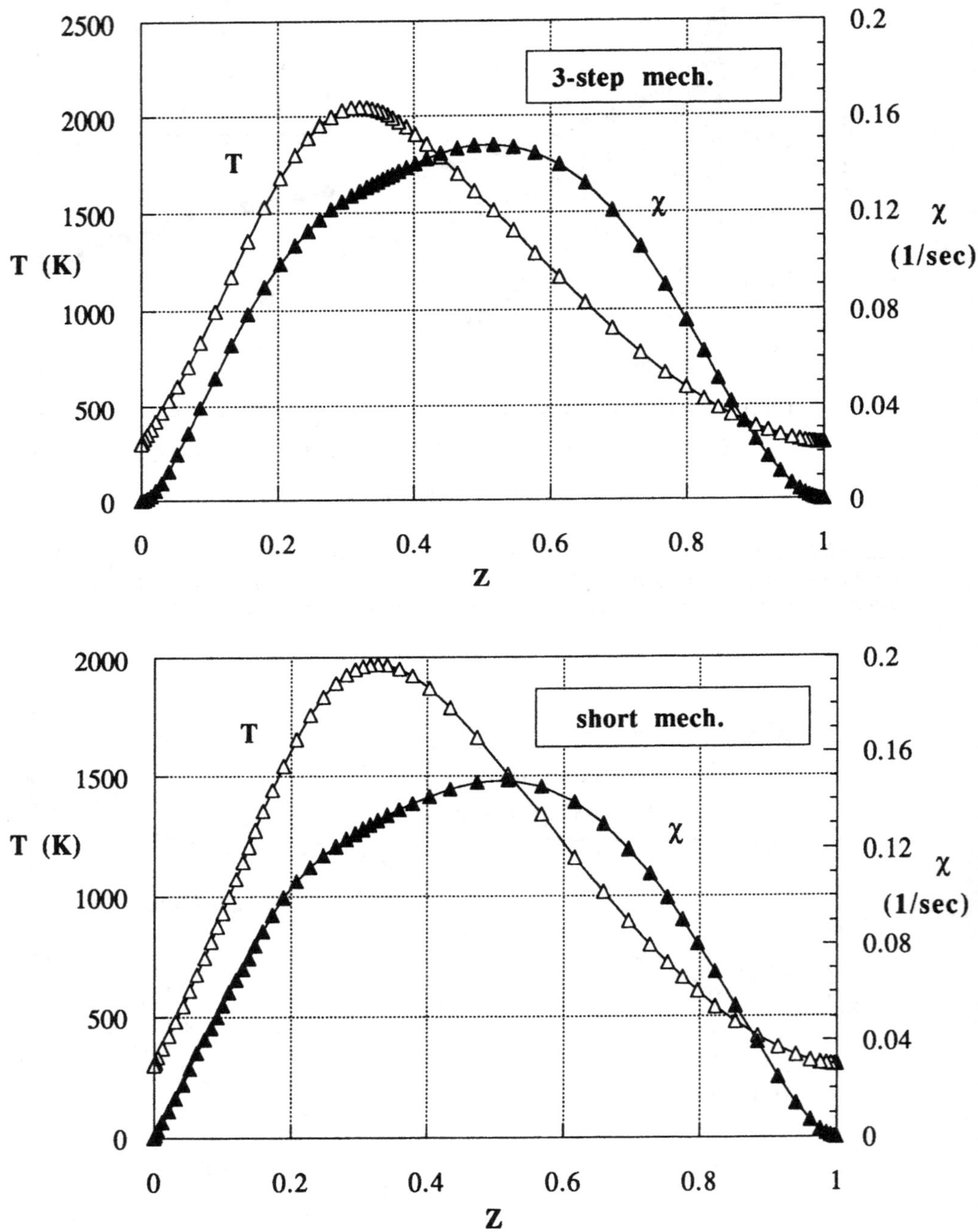

Fig. 12.5.a. Temperature T and scalar dissipation rate χ as functions of the mixture fraction Z for an opposed-jet diffusion flame at $p = 3\,\mathrm{bar}$, subject to $a = 2000\,\mathrm{s}^{-1}$. Top: results obtained with the short mechanism; bottom: results obtained with the 3-step mechanism.

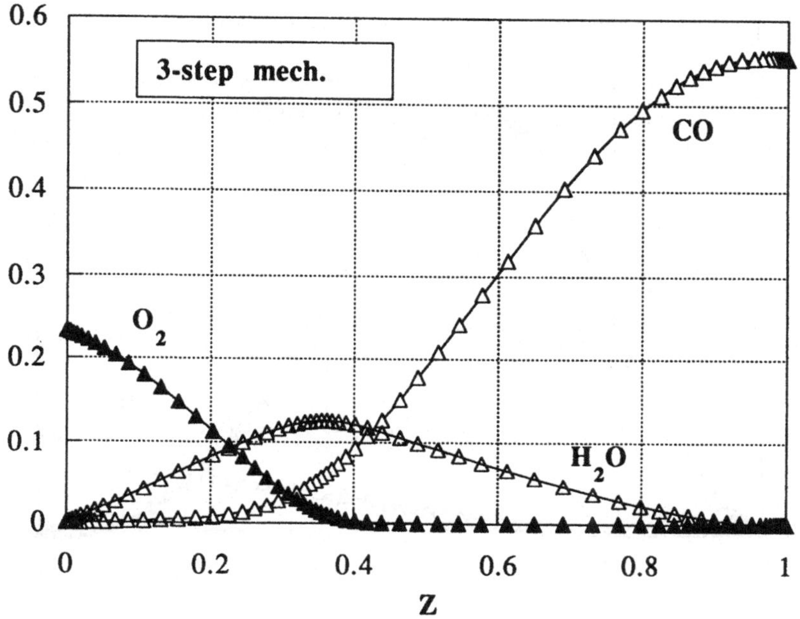

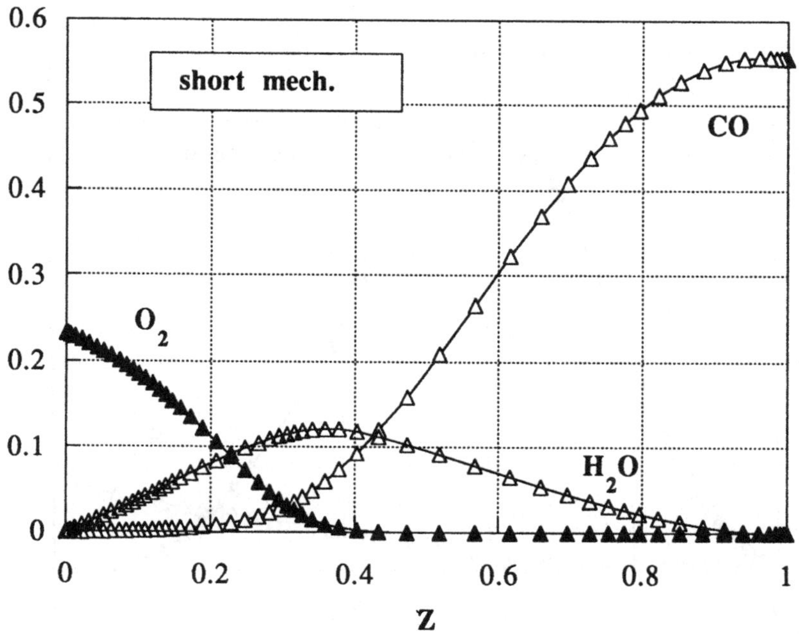

Fig. 12.5.b. As Fig. 12.5.a, but mass fractions of species CO, O_2 and H_2O.

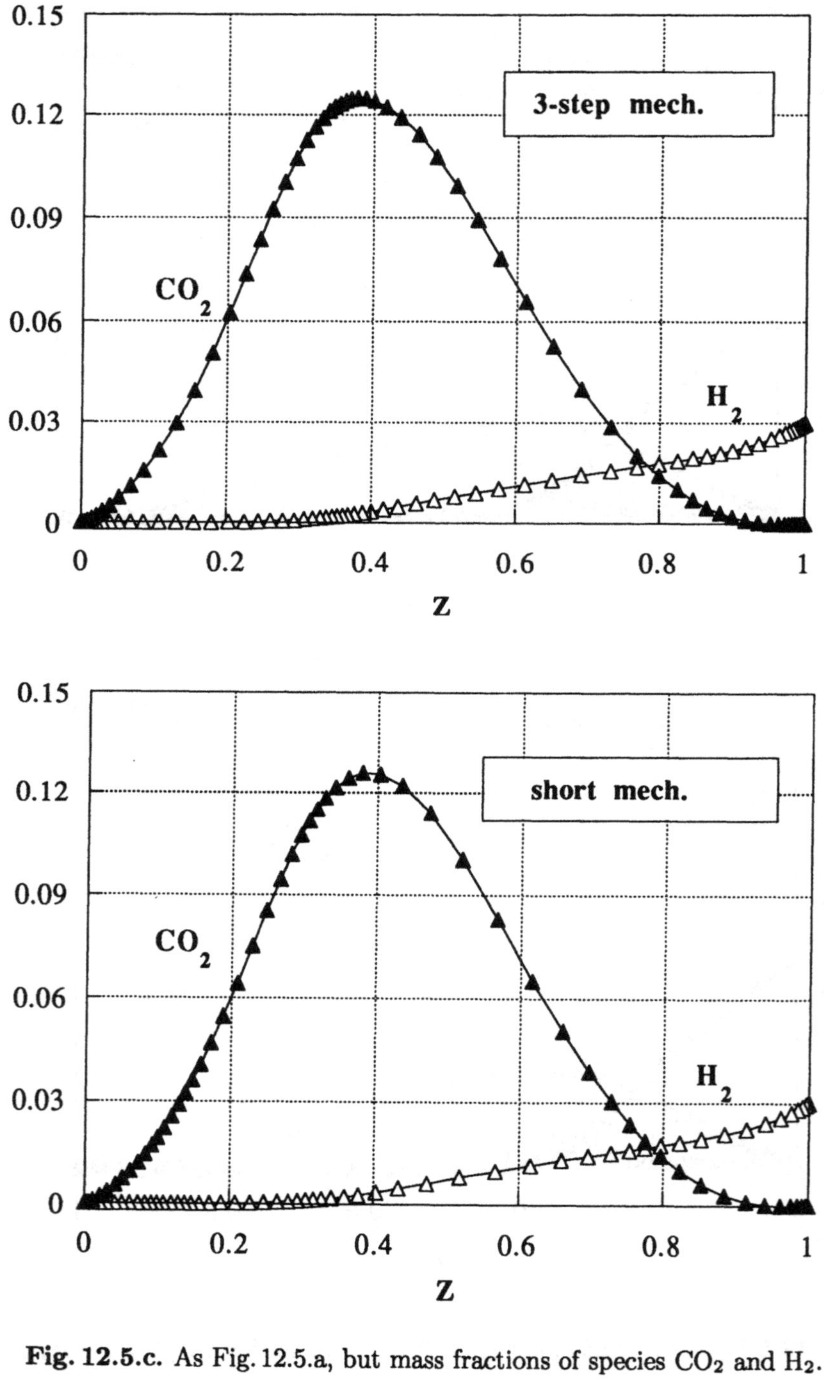

Fig. 12.5.c. As Fig. 12.5.a, but mass fractions of species CO_2 and H_2.

the reduced mechanism predicts slightly higher flame temperatures than the short mechanism.

Shown in Figs. 12.5.a–12.5.c is a comparison of flame structures calculated with the short mechanism and the 3-step mechanism for a pressure of 3 bar and a strain rate of $a = 2000\,\mathrm{s}^{-1}$. The various profiles in Fig. 12.5 are plotted versus the mixture fraction Z, which has been calculated as part of the solution to the problem by solving the Z-transport equation specified in Chap. 1. In each of Figs. 12.5.a–12.5.c the top picture represents the result obtained with the short mechanism, the bottom picture the result obtained with the three-step mechanism. Shown in Fig. 12.5.a are the profiles of temperature and scalar dissipation rate; the latter quantity has been calculated from the solution according to the second equality in (12.21). Shown in Figs. 12.5.b and 12.5.c are profiles of the mass fractions of those species that appear explicitly in the 3-step mechanism, with the exception of Y_{H}. It is seen from Fig. 12.5 that the agreement of the flame structure, including the H-atom concentration which is not shown, calculated with the short mechanism and the reduced mechanism is excellent. Agreement of comparable quality has been found for all pressures considered, and in the entire range of strain rates, even for strain rates with values relatively close to the extinction value a_q of the short mechanism.

12.6.2 Results of Steady-State Calculations

For the calculations, results of which are shown in Figs. 12.6 and 12.7, the boundary conditions were chosen to match those of the flames investigated experimentally by Drake and Blint [12.5], i.e., the temperature at the burner surface was taken as 395 K and 465 K, respectively, and the temperature of the air as 298 K. Calculations were performed with both the short mechanism and the three-step mechanism, and with and without taking thermal radiation into account. In Figs. 12.6 and 12.7 the top picture represents results obtained with the short mechanism, the bottom picture results obtained with the three-step mechanism. In all pictures, symbols not connected by a line represent experimental data by Drake and Blint [12.5], symbols connected by a line represent computational results obtained herein.

Shown in Fig. 12.6 are results obtained for a strain rate of $180\,\mathrm{s}^{-1}$. Temperature profiles are shown in Fig. 12.6.a, species mole fraction profiles in Fig. 12.6.b. It is seen that at this strain rate thermal radiation has only a small influence on the predicted maximum temperature, and virtually no influence on the concentrations of the major species. It should be noted, however, that thermal radiation at this value of the strain rate, and even more at lower values, has a pronounced effect on the peak values predicted for the concentrations of OH and NO_x; for details reference [12.8] should be consulted.

Furthermore, it is seen from Fig. 12.6 that the agreement between the computational and the experimental results is reasonable; the agreement between the results obtained with the short mechanism and the reduced mechanism is excellent. The slight discrepancies between the computed and measured concentrations shown in Fig. 12.6.b were discussed previously by Drake and

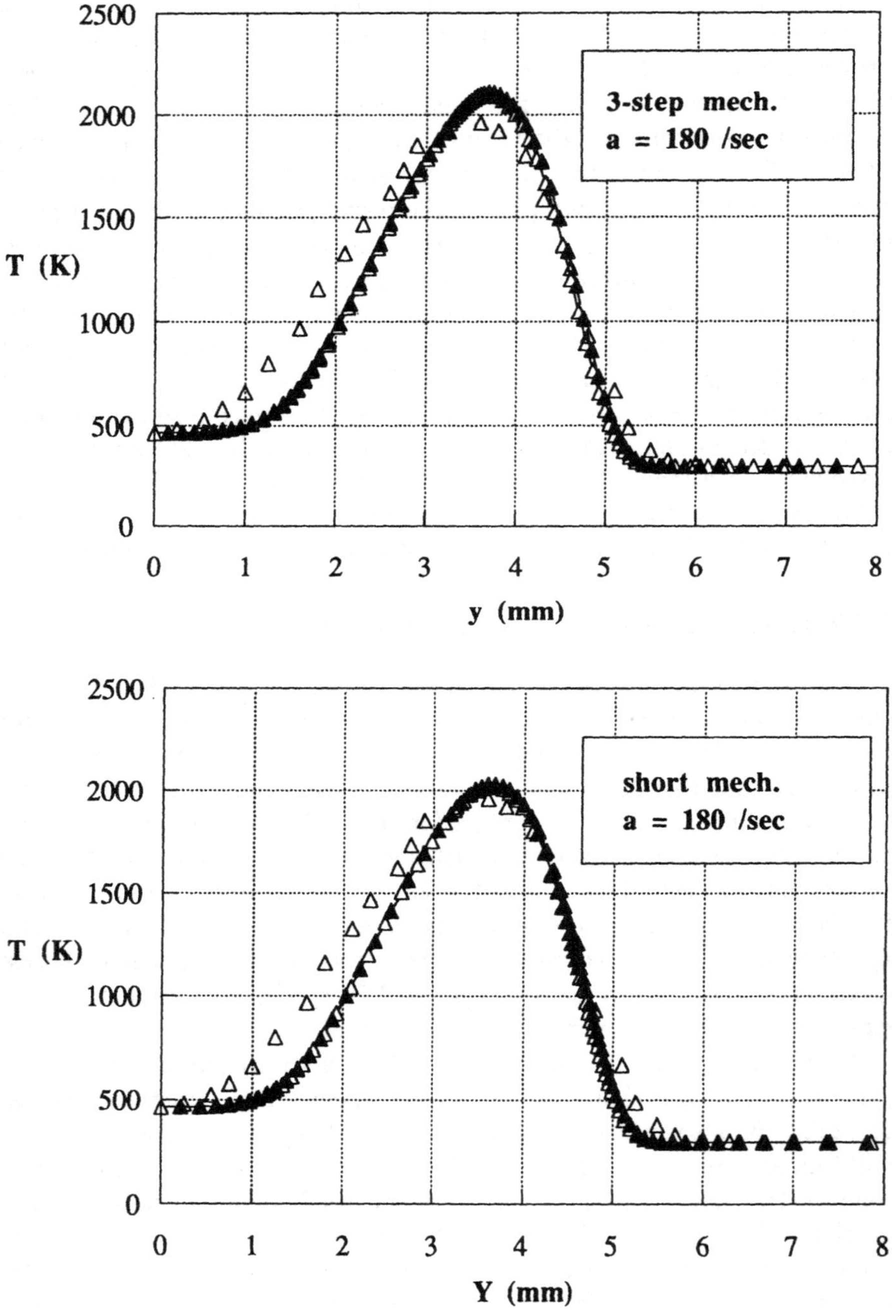

Fig. 12.6.a. Temperature T as a function of the distance from the burner surface for a flame subject to $a = 180\,s^{-1}$. Triangles: experimental data by Drake and Blint [12.5]; void rectangles: without radiation model; filled rectangles: with radiation model. Top: results obtained with the short mechanism; bottom: results obtained with the 3-step mechanism.

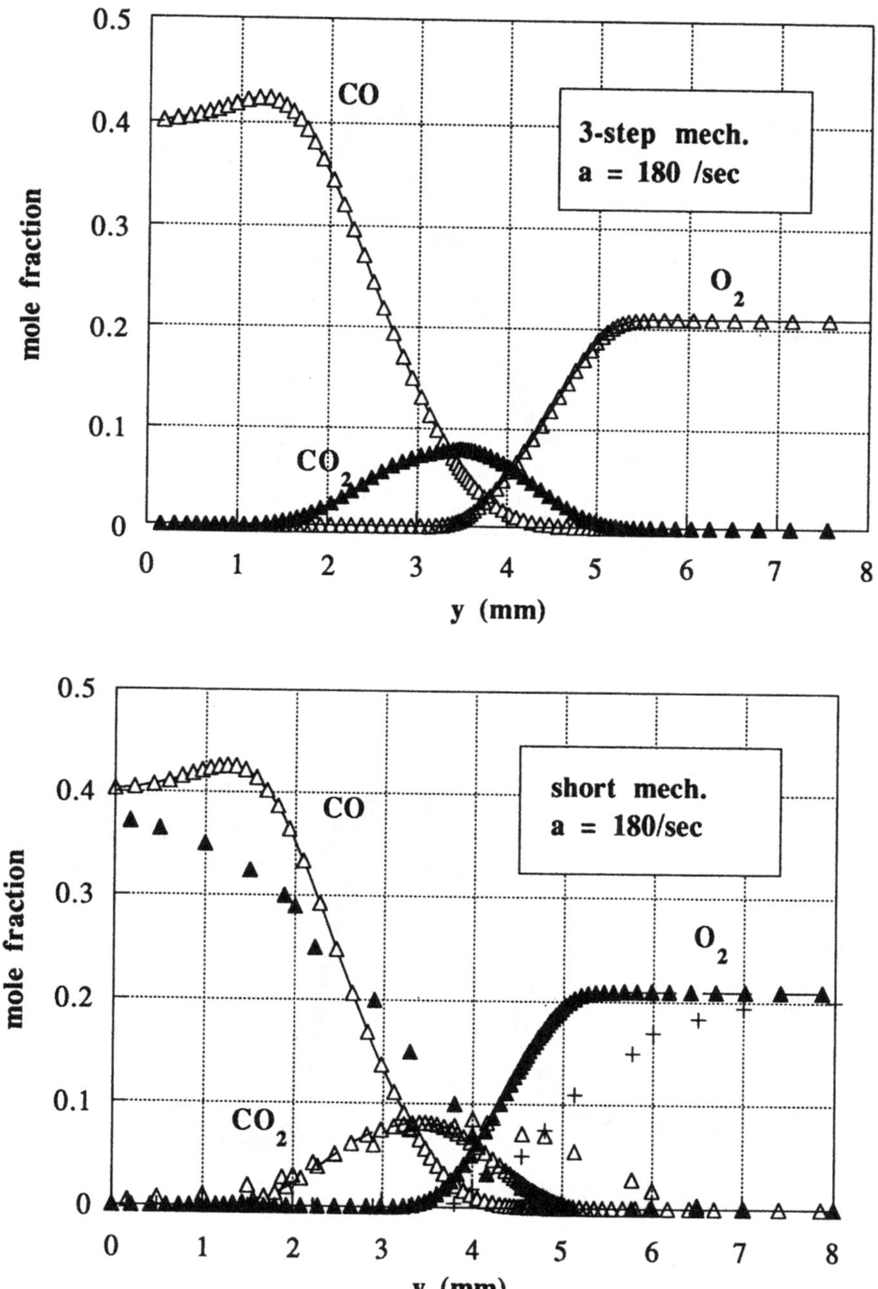

Fig. 12.6.b. Flame of Fig. 12.6.a, but mass fractions of species CO, O_2 and CO_2. Symbols not connected by line: experimental data by Drake and Blint [12.5]; symbols connected by line: computational results. For the species shown here, and at this particular strain rate, the computational results obtained with and without radiation model are virtually identical; this is not so, however, for the OH radical and NO_x [12.8]

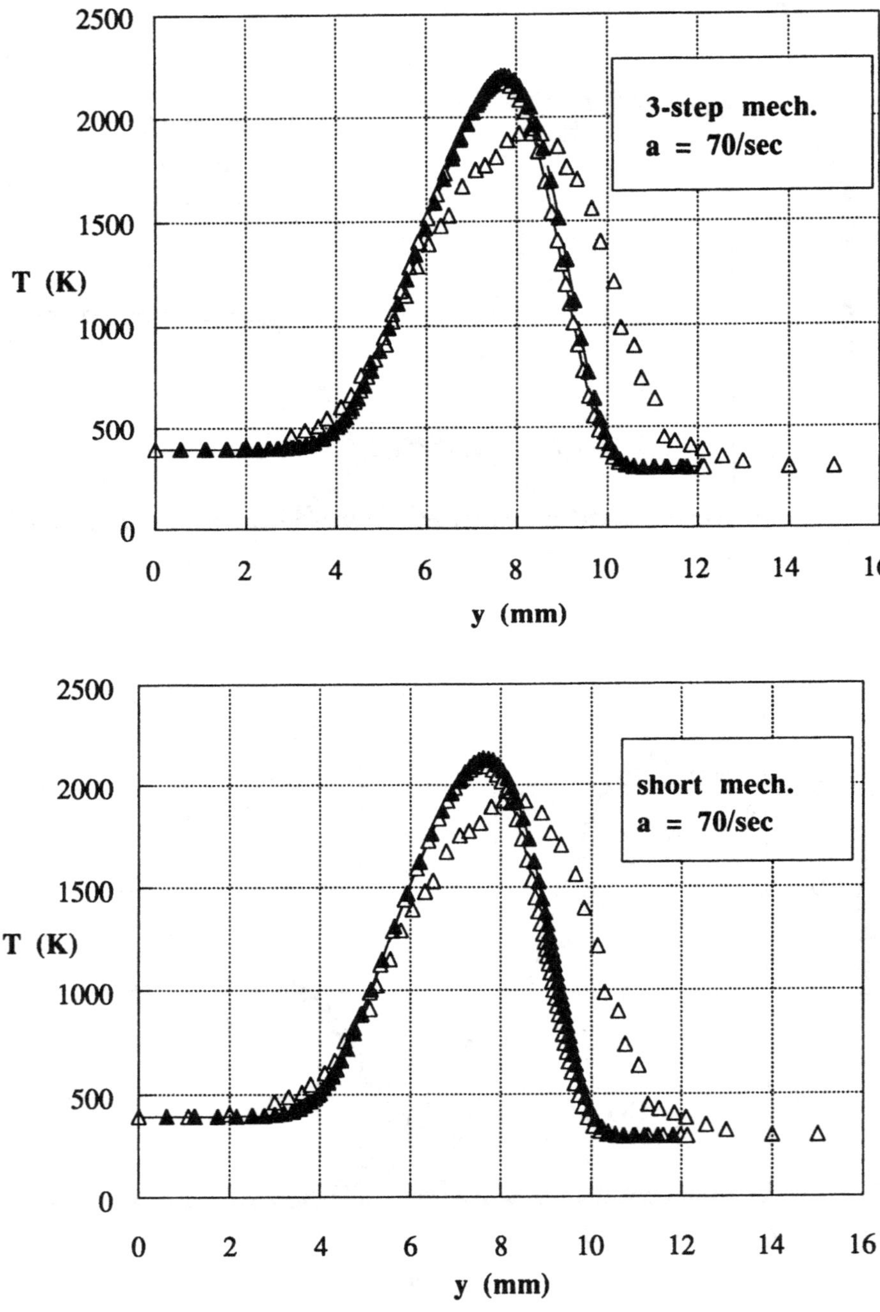

Fig. 12.7. As Fig. 12.6.a, but for strain rate of $a = 70\,\mathrm{s}^{-1}$.

Blint [12.5], who have also carried out numerical simulations, however, without taking effects of radiation into account.

Shown in Fig. 12.7 are temperature profiles for a strain rate of $70\,s^{-1}$. It is seen that at this lower value of the strain rate thermal radiation has a stronger effect on the maximum flame temperature which, in turn, has a strong influence on the predicted levels of NO_x [12.8]. The difference of the experimentally and numerically determined flame position must be contributed to deficiencies in the experimental technique or the flow-field model. Arguments along these lines can be found in Drake and Blint [12.5], and, therefore, need not to be repeated here.

12.6.3 Results of Time-Dependent Calculations

One of the important applications of systematically reduced reaction mechanisms is modelling of turbulent reacting flows. In a highly turbulent, chemically reacting flow, the magnitudes of the time scales of the fluctuating flow field vary rapidly, ranging over several orders of magnitude. The ability of reduced reaction mechanisms to respond accurately to such rapidly changing flow conditions is of critical importance to accurate simulations and, hence, predictions of chemically reacting turbulent flows. In the flamelet regime of non-premixed turbulent combustion, the mean scalar dissipation rate is one of the relevant variables that describe the effects of the turbulent flow on the combustion chemistry.

In this subsection we solve (12.19) and (12.20) to study the dynamic response of chemical kinetics subject to sinusoidal variations of the scalar dissipation rate at stoichiometric conditions, χ_{st}. Specifically, χ_{st} is varied according to

$$\chi_{st} = \chi_{st,\mathrm{mean}} + \Delta_\chi \sin(2\pi\Omega t)\,, \tag{12.25}$$

where $\chi_{st,\mathrm{mean}}$ denotes the time-independent mean value of χ_{st}; Δ_χ is the amplitude of the fluctuation, and Ω its frequency. In the calculations numerical values for $\chi_{st,\mathrm{mean}}$, Δ_χ and Ω are specified. Specifically, $\chi_{st,\mathrm{mean}} = 300\,s^{-1}$, $\Delta_\chi = 280\,s^{-1}$, and $\Omega = 1\,kHz$ or, alternatively, $5\,kHz$, are selected. The temperature at both the fuel and the oxidizer boundary are kept constant at a value of 300 K. As initial profiles of the calculations we use the steady-state solution for $\chi_{st} = \chi_{st,\mathrm{mean}}$.

Numerical calculations were performed with both the short mechanism and the three-step mechanism. Figures 12.8–12.10 pertain to $\Omega = 1\,kHz$, Figs. 12.11–12.13 to $\Omega = 5\,kHz$. Shown in Figs. 12.8 and 12.11 is a comparison of the histories of the maximum temperature in the flame, T_{max}, and of the maximum H-atom mass fraction in the flame, $Y_{\mathrm{H,max}}$, as calculated on the basis of the short mechanism and the three-step mechanism. It is seen that the results obtained from the two mechanisms agree reasonably well. Furthermore, inspection of Figs. 12.8 and 12.11 shows that, after an initial transition from the steady-state solution for $\chi_{st} = \chi_{st,\mathrm{mean}}$, the transient solutions settle down into a repeated pattern for both $\Omega = 1\,kHz$ and $\Omega = 5\,kHz$. A comparison

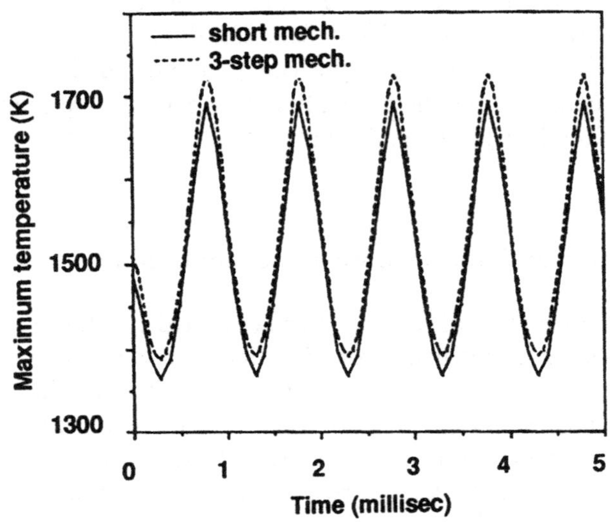

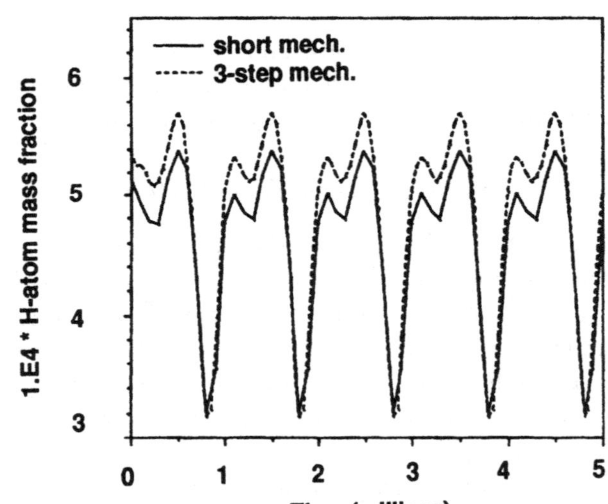

Fig. 12.8. History of maximum flame temperature T_{max} (top) and H-atom mass fraction Y_H (bottom). The scalar-dissipation rate varies sinusoidally with a frequency of 1 kHz according to $\chi_{st} = 300 + 280\, sin\,(2\pi\, 1000\, t/1\,\mathrm{s})\,[\mathrm{s}^{-1}]$. Solid lines: short mechanism; dashed lines: 3-step mechanism.

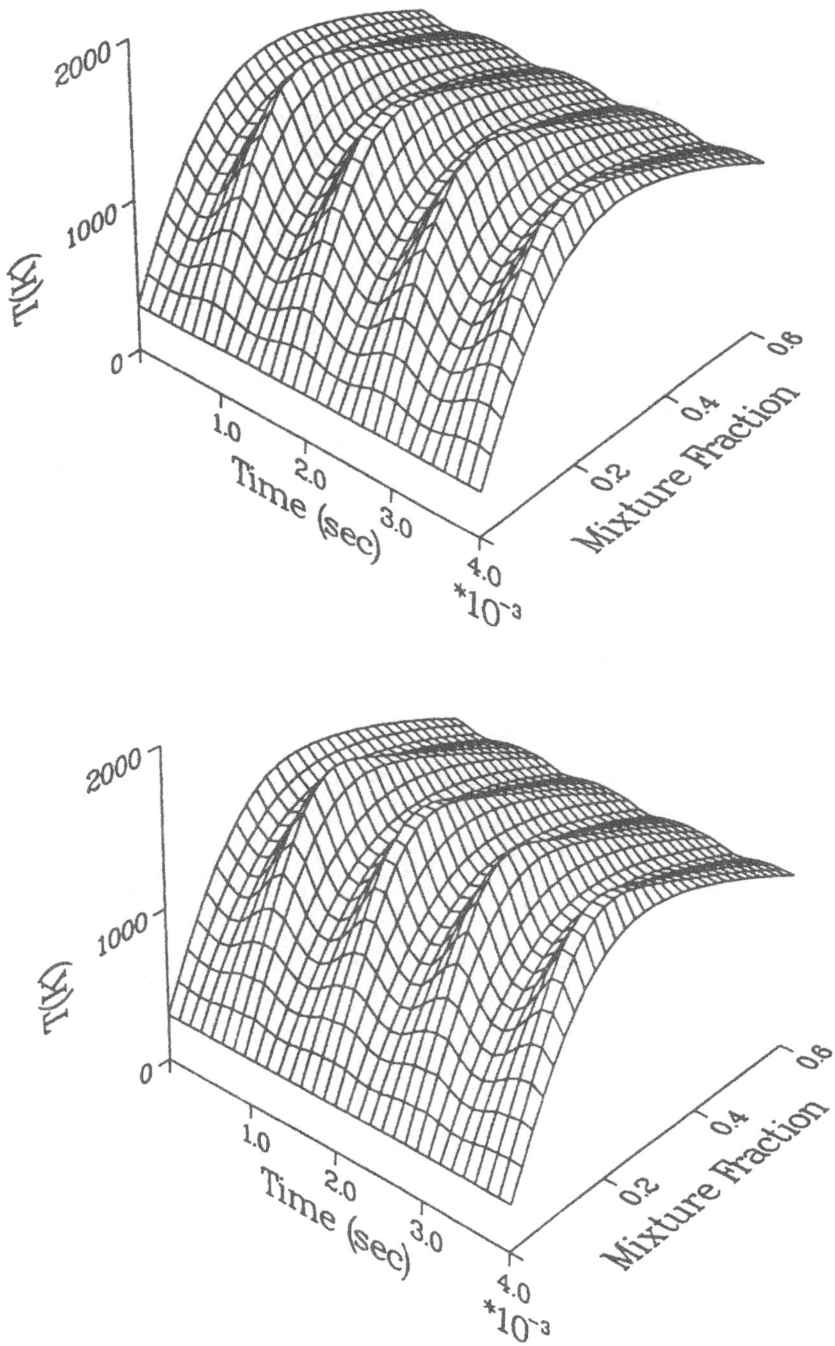

Fig. 12.9. Temperature T as a function of time t and mixture fraction Z. The scalar-dissipation rate varies sinusoidally with a frequency of 1 kHz according to $\chi_{st} = 300 + 280\, sin\, (2\pi\, 1000\, t/1\,\mathrm{s})\ [\mathrm{s}^{-1}]$. Top: short mechanism; bottom: 3-step mechanism.

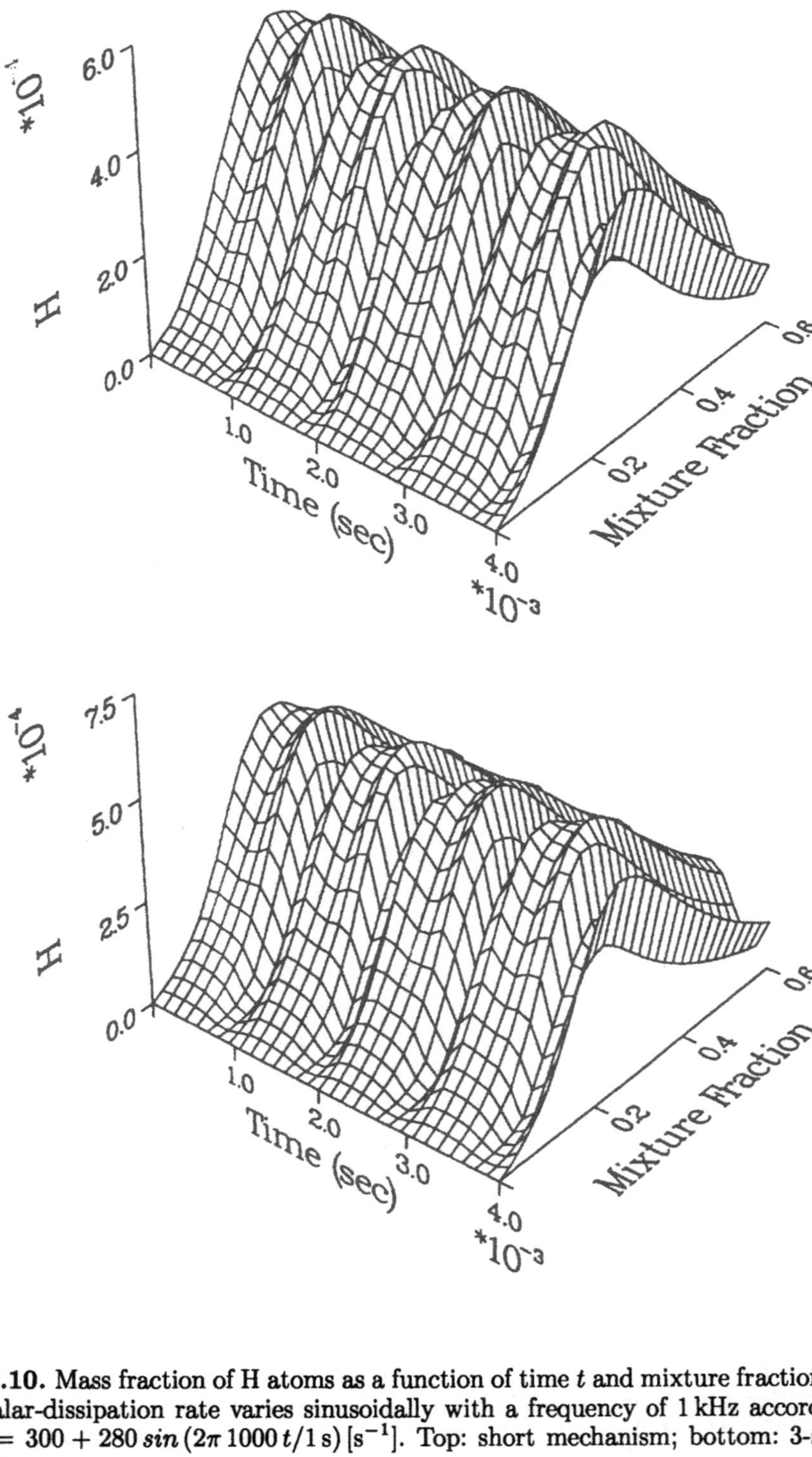

Fig. 12.10. Mass fraction of H atoms as a function of time t and mixture fraction Z. The scalar-dissipation rate varies sinusoidally with a frequency of 1 kHz according to $\chi_{st} = 300 + 280\, sin\,(2\pi\, 1000\, t/1\,\mathrm{s})\,[\mathrm{s}^{-1}]$. Top: short mechanism; bottom: 3-step mechanism.

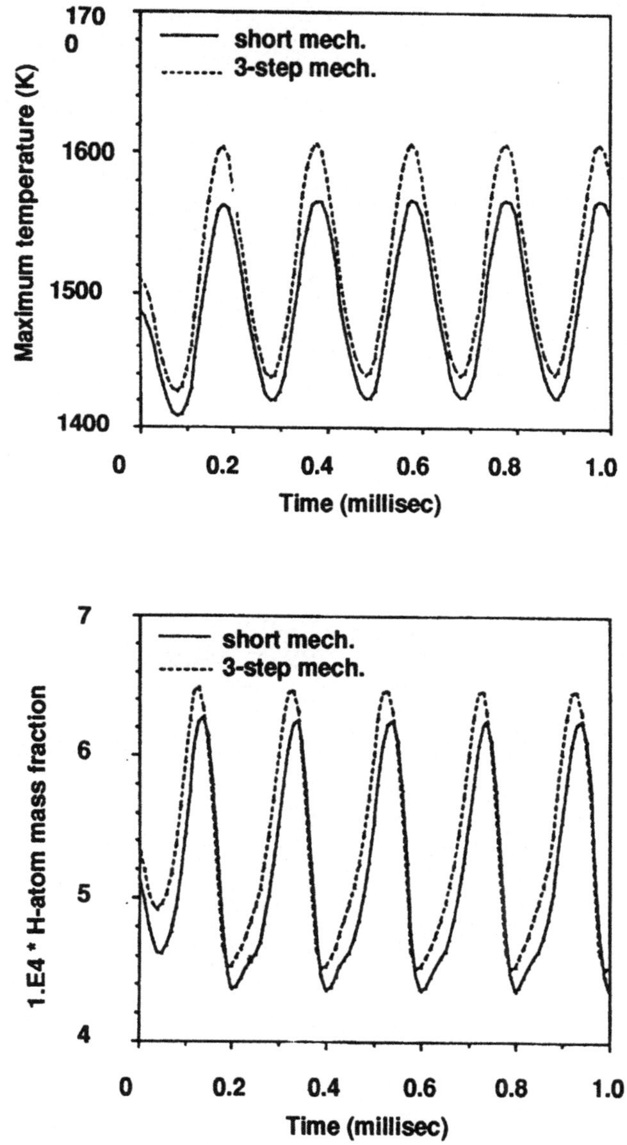

Fig. 12.11. History of maximum flame temperature $T_{\max}$ (top) and H-atom mass fraction Y_H (bottom). The scalar-dissipation rate varies sinusoidally with a frequency of 5 kHz according to $\chi_{st} = 300 + 280\, sin\,(2\pi\, 5000\, t/1\,\mathrm{s})\ [\mathrm{s}^{-1}]$. Solid lines: short mechanism; dashed lines: 3-step mechanism.

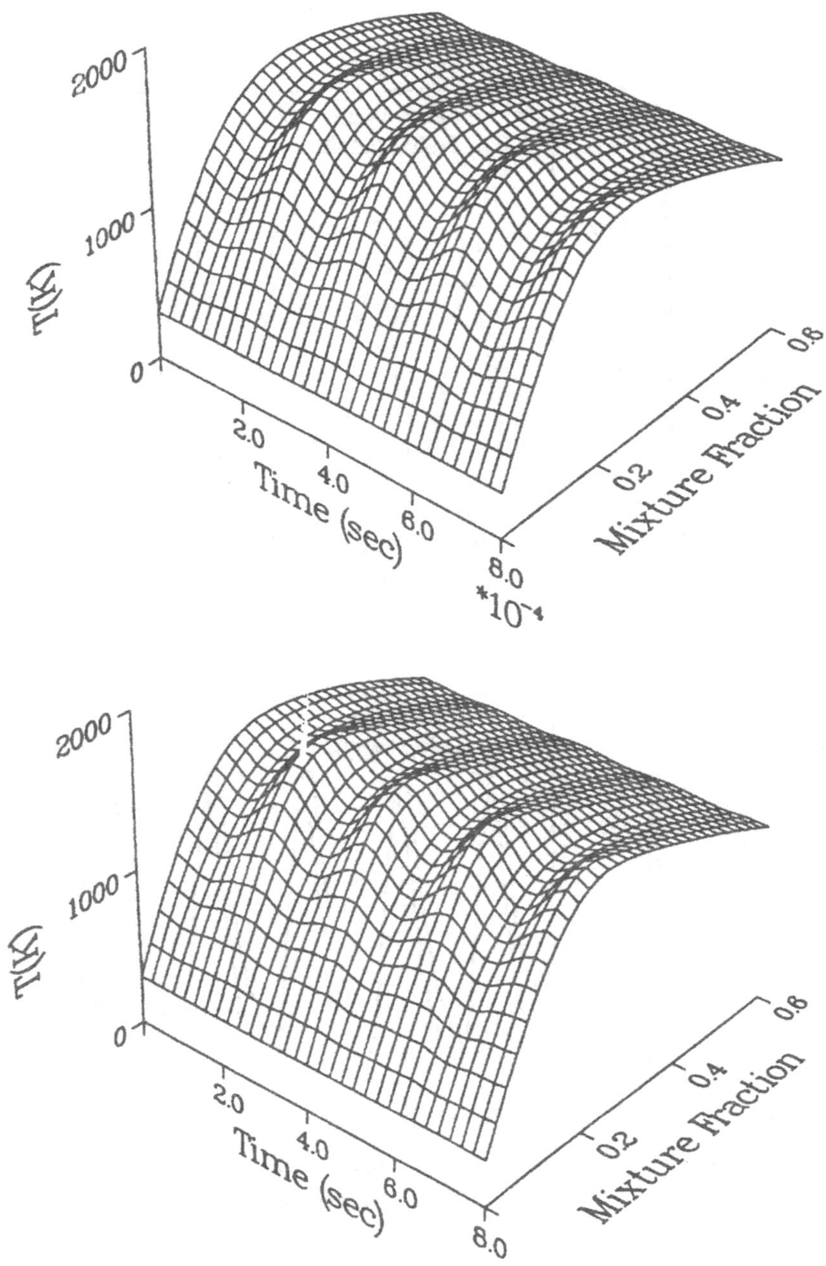

Fig. 12.12. Temperature T as a function of time t and mixture fraction Z. The scalar-dissipation rate varies sinusoidally with a frequency of 5 kHz according to $\chi_{st} = 300 + 280\,sin\,(2\pi\,5000\,t/1\,\mathrm{s})\,[\mathrm{s}^{-1}]$. Top: short mechanism; bottom: 3-step mechanism.

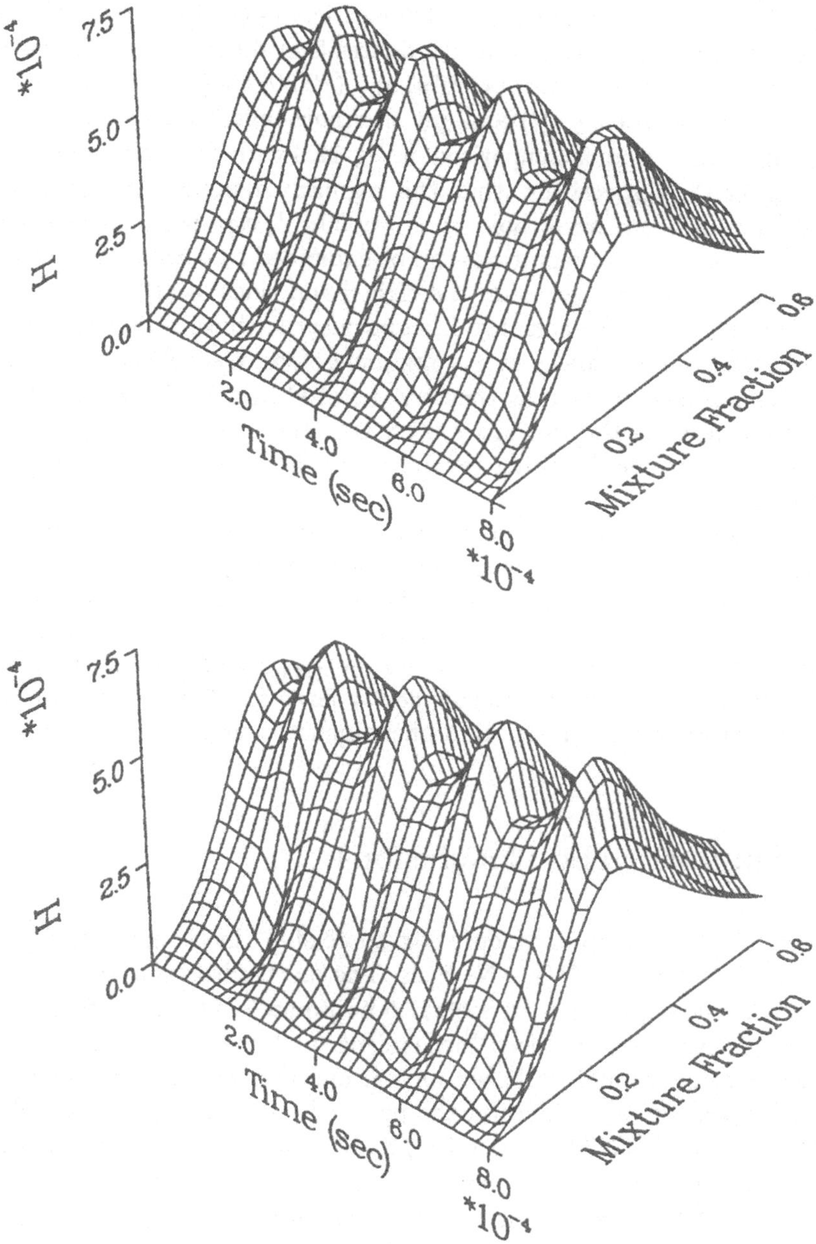

Fig. 12.13. Mass fraction of H atoms as a function of time t and mixture fraction Z. The scalar-dissipation rate varies sinusoidally with a frequency of 5 kHz according to $\chi_{st} = 300 + 280\,sin\,(2\pi\,5000\,t/1\,\mathrm{s})\,[\mathrm{s}^{-1}]$. Top: short mechanism; bottom: 3-step mechanism.

between Figs. 12.8 and 12.11 further shows that the transient temperature and concentrations are not symmetrical about their time-independent values for χ_{mean}. However, they become more symmetrical as the frequency is increased. Furthermore, it can be seen from the figures that due to the nonlinear nature of the chemical source terms in (12.19) and (12.20) there are phase differences in the profiles between $\chi_{st}(t)$ and any thermochemical variables.

It can be seen from a comparison of Fig. 12.9 with Fig. 12.12 and Fig. 12.10 with Fig. 12.13, respectively, that the regularly repeated patterns at $\omega = 1000\,s^{-1}$ are different from those at $\omega = 5000\,s^{-1}$, especially for radicals such as the H atom. These observations can be explained by considering the relative importance of the two terms on the right hand sides of (12.19) and (12.20). At high frequencies, the chemical kinetics may not have sufficient time to respond to the rapid changes of χ_{st} and, therefore, the solutions follow more or less a sinusoidal pattern as they are dictated by the changes in the scalar dissipation rate. As the frequency decreases, the effects of chemical kinetics become more profound leading to solutions that deviate significantly from a sinusoidal pattern.

Although in the present work only a limited number of unsteady flame structures could be studied, in general the response of the three-step mechanism to a time-varying scalar dissipation rate is essentially the same as that of the short mechanism. A more stringent test, which will be carried out in future studies, would be to select values for $\chi_{st,mean}$, Δ_χ and Ω such that at $\chi_{st}(t)$ attains temporary values sufficiently large for flame extinction so that transient extinction and re-ignition phenomena occur.

12.7 Summary and Conclusions

In Sect. 12.2 of this chapter we derived a three-step kinetic mechanism for non-premixed flames with $CO/H_2/N_2$ fuels. The mechanism was validated in the Tsuji counterflow geometry. In Sect. 12.6 we pointed out that in future studies both the derivation of reduced mechanisms and the generation of laminar-flamelet libraries should be based on the opposed-jet geometry rather than on the Tsuji geometry. Based on the opposed-jet geometry, for a molar fuel composition of 40/30/30 percent $CO/H_2/N_2$ and air as oxidizer, a flamelet library was computed for pressures ranging from 1 to 10 bars.

In Sect. 12.3 the most general set of similarity equations allowing for a time-varying strain rate was given; in Sect. 12.4 a model of thermal radiation was derived. Both the detailed mechanism of elementary reactions, from which the three-step mechanism was derived, and the three-step mechanism were used in numerical simulations of thermally radiating steady flames and non-radiating transient flames. For the study of the latter flames the scalar dissipation rate was varied sinusoidally with a frequency of 1 kHZ and, alternatively, 5 kHz. For all cases studied good agreement was found between the numerical results based on the detailed and the reduced mechanism.

Acknowledgement

Much of the work stemming from the Cambridge group was done whilst B. Rogg was hosted by CERMICS at Sophia-Antipolis, France.

References

[12.1] Vandooren, J., Peeters, J., van Tiggelen, P. J., Fifteenth Symposium (International) on Combustion, The Combustion Institute, p. 745–753, 1974.

[12.2] Safieh, H. Y., Vandooren, J., van Tiggelen, P. J., Nineteenth Symposium (International) on Combustion, The Combustion Institute, p. 117–126, 1982.

[12.3] Vandooren, J., Nelson da Cruz, F., van Tiggelen, P. J., Twentysecond Symposium (International) on Combustion, The Combustion Institute, p. 1587–1595, 1988.

[12.4] Rogg, B. and Williams, F. A., Twentysecond Symposium (International) on Combustion, The Combustion Institute, p. 1441–1451, 1988.

[12.5] Drake, M. C., Blint, R. J., Combust. Flame, **76**, p. 151, 1989.

[12.6] Chen, J.-Y., Dibble, R. W., Bilger, R. W., Twentythird Symposium (International) on Combustion, The Combustion Institute, p. 775–780, 1990.

[12.7] Mauß, F., Keller, D., Peters, N., Twentythird Symposium Symposium (International) on Combustion, to appear, The Combustion Institute, 1990.

[12.8] Liu, Y., Rogg, B., Modelling and Computation of Strained Laminar Diffusion Flames with Thermal Radiation, Paper presented at the 13th ICDERS conference, Nagoya (Japan), 28 July–02 August, 1991; also: Technical Report CUED/A-THERMO/TR46, Cambridge University Engineering Department, August 1991.

[12.9] Rogg, B., RUN-1DL: A Computer Program for the Simulation of One-Dimensional Chemically Reacting Flows, Technical Report CUED/A-THERMO/TR39, Cambridge University Engineering Department, April 1991, Appendix C of this book.

[12.10] Kee, R. J., Rupley, F. M., Miller, J. A., Chemkin II: A Fortran Chemical Kinetics Package for the Analysis of Gas-Phase Chemical Kinetics, Technical Report SAND89-8009, Sandia National Laboratories, Livermore (CA), 1989.

[12.11] Peters, N. and Kee, R. J., Combust. Flame, **68**, p. 17, 1987.

[12.12] Hindmarsh., A. C., LSODE, Lawrence Livermore Laboratory, Livermore, California, USA, 1978.

[12.13] Rogg, B., RUN-1DL: The Cambridge Laminar-Flamelet Code, Appendix C of this book.

13. Reduced Kinetic Mechanisms for Counterflow Methane–Air Diffusion Flames

H. K. Chelliah
Department of Mechanical and Aerospace Engineering,
Princeton University, Princeton, NJ 08544, USA

K. Seshadri
Department of Applied Mechanics and Engineering Sciences,
University of California, San Diego, La Jolla, CA 92093, U.S.A.

C. K. Law
Department of Mechanical and Aerospace
Engineering, Princeton University, Princeton, NJ 08544, USA

13.1 Introduction

Results of numerous studies describing the structure and mechanisms of extinction of laminar, methane-air diffusion flames, are available [13.1]–[13.11]. These studies include numerical calculations employing a detailed set of elementary chemical reactions [13.1]–[13.4], suitably reduced chemical kinetic mechanisms [13.1], [13.5]–[13.7], and asymptotic analysis using reduced kinetic mechanisms [13.8]–[13.11]. These previous studies have shown that the principal path of oxidation of methane is CH_4-CH_3-CH_2O-HCO-CO, H_2-CO_2,H_2O [13.1]–[13.7], and the structures of this flame can be predicted fairly accurately by including only the C_1 mechanism shown in Table 1 of Chap. 1. The results of previous numerical calculations also show, for example, that steady-state approximations are valid for CH_2O and HCO [13.1], but that for CH_3 is questionable [13.7]. Four step mechanisms were deduced previously [13.1], [13.5]–[13.7] by introducing steady-state approximations for all species, except CH_4, O_2, N_2, CO_2, H_2O, CO, H_2 and H. Comparison of the results for the structure of the stagnation point diffusion flame, calculated using the 4-step mechanism were found to agree reasonably well with the those calculated using the detailed mechanism [13.7], except for the profile for CH_3 [13.7].

In this chapter the structure and critical conditions of extinction of the stagnation point diffusion flames are calculated using the C_1 mechanism shown in Table 1 of Chap. 1. A 5-step mechanism is deduced here by introducing steady-state approximations for CH_2, CH, CH_2O, HCO, OH, O, HO_2 and H_2O_2, and the results of calculations using this mechanism are compared with those obtained using the detailed mechanism. To explore the impact of making a steady-state approximation for CH_3, this 5-step mechanism was further reduced to 4-steps by eliminating CH_3. The results of calculations using the 4-step are compared with those obtained using the detailed mechanism and the 5-step mechanism.

13.2 Chemical Kinetic Mechanism and Formulation of the Numerical Problem

The starting mechanism used here is essentially the C_1 mechanism shown in Table 1 of Chap. 1. Thus, the mechanism includes reactions 1–35, 37–40 involving 17 species. This starting C_1 mechanism is different from that employed previously in [13.1] and the essential differences are that reactions involving the species CH and CH_2 were neglected previously [13.1]. Also, in [13.1] the products of reaction 37 was presumed to be CH_3O and O and the associated rate data were different than that shown in Table 1 of Chap. 1. To evaluate the differences between these mechanisms, the burning velocity for a stoichiometric premixed flame was calculated using the starting mechanism with and without reactions involving CH and CH_2. The calculated value of the burning velocity excluding reactions involving CH and CH_2 was found to be considerably lower than that calculated including reactions involving these two species. Also, reaction 37 was found to have a significant influence on the calculated value of the burning velocity. However, the change in the calculated value of the burning velocity is negligibly small if reactions involving CH_3O, which were introduced in the previous mechanism [13.1], are also included in the present starting C_1 mechanism.

The governing equations for the numerical problem are described in detail in Chap. 1, and the notation introduced in Chap. 1 is used here. All calculations were performed in the axisymmetric configuration using the potential-flow and plug-flow boundary conditions. As discussed in Chap. 1, the mixture fraction Z is presumed to represent the normalized enthalpy based on the assumption of unity Lewis numbers, and is calculated as a function of the independent spatial coordinate, y, using the source free balance equation for Z (1.10) in Chap. 1. Boundary conditions are applied at $y = y_{-\infty}$ and $y = y_{\infty}$, where the subscripts $-\infty$ and ∞ identify conditions in the ambient oxidizer stream and fuel stream respectively. Since, the density ρ is constant in the vicinity of $y_{-\infty}$ and y_{∞}, the continuity equation implies that $2G = -dv/dy (= a)$ in the vicinity of each boundary. For potential flow, the boundary conditions at the ambient oxidizer stream are

$$\begin{aligned} &v = v_{-\infty}\,;\; G = a/2\,;\; T = T_{-\infty}\,;\; \rho = \rho_{-\infty}\,;\; Y_{O_2} = Y_{O_2,-\infty}\,; \\ &Y_{N_2} = Y_{N_2,-\infty}\,;\; Y_i = 0\,, i \neq O_2\,, N_2\,;\; Z = 0 \quad \text{at } y = y_{-\infty}\,, \end{aligned} \tag{13.1}$$

and at the ambient fuel stream are

$$v = v_\infty\,;\; G = \sqrt{(\rho_{-\infty}/\rho_\infty)}(a/2)\,;\; T = T_\infty\,;\; \rho = \rho_\infty\,;\; Y_F = Y_{F,\infty}\,;$$
$$Y_{N_2} = Y_{N_2,\infty}\,;\; Y_i = 0\,,\; i \neq F\,, N_2\,;\; Z = 1 \text{ at } y = y_\infty\,, \tag{13.2}$$

Here the subscript F represents the fuel and a the specified strain rate. For plug flow, the boundary conditions at the ambient oxidizer stream are

$$v = v_{-\infty};\; G = 0;\; T = T_{-\infty};\; \rho = \rho_{-\infty};\; Y_{O_2} = Y_{O_2,-\infty};$$
$$Y_{N_2} = Y_{N_2,-\infty}; Y_i = 0, i \neq O_2, N_2;\; Z = 0 \text{ at } y = y_{-\infty}, \tag{13.3}$$

and at the ambient fuel stream are

$$v = v_\infty;\; G = 0;\; T = T_\infty;\; \rho = \rho_\infty;\; Y_F = Y_{F,\infty};$$
$$Y_{N_2} = Y_{N_2,\infty}; Y_i = 0, i \neq F, N_2;\; Z = 1 \text{ at } y = y_\infty, \tag{13.4}$$

The transport model [13.12] and the numerical procedure [13.13] are similar to those described elsewhere.

13.3 Reduced Kinetic Mechanism

To deduce a reduced chemical kinetic mechanism, which is capable of describing fairly accurately the structure and critical conditions of extinction of the flame, it is necessary to identify those species for which steady-state approximations are valid. In Figs. 13.1–13.7 the terms accounting for the flux rates of convection, diffusion, chemical production and consumption are shown as a function of the mixture fraction Z for the species CH_3, CH_2, CH, CH_2O, H, OH and O. The calculations were performed using the starting mechanism together with the boundary conditions (13.1) and (13.2), for $p = 1\,\text{atm}$, $a = 300\,\text{s}^{-1}$, $Y_{O_2,-\infty} = 1 - Y_{N_2,-\infty} = 0.233$, $Y_{F,\infty} = 1.0$ and $T_\infty = T_{-\infty} = 298\,\text{K}$. Figure 13.2 shows that for the radical CH_2 the convective and diffusive terms are negligibly small, and the chemical production rates are nearly equal to the consumption rates everywhere. Hence, steady-state approximation is valid for CH_2 to a high degree of accuracy. Similarly, Figs. 13.3, 13.6 and 13.7 show that steady-state approximations can be introduced for CH, OH, and O. Also, Fig. 13.4 shows that it is reasonable to introduce steady-state approximation for CH_2O. However, Figs. 13.1 and 13.5 show that steady-state approximations for CH_3 and H are not justified. In Figs. 13.8 and 13.9 data similar to Figs. 13.1–13.7 are shown for CH_3 and CH_2O at a near extinction strain rate of $a = 518\,\text{s}^{-1}$, again with $p = 1\,\text{atm}$, $Y_{O_2,-\infty} = 1 - Y_{N_2,-\infty} = 0.233$, $Y_{F,\infty} = 1.0$ and $T_\infty = T_{-\infty} = 298\,\text{K}$, and they show that steady-state approximation is reasonably valid for CH_2O but not for CH_3. Plots similar to Figs. 13.1–13.9 were also prepared for HCO, HO_2 and H_2O_2 and they show that steady-state approximations are valid for these species.

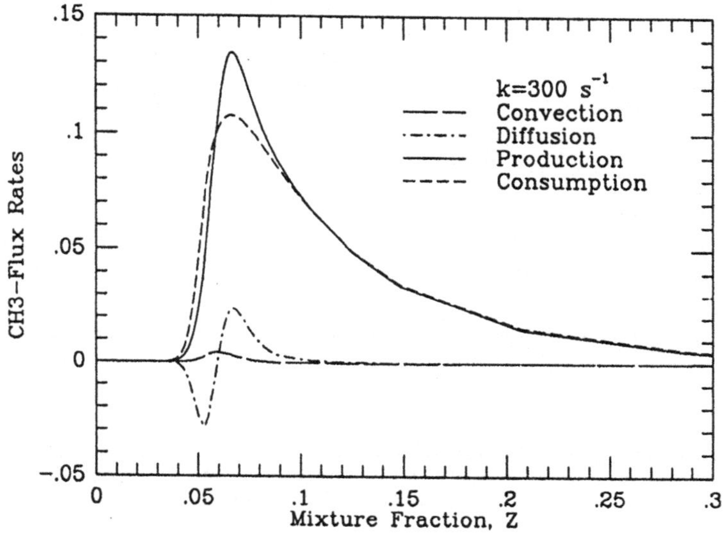

Fig. 13.1. Flux of convection, diffusion, production and negative consumption of CH_3 calculated using the starting mechanism with conditions $k = 300\,s^{-1}$, $p = 1$ atm, $T_{-\infty} = T_\infty = 298$ K, $Y_{F,\infty} = 1.0$ and $Y_{O_2,-\infty} = 1 - Y_{N_2,-\infty} = 0.233$.

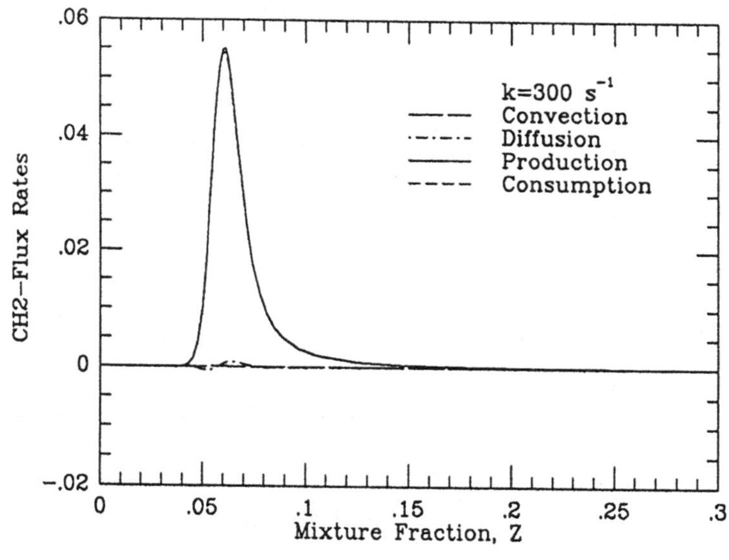

Fig. 13.2. Flux of convection, diffusion, production and negative consumption of CH_2 calculated using the starting mechanism with conditions $k = 300\,s^{-1}$, $p = 1$ atm, $T_{-\infty} = T_\infty = 298$ K, $Y_{F,\infty} = 1.0$ and $Y_{O_2,-\infty} = 1 - Y_{N_2,-\infty} = 0.233$.

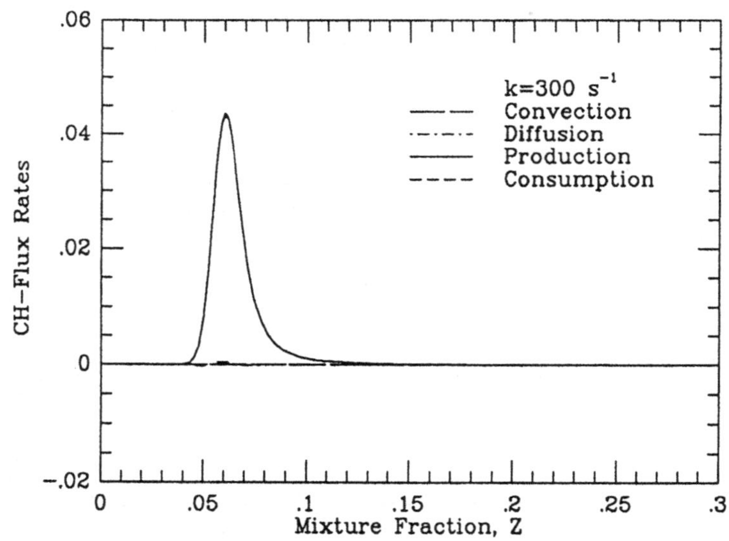

Fig. 13.3. Flux of convection, diffusion, production and negative consumption of CH calculated using the starting mechanism with conditions $k = 300\,\mathrm{s}^{-1}$, $p = 1\,\mathrm{atm}$, $T_{-\infty} = T_{\infty} = 298\,\mathrm{K}$, $Y_{F,\infty} = 1.0$ and $Y_{O_2,-\infty} = 1 - Y_{N_2,-\infty} = 0.233$.

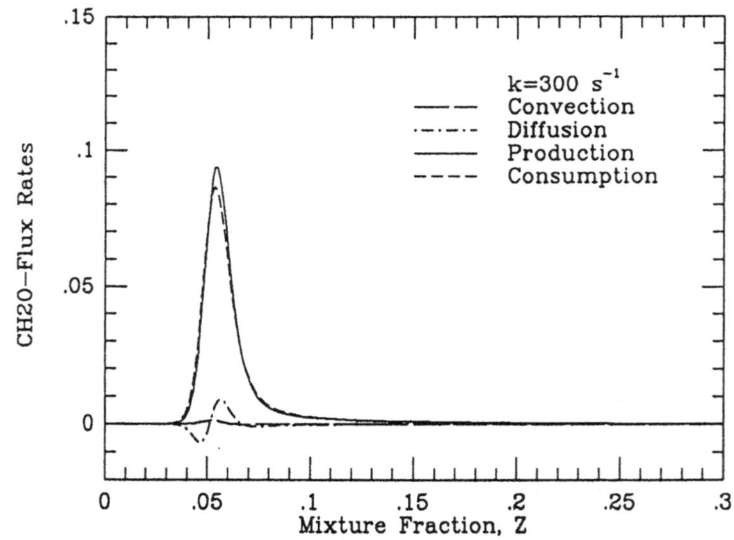

Fig. 13.4. Flux of convection, diffusion, production and negative consumption of CH_2O calculated using the starting mechanism with conditions $k = 300\,\mathrm{s}^{-1}$, $p = 1\,\mathrm{atm}$, $T_{-\infty} = T_{\infty} = 298\,\mathrm{K}$, $Y_{F,\infty} = 1.0$ and $Y_{O_2,-\infty} = 1 - Y_{N_2,-\infty} = 0.233$.

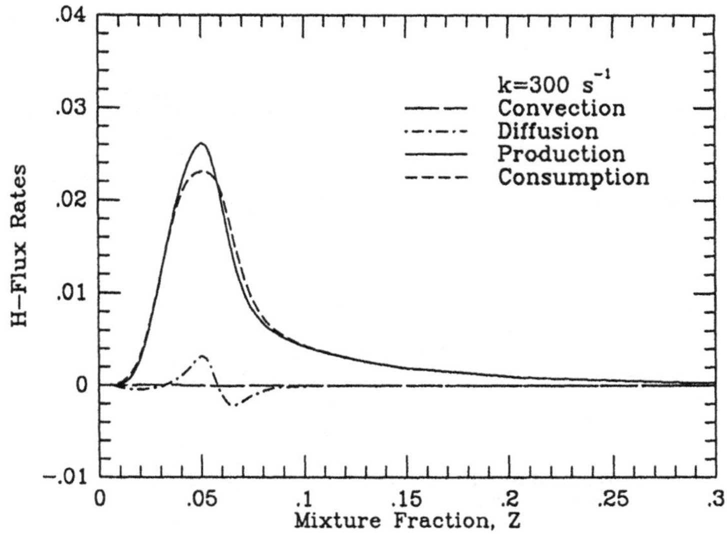

Fig. 13.5. Flux of convection, diffusion, production and negative consumption of H calculated using the starting mechanism with conditions $k = 300\,\mathrm{s}^{-1}$, $p = 1\,\mathrm{atm}$, $T_{-\infty} = T_{\infty} = 298\,\mathrm{K}$, $Y_{F,\infty} = 1.0$ and $Y_{O_2,-\infty} = 1 - Y_{N_2,-\infty} = 0.233$.

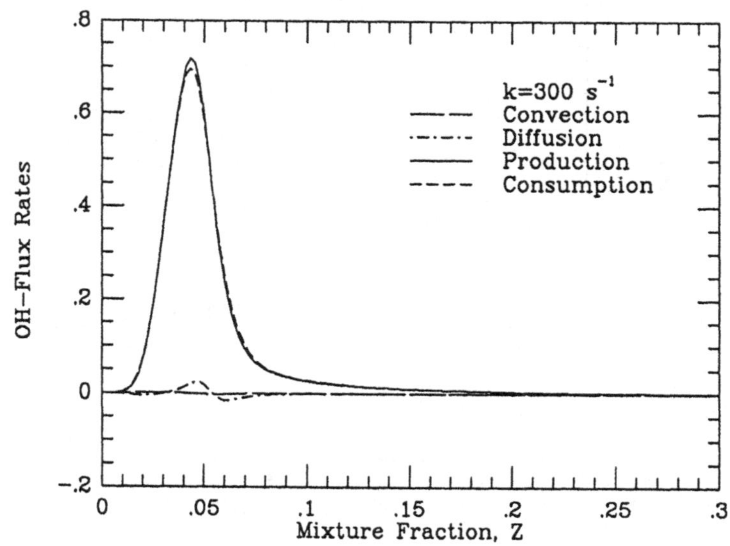

Fig. 13.6. Flux of convection, diffusion, production and negative consumption of OH calculated using the starting mechanism with conditions $k = 300\,\mathrm{s}^{-1}$, $p = 1\,\mathrm{atm}$, $T_{-\infty} = T_{\infty} = 298\,\mathrm{K}$, $Y_{F,\infty} = 1.0$ and $Y_{O_2,-\infty} = 1 - Y_{N_2,-\infty} = 0.233$.

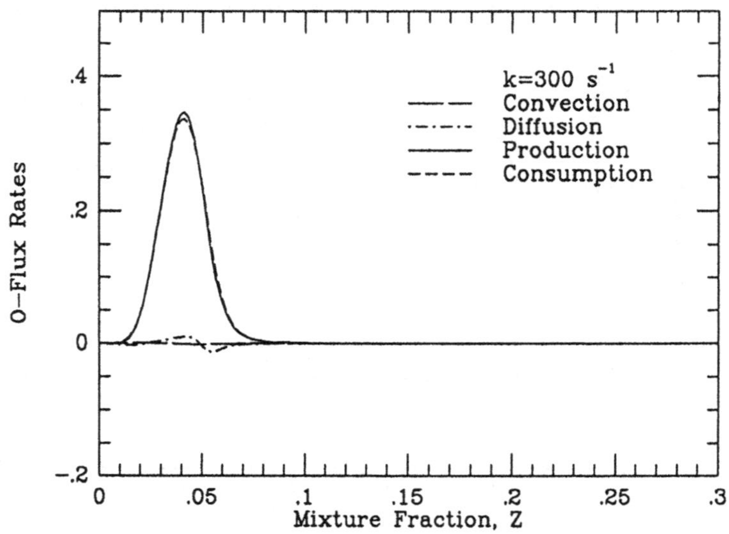

Fig. 13.7. Flux of convection, diffusion, production and negative consumption of O calculated using the starting mechanism with conditions $k = 300\,\mathrm{s}^{-1}$, $p = 1\,\mathrm{atm}$, $T_{-\infty} = T_{\infty} = 298\,\mathrm{K}$, $Y_{F,\infty} = 1.0$ and $Y_{O_2,-\infty} = 1 - Y_{N_2,-\infty} = 0.233$.

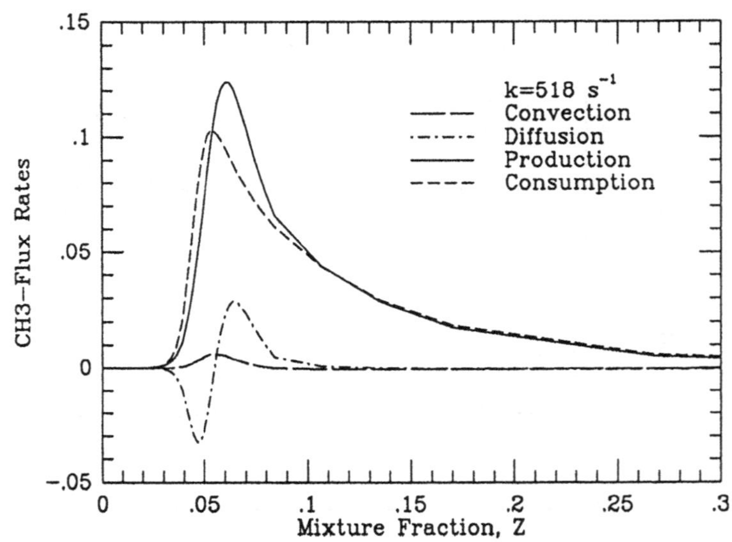

Fig. 13.8. Flux of convection, diffusion, production and negative consumption of CH_3 calculated using the starting mechanism with conditions $k = 518\,\mathrm{s}^{-1}$, $p = 1\,\mathrm{atm}$, $T_{-\infty} = T_{\infty} = 298\,\mathrm{K}$, $Y_{F,\infty} = 1.0$ and $Y_{O_2,-\infty} = 1 - Y_{N_2,-\infty} = 0.233$.

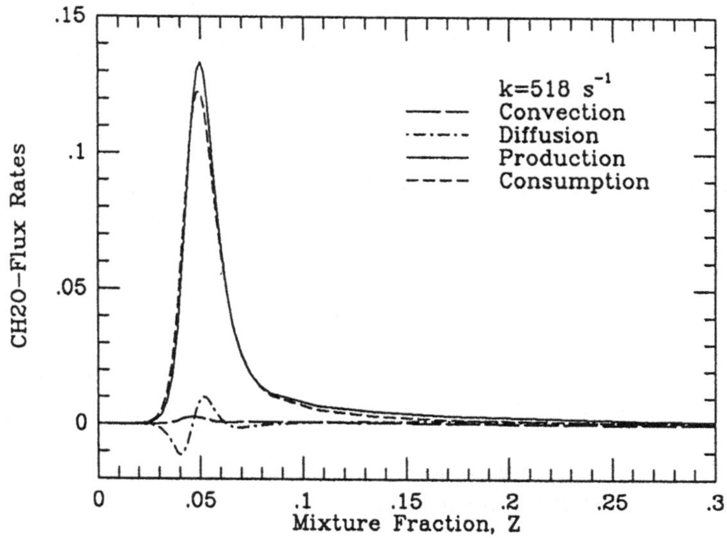

Fig. 13.9. Flux of convection, diffusion, production and negative consumption of CH_2O calculated using the starting mechanism with conditions $k = 518\,s^{-1}$, $p =$ 1 atm, $T_{-\infty} = T_\infty = 298\,K$, $Y_{F,\infty} = 1.0$ and $Y_{O_2,-\infty} = 1 - Y_{N_2,-\infty} = 0.233$.

By introducing steady-state approximations for the species CH_2, CH, CH_2O, HCO, OH, O, HO_2 and H_2O_2, a 5-step mechanism can be deduced from the starting mechanism and the resulting mechanism can be represented by

$$
\begin{aligned}
CH_4 + H &\rightleftharpoons CH_3 + H_2, && \text{Ia}\\
CH_3 + H_2O + H &\rightleftharpoons CO + 3H_2, && \text{Ib}\\
CO + H_2O &\rightleftharpoons CO_2 + H_2, && \text{II}\\
H + H + M &\rightleftharpoons H_2, && \text{III}\\
O_2 + 3H_2 &\rightleftharpoons 2H + 2H_2O, && \text{IV}
\end{aligned}
$$

The reaction rates for the overall steps Ia–IV can be expressed in terms of elementary reaction rates as

$$
\begin{aligned}
\omega_{Ia} &= -\omega_{34} + \omega_{38} + \omega_{39} + \omega_{40},\\
\omega_{Ib} &= \omega_{33} + \omega_{35} + \omega_{37},\\
\omega_{II} &= \omega_{18} - \omega_{20} + \omega_{28}, \qquad\qquad (13.5)\\
\omega_{III} &= \omega_5 + \omega_{12} + \omega_{15} + \omega_{16} + \omega_{17} + \omega_{21} + \omega_{22} + \omega_{23} - \omega_{32} + \omega_{34},\\
\omega_{IV} &= \omega_1 + \omega_6 + \omega_9 - \omega_{12} + \omega_{13} - \omega_{17} - \omega_{20} + \omega_{25} + \omega_{27} + \omega_{28} + \omega_{37}.
\end{aligned}
$$

A 4-step mechanism can be deduced by introducing a steady-state approximation for CH_3 and can be represented as

$$\mathrm{CH_4 + 2H + H_2O \rightleftharpoons CO + 4H_2}, \qquad \mathrm{I'}$$
$$\mathrm{CO + H_2O \rightleftharpoons CO_2 + H_2}, \qquad \mathrm{II'}$$
$$\mathrm{H + H + M \rightleftharpoons H_2}, \qquad \mathrm{III'}$$
$$\mathrm{O_2 + 3H_2 \rightleftharpoons 2H + 2H_2O}, \qquad \mathrm{IV'}$$

The reaction rates for the overall steps I'–IV' can be expressed in terms of the elementary rates as

$$\begin{aligned}
\omega_{\mathrm{I'}} &= -\omega_{34} + \omega_{38} + \omega_{39} + \omega_{40}, \\
\omega_{\mathrm{II'}} &= \omega_{18} - \omega_{20} + \omega_{28}, \\
\omega_{\mathrm{III'}} &= \omega_5 + \omega_{12} + \omega_{15} + \omega_{16} + \omega_{17} + \omega_{21} + \omega_{22} + \omega_{23} - \omega_{32} + \omega_{34}, \\
\omega_{\mathrm{IV'}} &= \omega_1 + \omega_6 + \omega_9 - \omega_{12} + \omega_{13} - \omega_{17} - \omega_{20} + \omega_{25} + \omega_{27} + \omega_{28} + \omega_{37}.
\end{aligned} \tag{13.6}$$

The predictions using these reduced mechanism are compared with those calculated using the starting mechanism.

13.4 Steady-State Assumptions

The overall reaction rates for the reduced chemical kinetic mechanism contain the species CH_2, CH, CH_2O, HCO, OH, O, HO_2 and H_2O_2 which are assumed to be in steady-state. Therefore, it is necessary to obtain explicit algebraic expressions for the steady-state concentrations of these species in terms of the species appearing in the reduced mechanism.

Introducing truncations as discussed below, the steady-state concentrations of CH_2, CH, O, CH_2O, HCO, HO_2 and H_2O_2 can be written as

$$\begin{aligned}
[\mathrm{CH_2}] &= \frac{k_{33f}[\mathrm{CH_3}]}{k_{25f}[\mathrm{H}] + (k_{27f} + k_{28f})[\mathrm{O_2}]}, \\
[\mathrm{CH}] &= \frac{k_{25f}[\mathrm{H}][\mathrm{CH_2}]}{k_{19f}[\mathrm{O_2}] + k_{20f}[\mathrm{CO_2}]}, \\
[\mathrm{O}] &= \frac{k_{1f}[\mathrm{O_2}][\mathrm{H}] + k_{2b}[\mathrm{OH}][\mathrm{H}] + k_{4f}[\mathrm{OH}]^2 + k_{19f}[\mathrm{CH}][\mathrm{O_2}]}{k_{1b}[\mathrm{OH}] + k_{2f}[\mathrm{H_2}] + k_{4b}[\mathrm{H_2O}] + k_{26f}[\mathrm{CH_2}] + k_{35f}[\mathrm{CH_3}] + k_{38f}[\mathrm{CH_4}]}, \\
[\mathrm{CH_2O}] &= \frac{(k_{35f}[\mathrm{O}] + k_{37f}[\mathrm{O_2}])[\mathrm{CH_3}]}{k_{29f}[\mathrm{H}] + k_{30f}[\mathrm{O}] + k_{31f}[\mathrm{OH}] + k_{32f}[\mathrm{M}]}, \\
[\mathrm{HCO}] &= \frac{(k_{29f}[\mathrm{H}] + k_{31f}[\mathrm{OH}])[\mathrm{CH_2O}] + (k_{19f}[\mathrm{O_2}] + k_{20f}[\mathrm{O_2}])[\mathrm{CH}]}{k_{21f}[\mathrm{H}] + k_{22f}[\mathrm{OH}] + k_{23f}[\mathrm{O_2}] + k_{24f}[\mathrm{M}]}, \\
[\mathrm{HO_2}] &= \frac{k_{5f}[\mathrm{H}][\mathrm{O_2}][\mathrm{M}] + k_{23f}[\mathrm{HCO}][\mathrm{O_2}]}{(k_{6f} + k_{7f} + k_{9f})[\mathrm{H}] + k_{8f}[\mathrm{OH}] + k_{10f}[\mathrm{O}]}, \\
[\mathrm{H_2O_2}] &= \frac{(k_{11f}[\mathrm{HO_2}] + k_{14b}[\mathrm{H_2O}])[\mathrm{HO_2}] + k_{12f}[\mathrm{OH}]^2[\mathrm{M}]}{k_{12b}[\mathrm{M}] + k_{13f}[\mathrm{H}] + k_{14f}[\mathrm{OH}]}.
\end{aligned} \tag{13.7}$$

Here $[i]$ represents the molar concentration of species i and [M] is the molar concentration of the third body. In deducing the steady state expressions in (13.7) the influences of reactions 25b and 26 were neglected for [CH_2]; reactions 9,10,17 and 30 for [O]; reactions 30, 32, and 24b for [HCO]; and reactions 14f, 14b, 5b and 11 for [HO_2]. The steady-state concentrations for the various species shown in (13.7) involves [OH] for which algebraic relation must be deduced. Steady-state expressions for [OH] is algebraically complicated [13.15], and truncated expressions for [OH] have been previously suggested [13.15]. However, it has been shown by Bilger et al. [13.7], that truncated expressions for [OH] may give a better peak value, but overpredict the [H_2] on the oxidizer side. Therefore, as suggested in [13.7], the value of [OH] is calculated by assuming that reaction 3 is in partial equilibrium, yielding

$$[\mathrm{OH}] = \frac{k_{3b}[\mathrm{H_2O}][\mathrm{H}]}{k_{3f}[\mathrm{H_2}]}. \tag{13.8}$$

The use of this result will make the OH steady-state approximation redundant and will introduce changes to the representation of the overall reduced mechanism. However, such changes were found to have negligible effects on the extinction predictions. Once the value of [OH] is determined from (13.8), the steady-state concentrations of other species can be computed in the sequence they appear in (13.7). Numerical calculations using the 4-step mechanism given by steps I'–IV' requires an algebraic expression for the steady-state concentration of CH_3, which neglecting reactions 33b, 37 and 39 can be written as

$$[\mathrm{CH_3}] = \frac{(k_{38f}[\mathrm{H}] + k_{40f}[\mathrm{OH}])[\mathrm{CH_4}]}{(k_{33f} + k_{34f})[\mathrm{H}] + k_{35f}[\mathrm{O}] + k_{38b}[\mathrm{H_2}] + k_{40b}[\mathrm{H_2O}]}. \tag{13.9}$$

Introducing (13.9) into $(13.7)_3$ yields a quadratic equation for [O], the solution to which, neglecting reactions 19, 26, and 39 can be written as

$$[\mathrm{O}] = \frac{\beta + \sqrt{\beta^2 + 4\alpha\gamma}}{2\alpha}, \tag{13.10}$$

where

$$\begin{aligned}
\alpha &= k_{35f}(k_{1b}[\mathrm{OH}] + k_{2f}[\mathrm{H_2}] + k_{4b}[\mathrm{H_2O}]),\\
\beta &= (k_{1b}[\mathrm{OH}] + k_{2f}[\mathrm{H_2}] + k_{4b}[\mathrm{H_2O}])\\
&\quad (k_{1b}(k_{34f}[\mathrm{H}] + k_{38b}[\mathrm{H_2}] + k_{40b}[\mathrm{H_2O}] + k_{33f}[\mathrm{H}])\\
&\quad + k_{35f}\{k_{1f}[\mathrm{H}][\mathrm{O_2}] + k_{2b}[\mathrm{OH}][\mathrm{H}] + k_{4f}[\mathrm{OH}]^2\\
&\quad - (k_{38f}[\mathrm{H}] + k_{40f}[\mathrm{OH}])[\mathrm{CH_4}]\},\\
\gamma &= (k_{1f}[\mathrm{H}][\mathrm{O_2}] + k_{2b}[\mathrm{OH}][\mathrm{H}] + k_{4f}[\mathrm{OH}]^2)\\
&\quad (k_{34f}[\mathrm{H}] + k_{38b}[\mathrm{H_2}] + k_{40b}[\mathrm{H_2O}] + k_{33f}[\mathrm{H}]).
\end{aligned} \tag{13.11}$$

In numerical calculations using the 4-step reduced mechanism, [OH] is first calculated using (13.8), followed by (13.10) and (13.9) to obtain values of [O] and [CH_3], respectively. The steady-state concentrations of other species are then computed in the sequence they appear in (13.7).

13.5 Comparisons Between Starting Mechanism, 5-Step Mechanism and 4-Step Mechanism

13.5.1 Flame Structure

Numerical calculations were performed using the starting mechanism, the 5-step mechanism and the 4-step mechanism for wide range of strain rates. The flame structure results with boundary conditions given by (13.1) and (13.2) and with $a = 300\,\mathrm{s}^{-1}$, $p = 1\,\mathrm{atm}$, $T_{-\infty} = T_{\infty} = 298\,\mathrm{K}$, $Y_{O_2,-\infty} = 0.233$ and $Y_{F,\infty} = 1.0$ are plotted in Figs. 13.10–13.11 using the mixture fraction Z as the independent variable. For this intermediate strain rate case, Fig. 13.10 shows that the profiles of T predicted by the various mechanisms agree well. It is also seen that the profiles of CH_4, O_2, H_2O, H_2 and CO predicted by the 5-step mechanism agree better with the results obtained using the starting mechanism, than those obtained using the 4-step mechanism. In Fig. 13.10, it is seen that the 4-step mechanism predicts higher values of O_2 around the position where the maximum value of T occurs, implying a higher oxygen leakage at $a = 300\,\mathrm{s}^{-1}$. Figure 13.11 shows that the reduced mechanisms predict higher values of H when compared to those predicted by the starting mechanism. Also, Fig. 13.11 shows that the profile of CH_3 calculated using the 5-step mechanism agrees reasonably well with that calculated using the starting mechanism, whereas the profile calculated using the 4-step mechanism differs considerably from that calculated using the starting mechanism.

In Figs. 13.12–13.13, the structure of the flame, calculated using the starting mechanism, 5-step mechanism, and 4-step mechanism, at conditions close to flame extinction are compared. These calculations were performed using boundary conditions (13.1) and (13.2) with $p = 1\,\mathrm{atm}$, $T_{-\infty} = T_{\infty} = 298\,\mathrm{K}$, $Y_{O_2,-\infty} = 0.233$ and $Y_{F,\infty} = 1.0$. Since, the critical value of strain rate at flame extinction, a_{ext}, calculated using various mechanism are different, the results shown in Figs. 13.12–13.13 are at a value of $a_{\mathrm{ext}} = 518$, 561, and $547\,\mathrm{s}^{-1}$ using the starting mechanism, the 5-step mechanism and the 4-step mechanism, respectively. Figure 13.12 shows that the predictions with the 5-step mechanism agree better with the starting mechanism than those with the 4-step mechanism. The higher flame temperature, T_f, and the lower oxygen leakage seen with the 4-step mechanism may indicate that the prediction with the 4-step mechanism is not quite close to the extinction condition. However, a comparison of the T_f with strain rate (see Fig. 13.16) indicates that the turning point occurs at a higher T_f with the 4-step mechanism. Furthermore, as seen from Fig. 13.13 the CH_3 profile predicted by the 4-step mechanism is extremely poor, while that with the 5-step mechanism is reasonably good. Interestingly, the 4-step mechanism predicts the H-atom peak value better than the 5-step mechanism, but the peak is shifted towards the oxidizer side.

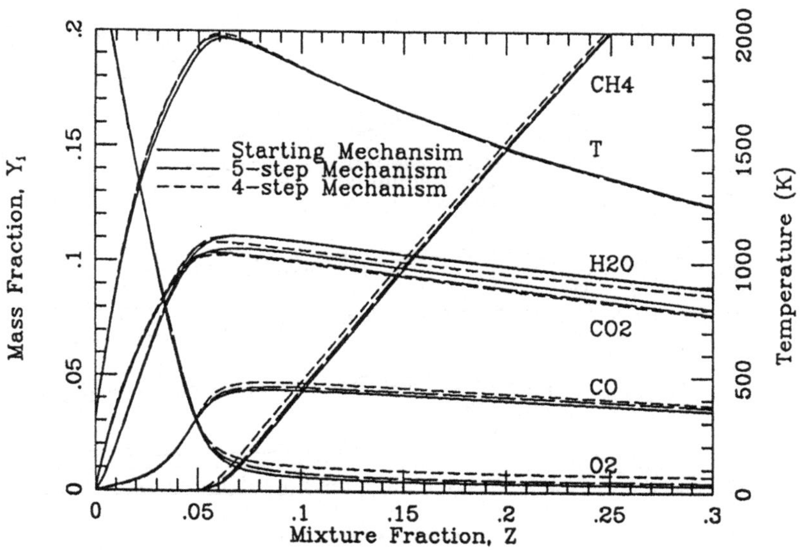

Fig. 13.10. Profiles of temperature (T) and of mass fractions of CH_4, O_2, H_2O, CO_2 and CO using starting, 5-step and 4-step mechanisms with conditions $k = 300\,s^{-1}$, $p = 1\,atm$, $T_{-\infty} = T_{\infty} = 298\,K$, $Y_{F,\infty} = 1.0$ and $Y_{O_2,-\infty} = 1 - Y_{N_2,-\infty} = 0.233$.

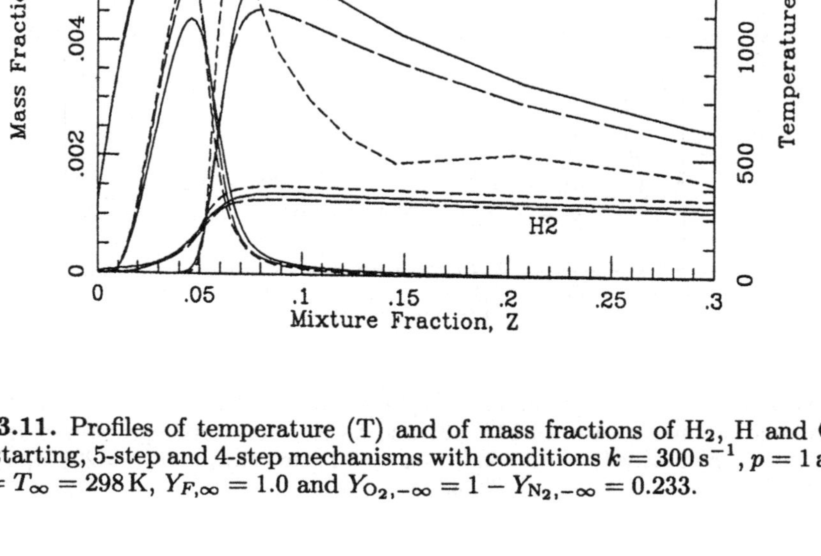

Fig. 13.11. Profiles of temperature (T) and of mass fractions of H_2, H and CH_3 using starting, 5-step and 4-step mechanisms with conditions $k = 300\,s^{-1}$, $p = 1\,atm$, $T_{-\infty} = T_{\infty} = 298\,K$, $Y_{F,\infty} = 1.0$ and $Y_{O_2,-\infty} = 1 - Y_{N_2,-\infty} = 0.233$.

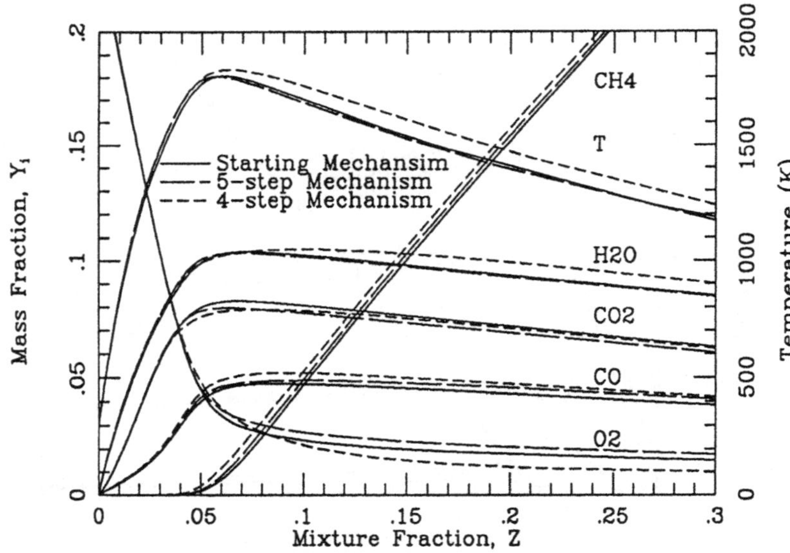

Fig. 13.12. Profiles of temperature (T) and of mass fractions of CH_4, O_2, H_2O, CO_2 and CO at near extinction strain rates of 518, 561 and 547 s^{-1} for starting, 5-step and 4-step mechanisms, respectively, with conditions $p = 1\,\text{atm}$, $T_{-\infty} = T_{\infty} = 298\,\text{K}$, $Y_{F,\infty} = 1.0$ and $Y_{O_2,-\infty} = 1 - Y_{N_2,-\infty} = 0.233$.

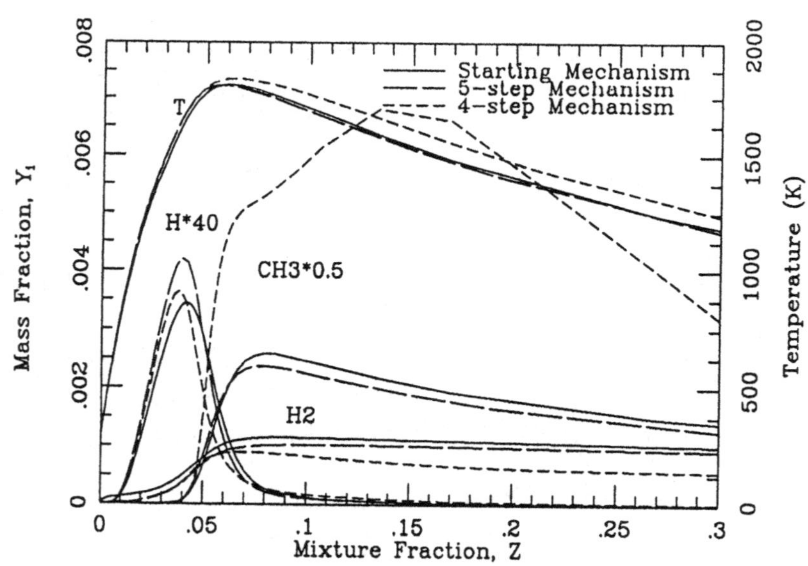

Fig. 13.13. Profiles of temperature (T) and of mass fractions of H_2, H CH_3 at near extinction strain rates of 518, 561 and 547 s^{-1} for starting, 5-step and 4-step mechanisms, respectively, with conditions $p = 1\,\text{atm}$, $T_{-\infty} = T_{\infty} = 298\,\text{K}$, $Y_{F,\infty} = 1.0$ and $Y_{O_2,-\infty} = 1 - Y_{N_2,-\infty} = 0.233$.

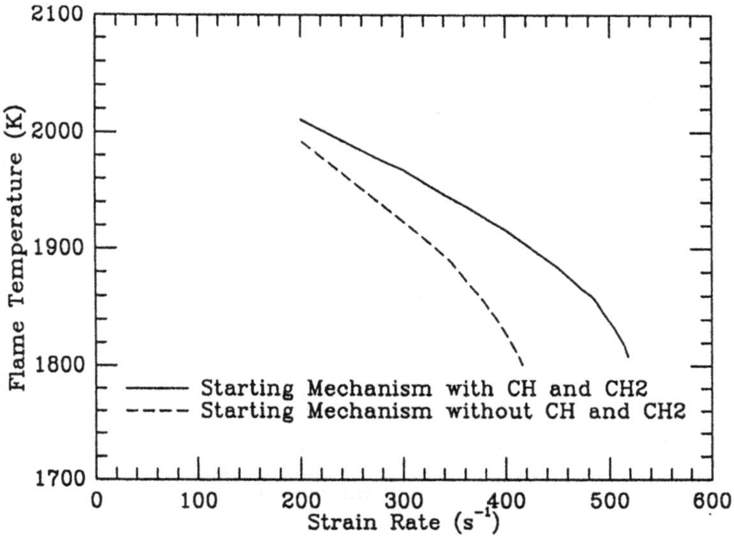

Fig. 13.14. The maximum flame teperature as a function of the strain rate, k, using the starting mechanism involving CH and CH_2 species and neglecting CH and CH_2 species, with conditions $p = 1\,\mathrm{atm}$, $T_{-\infty} = T_\infty = 298\,\mathrm{K}$, $Y_{F,\infty} = 1.0$ and $Y_{O_2,-\infty} = 1 - Y_{N_2,-\infty} = 0.233$.

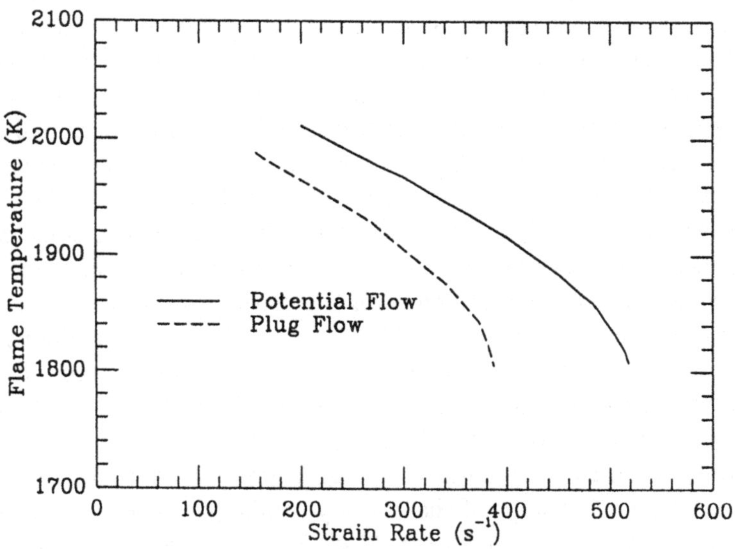

Fig. 13.15. The maximum flame teperature as a function of the strain rate, k, using the potential-flow and plug-flow boundary conditions, with $p = 1\,\mathrm{atm}$, $T_{-\infty} = T_\infty = 298\,\mathrm{K}$, $Y_{F,\infty} = 1.0$ and $Y_{O_2,-\infty} = 1 - Y_{N_2,-\infty} = 0.233$.

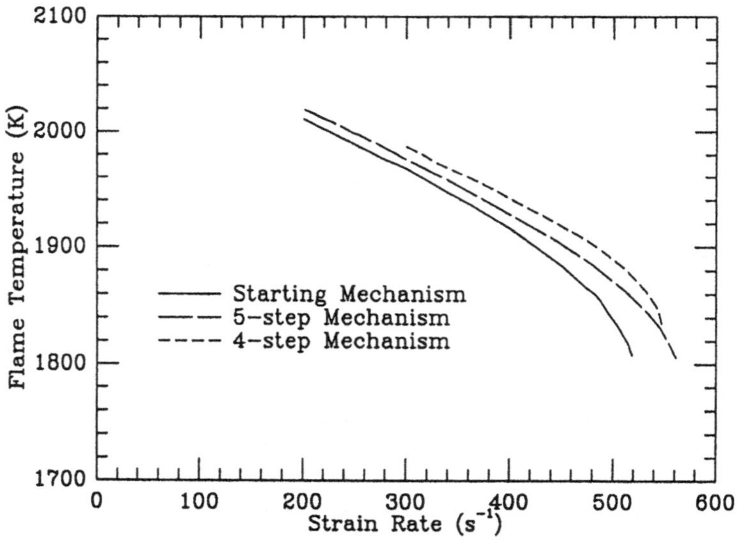

Fig. 13.16. The maximum flame teperature as a function of the strain rate, k, using starting, 5-step and 4-step mechanisms, with conditions $p = 1\,\text{atm}$, $T_{-\infty} = T_{\infty} = 298\,\text{K}$, $Y_{F,\infty} = 1.0$ and $Y_{O_2,-\infty} = 1 - Y_{N_2,-\infty} = 0.233$.

13.5.2 Critical Conditions of Extinction

Calculations were performed using the starting mechanism, the 5-step mechanism, and the 4-step mechanism to determine the critical conditions of flame extinction. To explore the influence of reactions involving the species CH and CH_2 on the critical conditions of extinction, calculations were performed using the starting mechanism with and without reactions involving these species, and the results of maximum flame temperature, T_f, as a function of strain rate, a, are shown in Fig. 13.14. The calculations were performed using boundary conditions (13.1) and (13.2). The value of a at extinction, calculated including the species CH and CH_2 is $518\,\text{s}^{-1}$ and is higher than that calculated by excluding these species ($420\,\text{s}^{-1}$). Hence, reactions involving the species CH and CH_2 exert a considerable influence on the critical conditions of extinction of diffusion flames and this increase of the extinction strain rate is consistent with the increase of burning velocity observed for premixed flames. Figure 13.15 shows plots of the T_f as a function of a calculated using the potential-flow boundary conditions (13.1) and (13.2) and the plug-flow boundary conditions (13.3) and (13.4). In the plug-flow configuration, the characteristic strain rate is calculated from the slope of the velocity profile (dv/dy) in the outer oxidizer flow, near the thermal mixing layer [13.4]. Figure 13.15 shows that the value of a at extinction calculated using the plug-flow boundary conditions ($387\,\text{s}^{-1}$) is lower than that calculated using the potential-flow boundary conditions ($518\,\text{s}^{-1}$). These predictions are consistent with previous results in [13.4].

Finally Fig. 13.16 shows the value of T_f as a function of a calculated using the various mechanisms, using the boundary conditions (13.1) and (13.2).

Predictions using the 5-step mechanism appear to agree better with that calculated using the starting mechanism. Although the value of T_f at extinction calculated using the 5-step mechanism is nearly the same as that calculated using the starting mechanism, the calculated value of a_{ext} is somewhat higher. At any given value of a, Fig. 13.16 shows that the calculated value of T_f using the 4-step mechanism is higher than those calculated using the 5-step mechanism and the starting mechanism. However, the 4-step mechanism is seen to predict a higher value of T_f at extinction. Table 13.1 shows a comparison of all the extinction conditions predicted here with the previously published [13.4]. The stoichiometric scalar dissipation rate at extinction, $(\chi_{st})_{ext}$, has been extimated at Z_{st}=0.051.

Table 13.1. Comparison of the Extinction Conditions.

Present	$a_{ext}(s^{-1})$	$\chi_{st}(s^{-1})$
Plug Flow-Starting	387	17.10
Potential Flow-Starting	518	17.14
Potential Flow-5-Step	561	18.55
Potential Flow-4-Step	547	17.97
Previous [13.4]	$a_{ext}(s^{-1})$	$\chi_{st}(s^{-1})$
Experiments	380	-
Plug Flow	391	14.74
Potential Flow	509	14.72

In addition to kinetic effects, one of the reasons for the large differences seen in present and previous estimates of $(\chi_{st})_{ext}$ is the use of different methods to evaluate Z.

13.6 Conclusions

Comparisons of the structure and the extinction conditions of methane-air diffusion flames have been made using a starting mechanism, a 5-step mechanism and a 4-step mechanism. Based on the C_1 mechanism employed, it is shown that the CH and CH_2 reaction paths can affect the extinction conditions considerably. Although the 4-step mechanism predicts the extinction strain rate slightly better than the 5-step mechanism, the predictions of the structure and extinction flame temperature with the 5-step mechanism is seen to agree better with the starting mechanism. The observed differences between the 5-step and the 4-step mechanisms can be attributed to the breakdown of the steady-state approximation for CH_3 in deducing the 4-step mechanism. Because of the importance of CH_3 in determining the concentrations of CH and CH_2, the 5-step mechanism is a better candidate, for example, in predicting prompt NO_x formation in methane-air flames.

Acknowledgements

The work at Princeton University was supported by the U.S. National Science Foundation and by the U.S. Air Force Office of Scientific Research, while that at University of California at San Diego by the U.S. Army Research Office.

References

[13.1] Smooke, M. D., Giovangigli, V., in Reduced Kinetic Mechanisms and Asymptotic Approximations for Methane-Air Flames (M.D. Smooke Ed.), Lecture Notes in Physics, **384**, Chapter 2, p. 29, 1991.

[13.2] Dixon-Lewis, G., David, T., Haskell, P. H., Fukutani, S., Jinno, H., Miller, J. A., Kee, R. J., Smooke, M. D., Peters, N., Effelsberg, E., Warnatz, J., Behrendt, F., Twentieth Symposium (International) on Combustion, The Combustion Institute, p. 1893–1904, 1985.

[13.3] Smooke, M. D., Puri, I. K., Seshadri, K., Twenty first Symposium (International) on Combustion, The Combustion Institute, p. 1783–1792, 1988.

[13.4] Chelliah, H. K., Law, C. K., Ueda, T., Smooke, M. D., Williams, F. A., Twenty-Third Symposium (International) on Combustion, The Combustion Institute, 1991.

[13.5] Peters, N., Kee, R. J., Combustion and Flame, **68**, p. 17–29, 1987.

[13.6] Bilger, R. W., Starner, S. H., Kee, R. J., Combustion and Flame, **80**, p. 185, 1990.

[13.7] Bilger, R. W., Esler, M. B., Starner, S. H., in Reduced Kinetic Mechanisms and Asymptotic Approximations for Methane-Air Flames (M. D. Smooke Ed.), Lecture Notes in Physics, **384**, Chapter 5, p. 86, 1991.

[13.8] Seshadri, K., Peters, N., Combustion and Flame, **73**, p. 23–44, 1988.

[13.9] Treviño, C., Williams, F. A., Dynamics of Reactive Systems, Part I: Flames, (A. L. Kuhl, J. R. Bowen, J.-C. Leyer, and A. Borisov, Eds.), Progress in Astronautics and Aeronautics, **113**, p. 129, 1988.

[13.10] Chelliah, H. K., Williams, F. A., Combustion and Flame, **80**, p. 17–48, 1990.

[13.11] Chelliah, H. K., Treviño, C., Williams, F. A., in Reduced Kinetic Mechanisms and Asymptotic Approximations for Methane-Air Flames (M. D. Smooke Ed.), Lecture Notes in Physics, **384**, Chapter 7, p. 137, 1991.

[13.12] Kee, R. J., Warnatz, J. and Miller, J. A., Sandia Report SAND 83-8209, UC-32, Livermore, 1983.

[13.13] Smooke, M. D., J. Computational Physics, No. 1, **48**, p. 72, 1982.

[13.14] Peters, N., Numerical Solution of Combustion Phenomena, Lecture Notes in Physics, **241**, p. 90, 1985.

[13.15] Peters, N., in Reduced Kinetic Mechanisms and Asymptotic Approximations for Methane-Air Flames (M. D. Smooke, Ed.), Lecture Notes in Physics, **384**, Chapter 3, p. 48, 1991.

14. Reduced Kinetic Mechanisms for Acetylene Diffusion Flames

R.P. Lindstedt
Department of Mechanical Engineering, Imperial College,
Exhibition Road, London SW7 2BX, England

F. Mauss
Institut für Technische Mechanik, RWTH Aachen,
W-5100 Aachen, West Germany

14.1 Introduction

Previous studies of acetylene combustion have predominantly focussed on premixed flames, e.g. Frenklach and Warnatz [14.1] and Miller and Melius [14.2], while little attention has been given to non–premixed combustion or a systematic reduction of the chemical kinetics for either case. For premixed flames the latter is the topic of Chap. 7 of the current book while the present study is using the planar counterflow geometry to investigate the structure of acetylene–air diffusion flames. The counterflow geometry forms an ideal and computationally efficient configuration for theoretical investigations of chemical kinetics in diffusion flames and many past studies have been performed for alkane fuels. Among these are the numerical studies of the structure of counterflow methane–air and propane–air diffusion flames with detailed [14.3,14.4] and simplified [14.3 – 14.5] chemistry. However, previous studies have not analysed diffusion flames with alkene or alkyne fuels. This is in part a reflection of the uncertainties surrounding the chemistry of such flames. Among the additional problems encountered is the formation of soot and cyclic compounds such as benzene. The former is particularly important at low rates of strain and as a consequence experimental flames are under such conditions strongly non-adiabatic due to radiation from soot particles. Recently soot models have been proposed [14.6] which in principle can be applied to obtain a first approximation of such effects in studies of the kind present here. However, the scope of the present study is limited to an investigation of the primary reaction channels including the formation of C_3 species but excluding the formation of aromatics or soot. Acetylene flames are here computed using strain rates from $10/s$ to extinction at pressures of 1, 2, 5 and 10 bar. Using the results of these computations the behaviour of non–premixed acetylene flames is analysed and the most important reaction paths identified as functions of rate of strain and pressure. These computations subsequently form the basis for the formulation of a reduced reaction mechanism incorporating five reaction steps. Derivations of such reduced mechanisms have already been achieved for methane [14.3,14.4] through the systematic application of partial equilibrium and steady state assumptions [14.7,14.8]. The mechanism resulting from the present work can either be used as a starting point for the derivation of further simplifications or as a basis for the inclusion of additional reaction steps describing the formation of higher hydrocarbons or soot.

14.2 Main Reaction Paths in Non-Premixed Acetylene Flames

The full mechanism used in the present study is shown in Table 1 of Chap. 1. The subset used as a starting mechanism here constitutes the full C_1 – C_2 mechanism with the addition of C_3H_2, C_3H_3 and C_3H_4 formation and consumption. Computations of flames in the range 1 to 10 bar for the entire strain rate range with this mechanism enabled a thorough investigation of different reaction paths. These computations showed that the overall characteristics of acetylene flames differ significantly from those observed in alkane flames. This is in part due to short reaction paths, see below, and the large energy release associated with acetylene. The most noticeable differences are the large amounts of CO generated, the much higher flames temperatures and the significantly greater resistance to extinction via strain.

The dominant reaction path for C_2H_2 oxidation in diffusion flames is under most conditions the very short chain constituted by O atom attack via w_{46} leading to the formation of CHCO which rapidly also reacts with an O-atom to form CO and H via w_{44}.

(46) $$C_2H_2 + O = CHCO + H$$

(44) $$CHCO + O = 2CO + H$$

The competing path via CH_2 radical formation e.g. w_{45} and leading to CH via w_{25} is also important but remains slower under most conditions.

(45) $$C_2H_2 + O = CH_2 + CO$$

(25) $$CH_2 + H = CH + H_2$$

The hydrogen attack on CH_2 to form CH and H_2 via w_{25} remains the dominant reaction for CH_2 destruction under all conditions for non-premixed acetylene flames. The key reactions involving the CH radical are discussed below.

The reaction of the H radical with acetylene is not important at low rates of strain but becomes a major path at high strain rates at all pressures and must therefore be taken into account.

(−41) $$C_2H_2 + H = C_2H + H_2$$
(42) $$C_2H + O_2 = CHCO + O$$

The C_2H formed via w_{41} is rapidly converted to CHCO via w_{42} and reacts further predominantly via w_{44} under all conditions. However, the hydrogen reaction with CHCO via w_{43} is of the same order and becomes increasingly important at high rates of strain and should be taken into account. This reaction increases the importance of the longer reaction chain via CH_2.

(43) $$CHCO + H = CH_2 + CO$$

A feature of the present scheme is that the formation of CHO is mainly controlled by the reactions of CH with CO_2 (w_{20}) leading directly to CHO and with H_2O (w_{90}, see below) leading to CHO via the formation of CH_2O. There are a number of reactions of importance in the breakdown of CH_2O depending on the combustion conditions. However, as this channel becomes more important at high rates of strain w_{29} is chosen as a representative reaction. This is not the only possible choice and reaction w_{31} with oxidation via the OH radical and the third body reaction w_{32} are particularly important at high pressures and low rates of strain. But as the overall importance of this reaction path increases with the rate of strain the following reactions are considered adequate.

(20) $$CH + CO_2 = CHO + CO$$
(90) $$CH + H_2O = CH_2O + H$$
(29) $$CH_2O + H = CHO + H_2$$

The dominant path for CHO destruction under all conditions is the third body reaction w_{24}.

(24) $$CHO + M = CO + H + M$$

The formation of CO_2 is as usual determined by w_{18} and similarly the dominant recombination reaction remains the formation of HO_2 via w_5. The role of the chain branching w_1 and the shuffle reactions w_2-w_4 is similar to that observed in other hydrocarbon flames and the findings resulting from the studies of atmospheric methane–air flames by Peters and Kee [14.3] and Bilger et al [14.4] appear equally valid for acetylene flames under the extended conditions of the present study.

In addition to the above reaction paths there are two short secondary chains which deserve some attention. The reaction of C_2H_2 with H_2 rapidly leads to the formation of C_2H_3 and an H atom via the revers of w_{49}. Subsequently, C_2H_3 decomposes back to acetylene and one further H atom via w_{51}. Consequently, the net effect of this brief chain is to maintain H radical concentration in the rich part of the flame.

(−49) $$C_2H_2 + H_2 = C_2H_3 + H$$
(51) $$C_2H_3 = C_2H_2 + H$$

An exactly analogous reaction path via CH_4 is formed by the reverse reactions of w_{38} and w_{34}.

$$(-38) \qquad CH_3 + H_2 = CH_4 + H$$
$$(-34) \qquad CH_4 = CH_3 + H$$

This latter reaction path is of less importance in the flames computed. In acetylene diffusion flames the reaction path via C_2H_4 is also of minor importance under most conditions but has some influence at high pressures via reaction w_{52} and the reverse of reaction w_{55}.

$$(-55) \qquad C_2H_2 + H_2 + M = C_2H_4 + M$$
$$(52) \qquad C_2H_4 + H = C_2H_3 + H_2$$

Reaction paths via C_2H_6 and C_2H_5 are insignificant and reactions involving H_2O_2 are only important at the lean edge of the flame.

The main interest in C_3H_3 chemistry is that it forms a possible reaction path to benzene e.g. [14.2] and may as a result be important in the formation of soot. For the diffusion flames considered in the present study the C_3 chain is initiated by reactions of acetylene with CH_2 and CH radicals. The chemistry of both of these species is to some extent uncertain along with the product distributions of some of the C_3 reactions. However, despite these reservations the reaction scheme listed in Chap. 1 was slightly extended to obtain an improved description of the CH, CH_2 and propargyl chemistry. The additional reactions are given in Tab. 1. Computations indicate, as outlined above, that the reactions of CH with H_2O and CO_2 dominate for CH destruction under all conditions. Similarily, the reaction of CH_2 with the H radical dominates destruction of CH_2. However, the reactions with acetylene of these two radicals via w_{48} and w_{91} form important secondary paths on the fuel rich side of the flames.

$$(91) \qquad C_2H_2 + CH_2 = C_3H_3 + H$$
$$(48) \qquad C_2H_2 + CH = C_3H_3$$

The reaction via CH_2 is infact the dominant formation path for C_3H_3 under most conditions. This indicates that a proper description of propargyl formation in acetylene diffusion flames also requires the consideration of the equivalent reaction with the singlet CH_2 isomer not considered in the present study. While the latter isomer is present in lower concentrations its addition reaction with acetylene is significantly faster [14.2]. The destruction paths for propargyl, in the absence of benzene formation reactions, are different depending on the rate of strain. At high rates of strain the increased oxygen penetration in the flame leads to the following two reactions having similar rates,

Table 14.1. Additional Reaction Steps

Additional Reaction Steps					
No	Reaction		A mole,cm^3,sec	n	E kJ/mole
88	$CH + O$	$\rightarrow CO + H$	5.70E+13	0	0
89	$CH + OH$	$\rightarrow CHO + H$	3.00E+13	0	0
90	$CH + H_2O$	$\rightarrow CH_2O + H$	1.17E+15	-0.75	0
91	$CH_2 + C_2H_2$	$\rightarrow C_3H_3 + H$	1.20E+13	0	27.62
92	$C_3H_3 + OH$	$\rightarrow C_3H_2 + H_2O$	2.00E+13	0	0
93	$C_3H_3 + H$	$\rightarrow C_3H_2 + H_2$	5.00E+13	0	12.55
94	$C_3H_2 + O_2$	$\rightarrow CHCO + CO + H$	5.00E+13	0	0
95	$C_3H_2 + OH$	$\rightarrow C_2H_2 + CHO$	5.00E+13	0	0

(62) $$C_3H_3 + O_2 = CHCO + CH_2O$$
(93) $$C_3H_3 + H = C_3H_2 + H_2$$

while at low rates of strain w_{93} dominates. The destruction of C_3H_2 is also dependent upon the rate of strain with w_{95} dominant at low strain rates.

(94) $$C_3H_2 + O_2 = CHCO + CO + H$$
(95) $$C_3H_2 + OH = C_2H_2 + CHO$$

Consequently, in the absence of reactions leading to higher hydrocarbons the net effect of the C_3H_2 path at low strain rates is the conversion of CH_2 to CHO. At high rates of strain the C_3 steps under consideration lead back to the shorter reaction path via CHCO not involving the CHO radical. Reactions via C_3H_4 are not important to the same extent for acetylene flames and can consequently be ignored.

14.3 A Reduced Five Step Mechanism

Following the above considerations and after neglecting the secondary reaction paths seventeen reactive species remain. The dominant reaction paths for these species can under the conditions encountered during the preliminary computations be represented by reactions 1–6, 8, 18, 20, 24, 25, 29, 41–46, 49, 51 and 90. These reactions lead to the following system of truncated balance equations.

$$
\begin{aligned}
L([\mathrm{H}]) &= -w_1 + w_2 + w_3 - w_5 - w_6 + w_{18} + w_{24} - w_{25} - w_{29} \\
&\quad + w_{41} - w_{43} + w_{44} + w_{46} + w_{90} - w_{49} + w_{51} \\
L([\mathrm{O}]) &= w_1 - w_2 + w_4 + w_{42} - w_{44} - w_{45} - w_{46} \\
L([\mathrm{H_2}]) &= -w_2 - w_3 + w_{25} + w_{29} - w_{41} + w_{49} \\
L([\mathrm{O_2}]) &= -w_1 - w_5 + w_8 - w_{42} \\
L([\mathrm{H_2O}]) &= w_3 + w_4 + w_8 - w_{90} \\
L([\mathrm{CO}]) &= -w_{18} + w_{20} + w_{24} + w_{43} + 2 \times w_{44} + w_{45} \\
L([\mathrm{CO_2}]) &= w_{18} - w_{20} \\
L([\mathrm{C_2H_2}]) &= w_{41} - w_{45} - w_{46} + w_{49} + w_{51} \\
0 = L([\mathrm{OH}]) &= w_1 + w_2 - w_3 - 2 \times w_4 + 2 \times w_6 - w_8 - w_{18} \\
0 = L([\mathrm{HO_2}]) &= w_5 - w_6 - w_8 \\
0 = L([\mathrm{C_2H}]) &= -w_{41} - w_{42} \\
0 = L([\mathrm{CHCO}]) &= w_{42} + w_{46} - w_{44} - w_{43} \\
0 = L([\mathrm{CH}]) &= w_{25} - w_{20} - w_{90} \\
0 = L([\mathrm{CHO}]) &= w_{20} + w_{29} - w_{24} \\
0 = L([\mathrm{CH_2O}]) &= w_{90} - w_{29} \\
0 = L([\mathrm{CH_2}]) &= w_{45} - w_{25} + w_{43} \\
0 = L([\mathrm{C_2H_3}]) &= -w_{49} - w_{51}
\end{aligned}
\tag{14.1}
$$

The convective-diffusive operator in the balance equations is here represented by $L([X_i])$. For all species the terms accounting for molecular transport and chemical production/consumption can readily be obtained from the solutions using the full mechanism and the validity of the steady state approximations explored. These computations show that steady-state approximations can be introduced for nine of the above species. The accuracy of these approximations depend to some extent on the combustion conditions and the position in the flame. Their validity must therefore be assessed by comparison of predictions obtained by detailed and reduced mechanisms for a wide range of conditions. The steady state species can be used to eliminate 9 of the 21 remaining reactions. The procedure whereby this is performed is to some extent arbitrary but parallell paths should be represented by characteristic reactions. The elimination of individual reactions is here performed in close accordance with the discussion in the previous section.

In the present work w_{46} is retained as characteristic of the fastest reaction path and w_{44} is eliminated via the steady state balance for CHCO. For the longer reaction path which passes via CH_2 reaction w_{45} is retained. At high rates of strain this path is becoming more important as a consequence of w_{43} approaching w_{44} in importance. As w_{45} is retained w_{43} must be eliminated via the steady state balance for CH_2.

The hydrogen attack on acetylene also becomes important at high rates of strain and the reaction sequence w_{41} followed by w_{42} and leading to w_{43}

and w_{44} is here represented by w_{42} and w_{41} is eliminated via the steady state balance for C_2H. This leads to the retention of w_{42}, w_{45} and w_{46} as representative reactions for the three main breakdown paths of C_2H_2 to CO.

Further down the reaction chain passing via CH_2 the reaction responsible for the conversion to CH (w_{25}) is eliminated by the balance equation for CH. The CH radical is predominantly oxidised to CHO via reactions with CO_2 and H_2O via w_{20} and w_{90}. The latter reaction leads to CH_2O which as discussed above has many parallell oxidation paths of which w_{29} is on balance the most important. This reaction is eliminated from the scheme using the steady state balance for CH_2O. Oxidation of CH with O_2 also becomes important at high rates of strain due to increased oxygen penetration in the flame but this step is here neglected. The final step leading to CO in this reaction path is the third body reaction of CHO via w_{24} and this step is eliminated using the steady state balance for this specie.

Due to the general importance of the O radical in acetylene flames and the role of the H radical at high rates of strain both of these species are retained. As a consequence one of the shuffle reactions can be eliminated by the steady state balance for OH. The choice made here is to eliminate w_4. This leads to the retention of w_1, w_2 and w_3. Similarly to previous work on methane flames [14.3,14.4] the HO_2 reactions are characterised by w_5 and w_8 while w_6 is eliminated by the balance equation for HO_2.

With the above steady state assumptions eight reacting species remain. Taking into account the atom-conservation conditions (for C, H and O) a total number of five global steps can be formulated to describe the chemistry. The system of balance equations for the major species is derived by eliminating the reaction rates in accordance with the above discussion. By this procedure the following system of balance equations for the remaining non-steady-state species H, O, H_2, O_2, H_2O, CO, CO_2 and C_2H_2 can be formulated:

$$\begin{aligned}&L([\mathrm{H}]) + \{3 \times L([\mathrm{CH}]) + L([\mathrm{CHO}]) - L([\mathrm{HO_2}]) + L([\mathrm{C_2H}]) + L([\mathrm{CHCO}]+)\}\\&\qquad\{L([\mathrm{C_2H_3}]) + 2 \times L([\mathrm{CH_2}])\}\\&= -w_1 + w_2 + w_3 + w_8 + w_{18} + 2 \times (w_{46} + w_{45} - w_5 - w_{20} - w_{49} - w_{90})\end{aligned}$$

$$\begin{aligned}&L([\mathrm{H_2}]) + \{-L([\mathrm{C_2H}]) + L([\mathrm{CH_2O}]) - L([\mathrm{CH}])\}\\&= -w_2 - w_3 + w_{20} + w_{42} + 2 \times w_{90} + w_{49}\end{aligned}$$

$$\begin{aligned}&L([\mathrm{H_2O}]) + \{0.5 \times L([\mathrm{OH}]) + L([\mathrm{HO_2}])\}\\&= 0.5 \times (w_1 + w_2 + w_3 + 2 \times w_5 - w_8 - w_{18}) - w_{90}\end{aligned}$$

$$
\begin{aligned}
&L([O]) + \{0.5 \times L([OH]) - L([CH_2]) - L([CH]) + L([HO_2]) - L([CHCO])\} \\
&= 0.5 \times (3 \times w_1 - w_2 - w_3 + 2 \times w_5 - 3 \times w_8 - w_{18}) + w_{20} + w_{90} \\
&- 2 \times (w_{45} + w_{46}) \\
&L([O_2]) \\
&= -w_1 - w_5 + w_8 - w_{42} \\
&L([CO_2]) \\
&= w_{18} - w_{20} \\
&L([CO]) + \{L([CH_2O]) - L([CHO]) + 2 \times L([CHCO]) + L([CH])+)\} \\
&\qquad\qquad \{L([CH_2])\} \\
&= -w_{18} + w_{20} + 2 \times (w_{42} + w_{45} + w_{46}) \\
&L([C_2H_2]) + \{L([C_2H]) + L([C_2H_3])\} \\
&= -w_{46} - w_{45} - w_{42}
\end{aligned}
\tag{14.2}
$$

The convective-diffusive operators in the curly, brackets correspond to those of steady state species and are neglected. It can readily be shown that the above balance equations are consistent with the following five step global mechanism.

(I)	$C_2H_2 + 2O = 2CO + 2H$
(II)	$CO + H_2O = CO_2 + H_2$
(III)	$2H + M = H_2 + M$
(IV)	$H_2 + O_2 = O + H_2O$
(V)	$2H_2 + O = 2H + H_2O$

The corresponding global reaction rates are,

$$
\begin{aligned}
w_I &= w_{45} + w_{46} + w_{42} \\
w_{II} &= w_{18} - w_{20} \\
w_{III} &= w_5 + w_{49} \\
w_{IV} &= w_1 + w_5 - w_8 + w_{42} \\
w_V &= 0.5 \times (-w_1 + w_2 + w_3 + w_8 + w_{18}) - w_{20} - w_{42} - w_{90} \,.
\end{aligned}
\tag{14.3}
$$

The above reactions have been cast in a form which encompasses the normal flame processes e.g. fuel breakdown, CO oxidation, recombination, chain branching and a shuffle reaction.

The truncated steady state assumptions for all the species e.g. (14.1) can in principle be used to derive algebraic relations for the concentrations of these species as required by the reaction rate expressions in (14.3). It is, however, frequently better to use an extended skeleton mechanism or even the complete reaction scheme to determine the concentrations of the steady state species. The actual procedure is documented elsewhere, e.g. Paczko et al [14.9], and is therefore not discussed in detail here. The steady state balances used in the present work to determine minor species are based on an extended skeleton mechanism.

For the methenyl radical the following approximation was used,

$$[\mathrm{CH}] = \frac{k_{25f}\,[\mathrm{CH_2}][\mathrm{H}] + k_{20b}\,[\mathrm{CHO}][\mathrm{CO}]}{k_{19f}\,[\mathrm{O_2}] + k_{20f}\,[\mathrm{CO_2}] + k_{90f}\,[\mathrm{H_2O}] + k_{25b}\,[\mathrm{H_2}]} \tag{14.4}$$

and the methylene radical was approximated as,

$$[\mathrm{CH_2}] = \frac{k_{45f}\,[\mathrm{C_2H_2}][\mathrm{O}] + k_{43f}\,[\mathrm{CHCO}][\mathrm{H}] + k_{25b}\,[\mathrm{CH}][\mathrm{H_2}]}{k_{25f}\,[\mathrm{H}] + k_{43b}\,[\mathrm{CO}] + k_{26f}\,[\mathrm{O}] + k_{33b}\,[\mathrm{H_2}] + \mathrm{W_1}} \tag{14.5}$$

where

$$\mathrm{W_1} = \{k_{27f} + k_{28f}\}\,[\mathrm{O_2}] + k_{91f}\,[\mathrm{C_2H_2}]$$

The steady state balance applied for ethynyl was modified by a correction factor,

$$\mathrm{C_{41}} = \frac{k_{41b}[\mathrm{H}]}{k_{41b}[\mathrm{H}] + k_{45f}\,[\mathrm{O}] + k_{46f}\,[\mathrm{O}] + k_{47f}\,[\mathrm{OH}]} \tag{14.6}$$

required to reduce the influence of reaction 41 under very fuel rich conditions. The modified expression may be written as:

$$[\mathrm{C_2H}] = \frac{k_{41b}\,[\mathrm{C_2H_2}][\mathrm{H}] \times \{1 - \mathrm{C_{41}}\} + k_{47f}\,[\mathrm{C_2H_2}][\mathrm{OH}] + k_{42b}\,[\mathrm{CHCO}][\mathrm{O}]}{k_{41f}\,[\mathrm{H_2}] + k_{47b}\,[\mathrm{H_2O}] + k_{42f}\,[\mathrm{O_2}]} \tag{14.7}$$

The errors in ethynyl concentrations in the unmodified steady-state expression occurs where the reaction rate is approaching zero. The modification is not required on grounds of accuracy in major species predictions. It does, however, affect minor species particularly at high pressures as discussed below.

The remaining steady state balances for carbon containing species are:

$$[\mathrm{CHCO}] = \frac{k_{42f}\,[\mathrm{C_2H}][\mathrm{O_2}] + k_{46f}\,[\mathrm{C_2H_2}][\mathrm{O}] + k_{43b}\,[\mathrm{CH_2}][\mathrm{CO}] + k_{44b}\,[\mathrm{CO}]^2[\mathrm{H}]}{k_{43f}\,[\mathrm{H}] + k_{44f}\,[\mathrm{O}] + k_{46b}\,[\mathrm{H}] + k_{42b}\,[\mathrm{O}]} \tag{14.8}$$

$$[\mathrm{CH_2O}] = \frac{k_{90f}\,[\mathrm{CH}][\mathrm{H_2O}]}{k_{29f}\,[\mathrm{H}] + k_{31f}\,[\mathrm{OH}] + k_{30f}\,[\mathrm{O}] + k_{32f}\,[\mathrm{M}]} \tag{14.9}$$

and finally the formyl radical,

$$[\mathrm{CHO}] = \frac{k_{19f}\,[\mathrm{CH}][\mathrm{O_2}] + k_{20f}\,[\mathrm{CH}][\mathrm{CO_2}] + k_{24b}\,[\mathrm{CO}][\mathrm{H}][\mathrm{M}] + \mathrm{W_2}}{k_{23f}\,[\mathrm{O_2}] + k_{22f}\,[\mathrm{OH}] + k_{95b}\,[\mathrm{C_2H_2}] + \mathrm{W_3}} \tag{14.10}$$

where

$$\mathrm{W_2} = \{k_{29f}\,[\mathrm{H}] + k_{30f}\,[\mathrm{OH}] + k_{31f}\,[\mathrm{OH}] + k_{32f}\,[\mathrm{M}]\}\,[\mathrm{CH_2O}]$$

$$\mathrm{W_3} = k_{21f}\,[\mathrm{H}] + k_{32b}\,[\mathrm{M}][\mathrm{H}] + k_{24f}\,[\mathrm{M}]$$

The steady-state approximations used for the hydrogen - oxygen system were also defined by the consideration of all relevant reactions giving the following expression for the hydroxyl radical:

$$[\mathrm{OH}] = \frac{k_{1f}\,[\mathrm{H}][\mathrm{O_2}] + k_{2f}\,[\mathrm{O}][\mathrm{H_2}] + 2\times k_{6f}\,[\mathrm{H}][\mathrm{HO_2}] + \mathrm{W_4}}{k_{1b}\,[\mathrm{O}] + k_{3f}\,[\mathrm{H_2}] + k_{8f}\,[\mathrm{HO_2}] + k_{18f}\,[\mathrm{CO}] + \mathrm{W_5}} \tag{14.11}$$

where,

$$\mathrm{W_4} = \{k_{3b}\,[\mathrm{H}] + 2\times k_{4b}\,[\mathrm{O}] + k_{8b}\,[\mathrm{O_2}] + k_{16b}\,[\mathrm{M}]\}\;[\mathrm{H_2O}] + k_{18b}\,[\mathrm{H}][\mathrm{CO_2}]$$

$$\mathrm{W_5} = \{2\times k_{4f} + 2\times k_{6b}\}\;[\mathrm{OH}] + \{k_{2b} + k_{16f}\,[M]\}\;[\mathrm{H}] + k_{31f}\,[\mathrm{CH_2O}]$$

The approximation used for hydrogen-peroxide was,

$$[\mathrm{H_2O_2}] = \frac{k_{12f}\,[\mathrm{OH}]^2[M] + k_{11f}\,[\mathrm{HO_2}]^2 + \{k_{13b}\,[\mathrm{OH}] + k_{14b}\,[\mathrm{HO_2}]\}\;[\mathrm{H_2O}]}{k_{13f}\,[\mathrm{H}] + k_{14f}\,[\mathrm{OH}] + k_{12b}\,[\mathrm{M}] + k_{11b}\,[\mathrm{O_2}]} \tag{14.12}$$

and for the HO_2 radical,

$$[\mathrm{HO_2}] = \frac{k_{5f}\,[\mathrm{H}][\mathrm{O_2}][\mathrm{M}] + k_{8b}\,[\mathrm{O_2}][\mathrm{H_2O}] + \mathrm{W_6}}{k_{8f}\,[\mathrm{OH}] + k_{5b}\,[\mathrm{M}] + k_{14b}\,[\mathrm{H_2O}] + k_{10f}\,[\mathrm{O}] + \mathrm{W_7}} \tag{14.13}$$

where

$$\mathrm{W_6} = \{k_{14f}\,[\mathrm{H_2O_2}] + k_{6b}\,[\mathrm{OH}]\}\;[\mathrm{OH}]$$

$$\mathrm{W_7} = \{k_{6f} + k_{7f} + k_{9f}\}\;[\mathrm{H}] + 2\times k_{11f}\,[\mathrm{HO_2}]$$

All the steady state approximations derived above are explicit but remain coupled and must be solved iteratively.

The above reduced mechanism assumes that as customary the steady state species are neglected in the elemental balance so that the mass fractions of the non-steady state species add up to unity. This assumption therefore implicitly ignores the elemental mass is stored in the steady state species.

14.4 Comparison of Detailed and Reduced Mechanisms

The first parameter to be studied with the reduced mechanism was the dependance on the peak temperature in the flame as a function of strain rate. Flames at atmospheric pressure where computed with the detailed mechanism at $a = 10, 40, 100, 400, 800, 1600, 3200, 10000, 15000/s$ with the last value very close to the extinction point. A comparison of the result obtained with the reduced mechanism is shown in Fig. 14.1. The agreement is generally very satisfactory with the extinction point for the reduced scheme occurring around $18000/s$. The same procedure was repeated at pressures of 2, 5 and 10 bar. The results for the highest pressure can be found in Fig. 14.2. While the results are generally satisfactory it can be noted that the agreement between the extinction points have deteriorated to some extent. A further indication

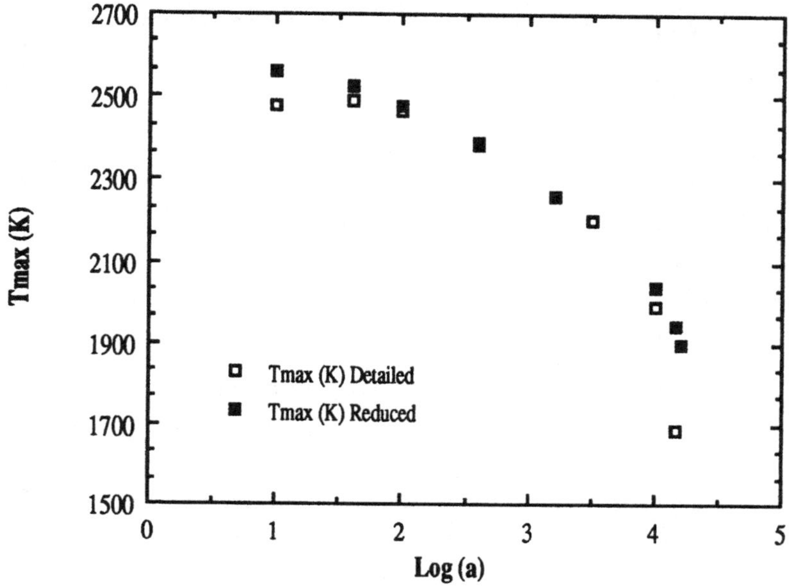

Fig. 14.1. Comparison of maximum flame temperatures obtained as a function of rate of strain with detailed and reduced chemistry at 1 bar pressure.

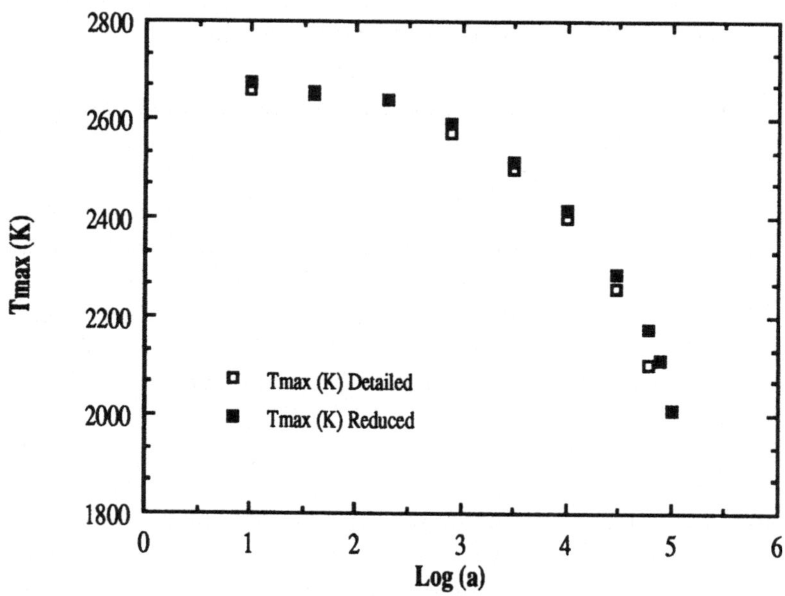

Fig. 14.2. Comparison of maximum flame temperatures obtained as a function of rate of strain with detailed and reduced chemistry at 10 bar pressure.

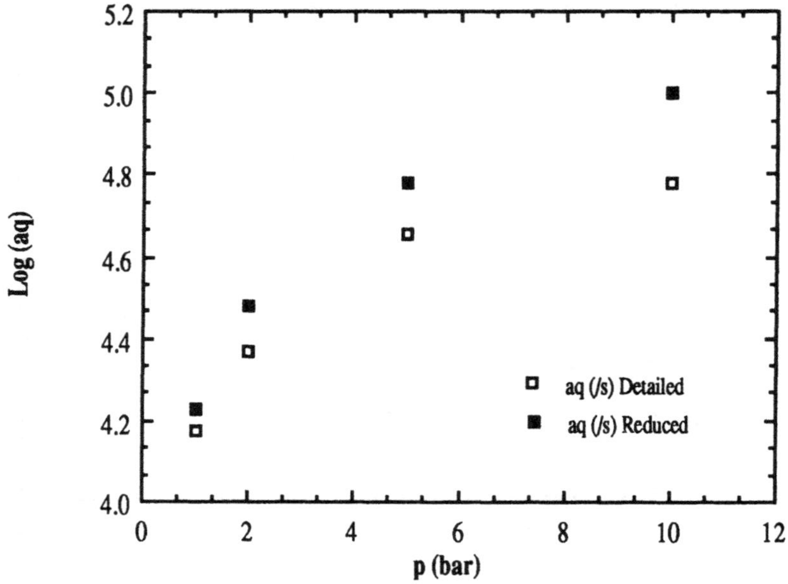

Fig. 14.3. The approximate extinction points as a function of pressure obtained with detailed and reduced chemistry.

of this can be seen in Fig. 14.3 where the extinction points have been plotted as a function of pressure. Part of the reason for this discrepancy is the poor steady state approximation for C_2H discussed above.

The temperature profiles obtained with the detailed and reduced mechanisms are very similar away from extinction as can be seen from Fig. 14.4 where the conditions correspond to a pressure of 5 bar and a strain rate of $6400/s$ and where the profiles have been plotted against the mixture fraction Z obtained assuming unity Lewis numbers as outlined in Chap. 1. In Fig. 14.5 the scalar dissipation rate as defined by Peters and Kee [14.3] is shown along with the temperature profile plotted against physical distance along the stagnation point streamline. As a consequence of the high rate of strain and the sharper gradients the scalar dissipation rates are significantly higher in these flames than in alkane flames.

The agreement for some of the major species, C_2H_2, O_2, CO and CO_2, predicted by the two mechanisms can be found in Fig. 14.6 where the conditions correspond to atmospheric pressure and a low rate of strain of $a = 40/s$. The agreement is very satisfactory with only a minor discrepancy in the CO–CO_2 balance leading to an under–prediction of CO with the reduced mechanism. Similar agreement is also obtained for a flame at more extreme conditions with a pressure of 10 bar and a strain rate of $10000/s$ as can be seen from Fig. 14.7. For this flame the CO levels are somewhat over–predicted with the reduced scheme. A comparison of the agreement for the other major species,

H_2 and H_2O, is shown in Fig. 14.8 at a pressure of 5 bar and $a = 6400/s$. The balance between all major species is satisfactory in view of the neglect of all intermediate species.

Regarding the predictions of the key radical concentration, e.g. H, O and OH, comparisons are shown in Fig. 14.9 and Fig. 14.10 obtained from the same computations as in Fig. 14.6 and Fig. 14.7. The agreement for the flame at the low strain rate is remarkably good while the highly strained high pressure flame shows more significant differences. The most important among these is the higher H radical concentrations observed on the rich side of the flame. It can also be noted that the O concentration at the lean edge of the flame exhibits deviations from the solution with the detailed mechanism. However, the agreement is generally acceptable.

The steady state predictions of some minor species, HO_2, CH and CH_2 are shown in Fig. 14.11 for the atmospheric pressure flame at $a = 40/s$. The agreement is very satisfactory with maximum errors in molefractions typically significantly less than 50%. For the 10 bar flames at $a = 10000/s$ the agreement is less satisfactory as can be seen from Fig. 14.12 where profiles of CH, C_2H and CHCO radicals are shown. The agreement deteriorates particularly on the fuel rich side of the flame where the correction factor C_{41} reduces C_2H levels excessively. However, in the absence of a limiter on reaction 41 the resulting C_2H concentrations are much too high. Due to the low reaction rates in this part of the flame the consequences of this discrepancy are less than anticipated. However, the reaction of C_2H with O_2 via w_{42} is part of the reduced reaction mechanism and an improved steady state approximation for this specie is desirable.

14.5 Conclusions

The major reaction paths in non-premixed acetylene diffusion flames exhibit very short reaction chains in which both the O and H radicals have a crucial role. The short reaction paths together with the high energy release associated with acetylene leads to significantly different extinction points compared to alkane flames. The formation of C_3 species which is a possible formation path to aromatic species [14.2] forms a secondary part of a secondary reaction chain in acetylene flames. The present work also indicates that the accuracy of predictions of species such as the propargyl radical is strongly dependent upon the chemistry of CH and CH_2 radicals.

It has been shown that non-premixed acetylene flames can be well represented by a five step reduced mechanism for a wide range of conditions. In particular major species and key radicals are well predicted along with peak temperatures. However, some discrepancies are observed at high rates of strain and high pressures. These problems may in part be attributed to the breakdown of the steady state approximation for ethynyl.

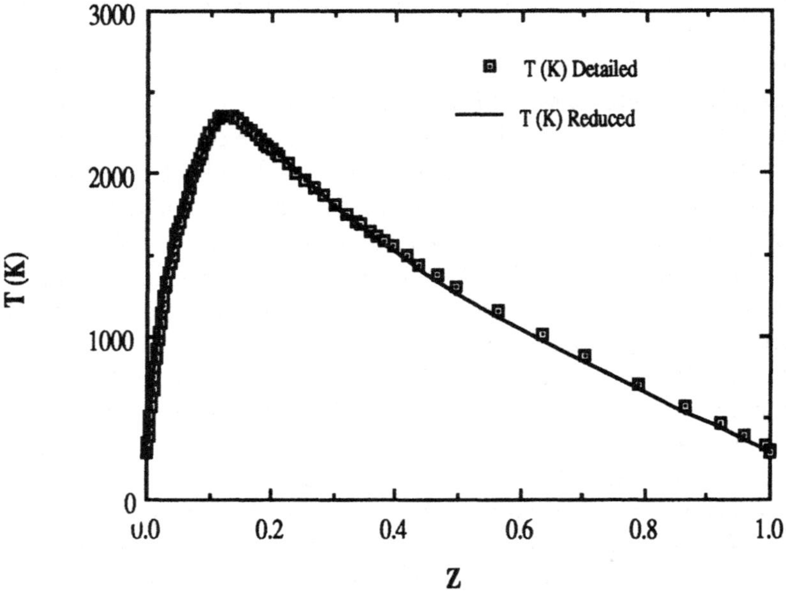

Fig. 14.4. Temperature profiles plotted against the mixture fraction Z for a flame with $a = 6400$/s and 5 bar pressure. The points correspond to the detailed chemistry prediction.

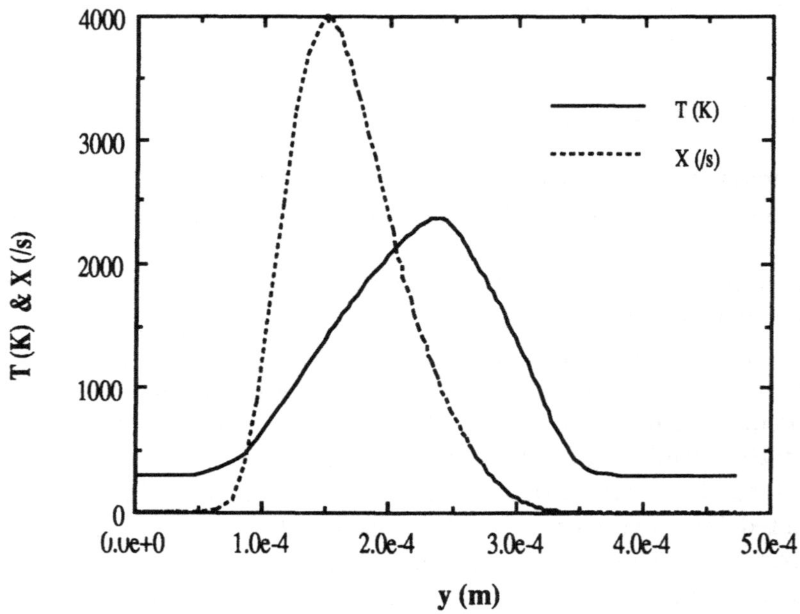

Fig. 14.5. Predicted values of the scalar dissipation rate and temperature vs. distance along the stagnation point stream line. Conditions identical to those in Fig. [14.4].

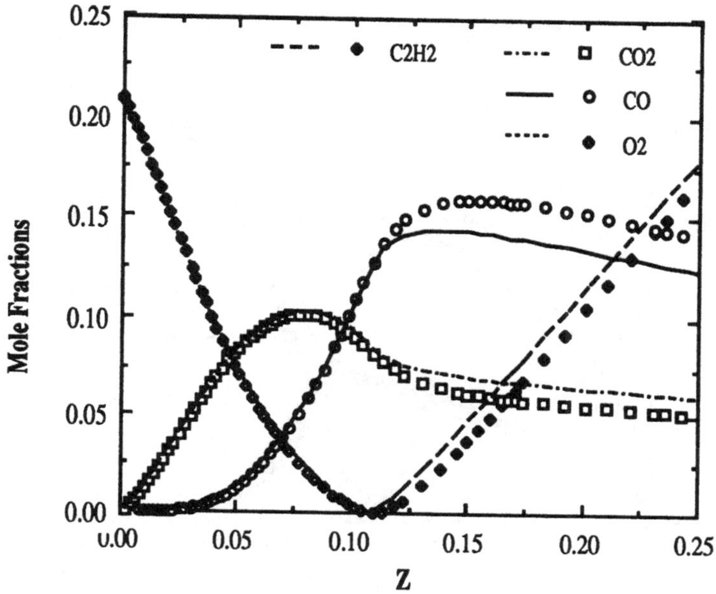

Fig. 14.6. Comparison of predictions of major species obtained with detailed (points) and reduced (lines) chemistry for a flame with $a = 40/\mathrm{s}$ and 1 bar pressure.

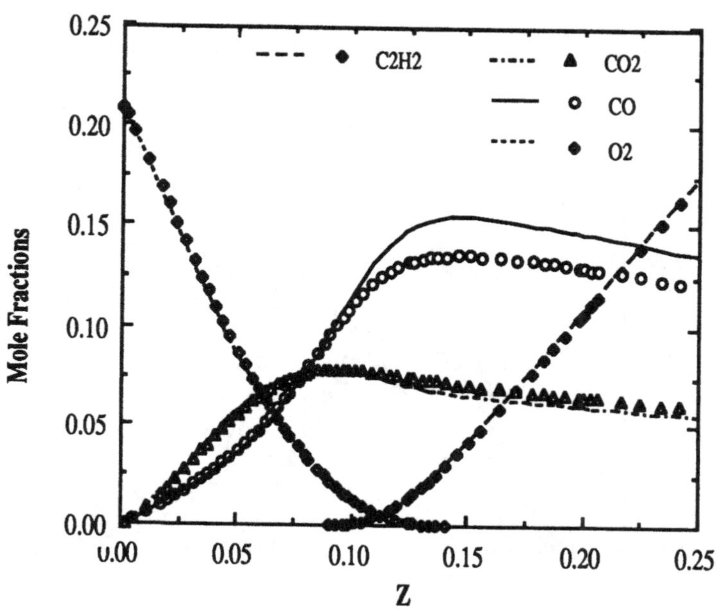

Fig. 14.7. Comparison of predictions of major species obtained with detailed (points) and reduced (lines) chemistry for a flame with $a = 10000/\mathrm{s}$ and 10 bar pressure.

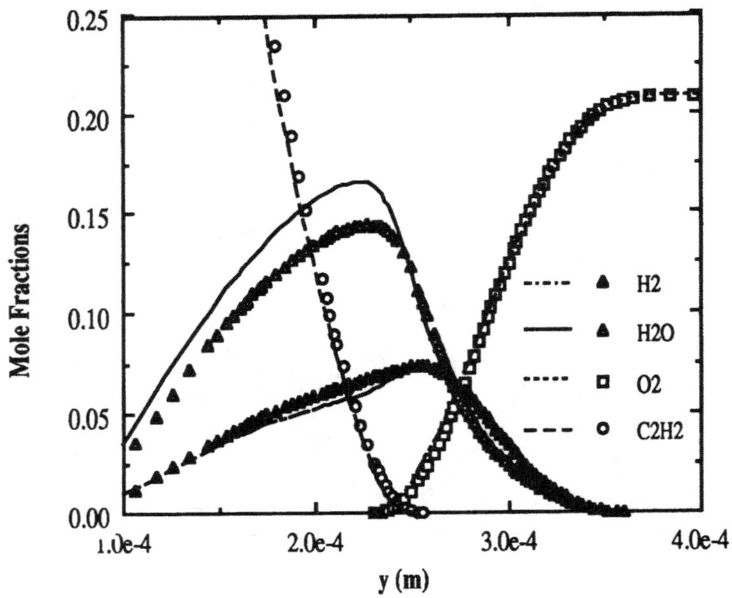

Fig. 14.8. Comparison of predictions of major species obtained with detailed (points) and reduced (lines) chemistry for a flame with $a = 6400$/s and 5 bar pressure.

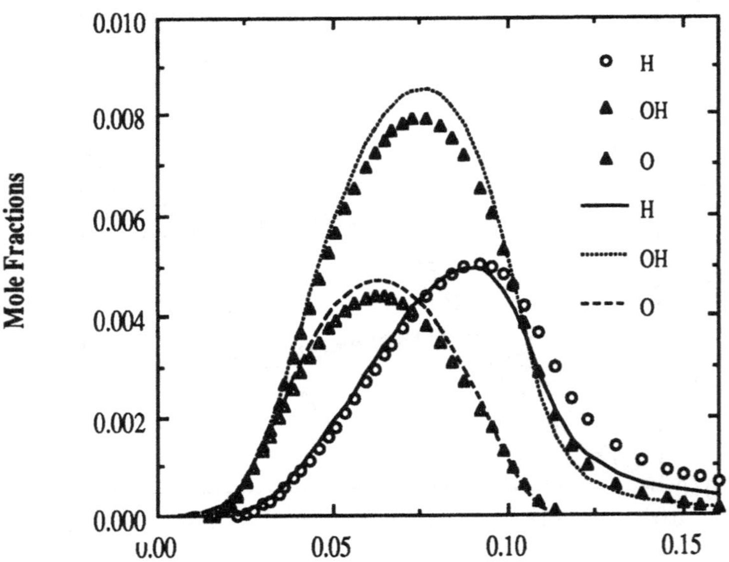

Fig. 14.9. Comparison of predictions of major radical species obtained with detailed (points) and reduced (lines) chemistry for a flame with $a = 40$/s and 1 bar pressure.

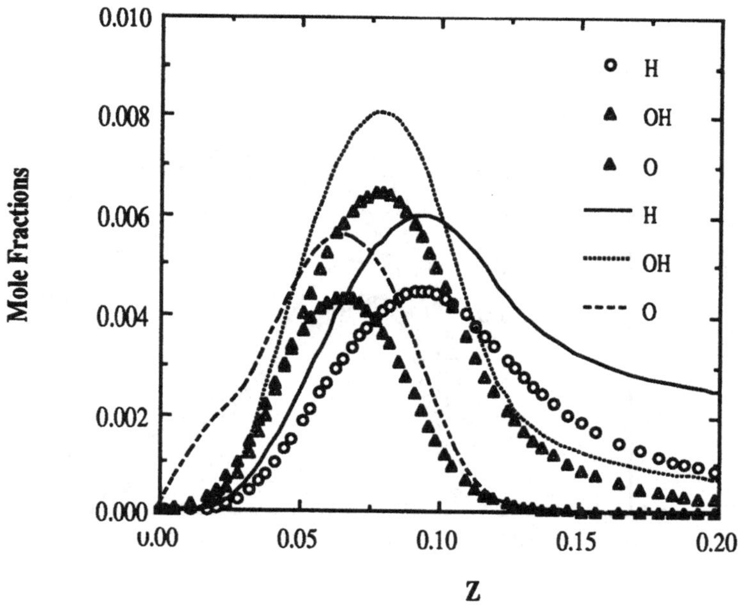

Fig. 14.10. Comparison of predictions of major radical species obtained with detailed (points) and reduced (lines) chemistry for a flame with $a = 10000/s$ and 10 bar pressure.

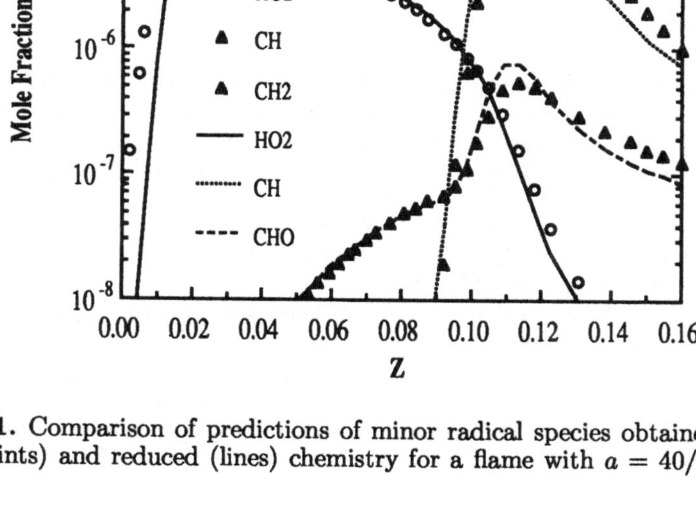

Fig. 14.11. Comparison of predictions of minor radical species obtained with detailed (points) and reduced (lines) chemistry for a flame with $a = 40/s$ and 1 bar pressure.

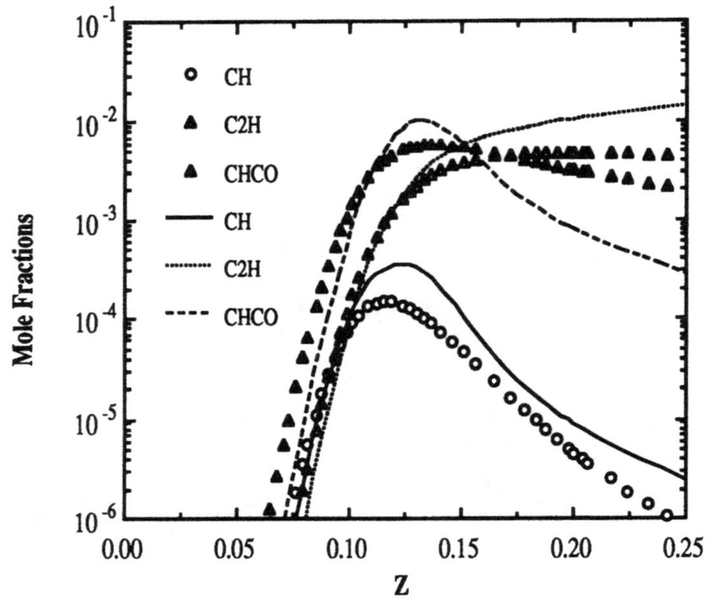

Fig. 14.12. Comparison of prediction of key radical species obtained with detailed (points) and reduced (lines) chemistry for a flame with a = 10000/s and 10 bar pressure.

Acknowledgement

The authors wish to thank the MoD at RAE Pyestock for in part supporting this work.

References

[14.1] Frenklach, M. and Warnatz, J., Comb. Sci. and Tech. **51**, 265–283, 1987
[14.2] Miller, J.A. and Melius, C.F., Submitted to Combustion and Flame 1990
[14.3] Peters, N. and Kee, R.J., Comb. and Flame **68**, 17–30, 1987
[14.4] Bilger, R.W., Stårner, S.H. and Kee, R.J., Comb. and Flame **80**, 135–149, 1990
[14.5] Jones, W.P. and Lindstedt, R.P., Comb. Sci. and Tech. **61** 31–49, 1988
[14.6] Leung, K.M., Lindstedt, R.P. and Jones, W.P., Comb. and Flame Comb. and Flame **87** pp. 289-305, 1992
[14.7] Peters, N. in: Numerical Simulation of Combustion Phenomena. Lecture Notes in Physics **241**, pp. 90–109, 1985
[14.8] Peters, N. in: Reduced Kinetic Mechanisms and Asymptotic Approximations for Methane-Air Flames. (M.D. Smooke, Ed.), Springer 1991
[14.9] Paczko, G., Lefdal, P.M., Peters, N., Twenty-First Symposium (International) on Combustion, The Combustion Institute, pp. 739–748, 1988

15. Reduced Kinetic Mechanisms for Propane Diffusion Flames

K.M. Leung and R.P. Lindstedt
Department of Mechanical Engineering, Imperial College,
Exhibition Road, London SW7 2BX, England

W.P. Jones
Department of Chemical Engineering and Chemical Technology,
Imperial College, Prince Consort Road, London SW7 2BY, England

15.1 Introduction

Propane is an important practical fuel and its high temperature combustion is characterised by the rapid decomposition into smaller C_1-C_3 intermediates [15.1]. This behaviour is similar to the combustion of more complex hydrocarbon fuels. From a modelling perspective, a propane combustion mechanism, compared to that of other higher hydrocarbons, requires the smallest number of species and reactions that are necessary for a thorough kinetic study of the C_1-C_3 species. Previous modelling studies of propane combustion with detailed [15.1,15.2] and simplified [15.3] chemistry have mainly focused on premixed flames with little attention given to non-premixed conditions [15.4–15.5]. There is also a lack of simplified mechanisms based on the systematic reduction of complex chemical mechanisms for non-premixed propane flames. The purpose of the present study is to formulate reduced reaction mechanisms based on the systematic theoretical investigation of propane-air diffusion flames using a planar counterflow geometry and the detailed chemistry defined in Chap. 1. Propane flames are here computed using rates of strain from 10/s to extinction at pressures ranging from 1 to 10 bar. The deduced mechanisms are also validated against the experimental results obtained by Tsuji and Yamaoka [15.6] for counterflow propane-air flames at strain rates of 150/s and 350/s. Using the results of the above computations the behaviour of non-premixed propane flames is analysed and the most important reaction paths indentified as functions of rate of strain and pressure. The validity of steady-state assumptions for intermediate species are also examined. Subsequent formulations of the reduced mechanisms are based on the use of steady-state assumptions for most of the species and the elimination of unimportant reactions from the original detailed mechanism.

15.2 Main Reaction Paths in Non-Premixed Propane Flames

The full mechanism used in the present study is shown in Table 1 of Chap. 1. The reactions (83)–(87), which involve CH_2OH and CH_3OH, are neglected in the present study as they are considered to be unimportant for non-premixed propane flames. Computations have been performed in the range from 1 to 10 bar for the entire strain rate range with this mechanism. These computations showed that the removal regions of propane and oxygen are well separated spatially even in highly strained flames at elevated pressures. For example, in a flame at 10 bar pressure and with a strain rate of 2000/s, which is close to the extinction point of 2300/s, the maximum rates of consumption of propane and oxygen occurs at mixture fractions of 0.094 and 0.085 respectively, with the mixture fraction (Z) obtained according to the definition in Chap. 1. Furthermore, this separation increases with decreasing strain rate and pressure.

Propane is rapidly consumed on the rich side of the flames to produce a large amount of C_1 and C_2 intermediates especially at low strain rate conditions. Therefore subsequent reactions proceed to a significant extent through the C_1- and C_2-chains. The attack by H and OH radicals are the major consumption paths for propane and the intermediate species with the exception of acetylene where the O atom attack is dominant as discussed in Chap. 14. The latter also becomes significant at high strain rate conditions because of the increased O_2 penetration to the rich side of the flame.

The dominant consumption path of C_3H_8 under all conditions is the H radical attack to form normal- and iso- radicals via reactions (77) and (78) respectively. The OH radical attack only plays a secondary role in the consumption of propane in diffusion flames. This is because of the low concentration of OH radicals in the rich side of the flames.

(77) $$C_3H_8 + H = n\text{-}C_3H_7 + H_2$$

(78) $$C_3H_8 + H = i\text{-}C_3H_7 + H_2$$

(81) $$C_3H_8 + OH = n\text{-}C_3H_7 + H_2O$$

(82) $$C_3H_8 + OH = i\text{-}C_3H_7 + H_2O$$

Both reactions (77) and (78) are of the same importance. Both propyl radicals decompose quickly to C_2H_4 and CH_3 via reactions (73) and (75).

(73) $$n\text{-}C_3H_7 = C_2H_4 + CH_3$$

(75) $$i\text{-}C_3H_7 = C_2H_4 + CH_3$$

The n-propyl radical also decomposes to form C_3H_6 via reaction (74) which also constitutes the major formation path for propene. However, the rate of this reaction is much slower than that of reaction (73) under all conditions as found in the present study.

(74) $$n\text{-}C_3H_7 = C_3H_6 + H$$

Propene is consumed via reactions (71) and (72). And with the current scheme the thermal decomposition reaction (71) is the dominant path under most conditions. This is due to the fast rate constant expression used for reaction (71) in the present mechanism. A slower rate constant such as that suggested recently by Rao and Skinner [15.7] will change the balance between these two paths.

(71) $$C_3H_6 = C_2H_3 + CH_3$$

(72) $$C_3H_6 + H = C_3H_5 + H_2$$

Since the decomposition of n-C_3H_7 via reaction (74) is much slower than via reaction (73), only a small amount of C_3H_6 are predicted. As a consequence, the C_3-chain via C_3H_6, C_3H_5 and C_3H_4 is of minor importance in the present study. On the other hand, the rapid formation of C_2H_4 and CH_3 on the rich side of the flames leads to significant accumulation of C_1 and C_2 intermediates. This is due to their relatively slow consumption paths in this part of the flames. This behavior is especially noticeable at low strain rate conditions.

Ethene is mainly consumed by H and OH radical attack via reactions (52) and (54) under all conditions. Both paths are of the same importance, but the reaction with the H radical is faster particularly at high rate of strain. The thermal decomposition of C_2H_4 into C_2H_2 and H_2 via reaction (55) becomes significant only at high pressure conditions.

(52) $$C_2H_4 + H = C_2H_3 + H_2$$

(54) $$C_2H_4 + OH = C_2H_3 + H_2O$$

(55) $$C_2H_4 + M = C_2H_2 + H_2 + M$$

The vinyl radical, C_2H_3, predominantly decomposes to form C_2H_2 via reaction (51).

(51) $$C_2H_3 = C_2H_2 + H$$

The dominant consumption path for acetylene in acetylene diffusion flames is the attack by the oxygen atom as discussed in Chap. 14. In the present study of propane diffusion flames reactions (45) and (46) with oxygen atom are still the major consumption paths for acetylene. However, the reaction with H radical also plays an important role for acetylene consumption in propane flames. The H radical attack is responsible for 30% of C_2H_2 consumption at rate of strain of 10/s and this increases up to 50% at rate of strain of 600/s at atmospheric pressure. The most important consumption paths for acetylene are therefore,

(−41) $$C_2H_2 + H = C_2H + H_2$$

(45) $$C_2H_2 + O = CH_2 + CO$$

(46) $$C_2H_2 + O = CHCO + H$$

(47) $$C_2H_2 + OH = C_2H + H_2O$$

The addition reactions of CH and CH_2 with C_2H_2 via reactions (48) and (91), see Tab. 1 of Chap. 14, are the major formation paths for C_3H_3 in propane flames. This is similar to the behaviour observed in acetylene flames. Under high pressure conditions these two reactions can proceed with rates comparable to the other consumption paths of C_2H_2. The main interest in these reactions is that the propargyl radical provides a possible reaction path to benzene formation as discussed in Chap. 14.

(48) $$C_2H_2 + CH = C_3H_3$$

(91) $$C_2H_2 + CH_2 = C_3H_3 + H$$

Methane is produced on the rich side by the recombination of CH_3 with H radicals via reaction (34) and with H_2 and H_2O via the reverse of reactions (38) and (40) respectively. These reactions form an important part of the kinetics on the rich side of the flame as they are responsible for the formation of both H and OH radicals in this part of the flame. Methane is subsequently consumed near the flame front via reactions (38) and (40).

(34) $$CH_3 + H = CH_4$$

(38) $$CH_4 + H = CH_3 + H_2$$

(40) $$CH_4 + OH = CH_3 + H_2O$$

The complex behaviour of the above reactions leads to difficulties in formulating steady state expressions for CH_4 as discussed below.

Ethane is mainly produced via the recombination reaction (36) of CH_3 radical and subsequently consumed by H and OH radical attack via reactions (59) and (61).

(36) $$CH_3 + CH_3 = C_2H_6$$

(59) $$C_2H_6 + H = C_2H_5 + H_2$$

(61) $$C_2H_6 + OH = C_2H_5 + H_2O$$

Ethyl radicals formed in the above reactions rapidly undergoes thermal decomposition into C_2H_4 and H via reaction (58).

(58) $$C_2H_5 = C_2H_4 + H$$

The remaining reaction paths leading from C_2-chain to C_1-chain are similar to those of non-premixed acetylene flames which are discussed further in Chap. 14 of the current book.

15.3 Comparison with Measurements

The results from computations of counterflow propane-air flames at rates of strain of 150/s and 350/s and 1 bar pressure with the present kinetic mechanism are here compared with the experimental results of Tsuji and Yamaoka

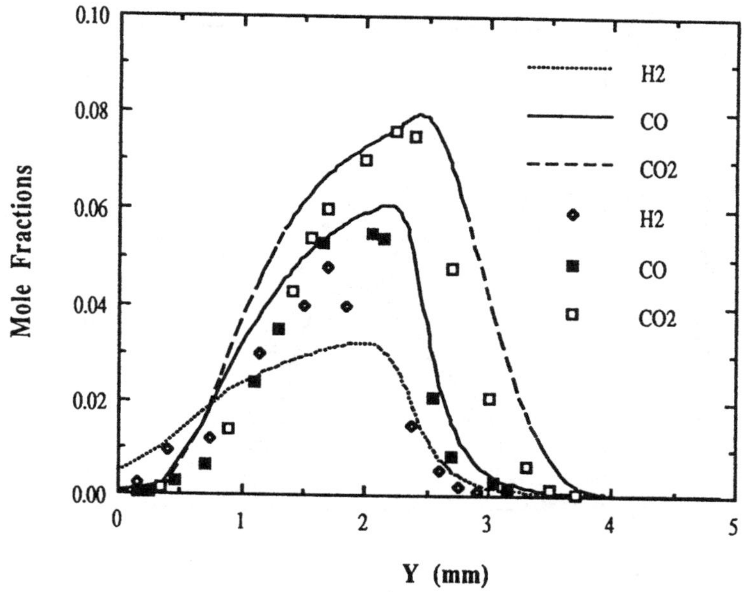

Fig. 15.1. Comparison of predictions of major species with detailed mechanism (lines) and experimental results (points) for a propane-air diffusion flame with $a = 150/s$ and 1 bar pressure.

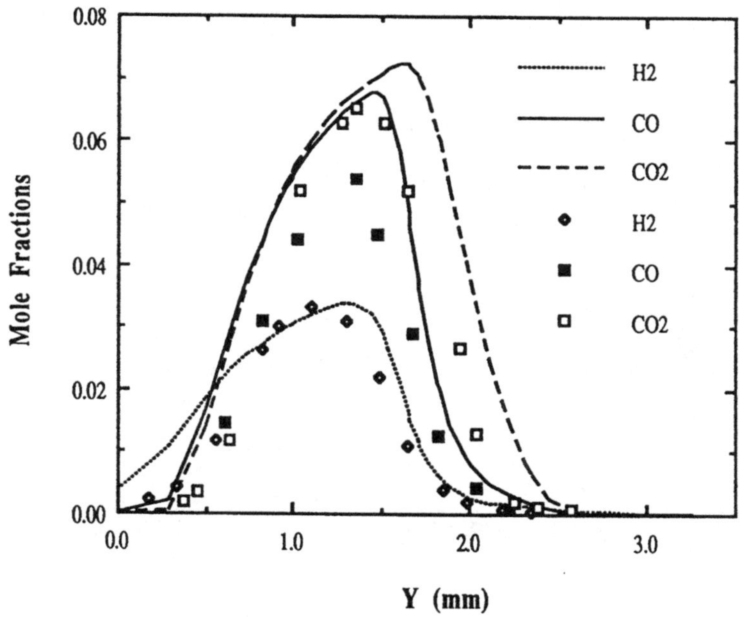

Fig. 15.2. Comparison of predictions of major species with detailed mechanism (lines) and experimental results (points) for a propane-air diffusion flame with $a = 350/s$ and 1 bar pressure.

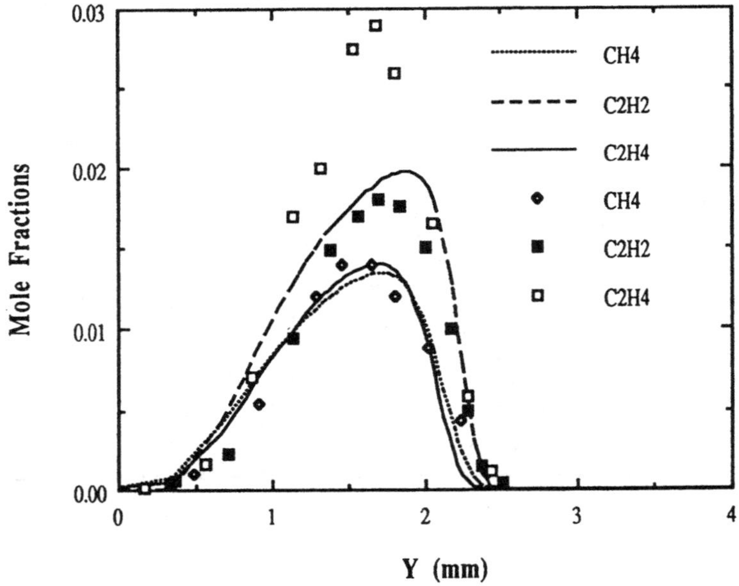

Fig. 15.3. Comparison of predictions of major intermediate species with detailed mechanism (lines) and experimental results (points) for a propane-air diffusion flame with $a = 150/\mathrm{s}$ and 1 bar pressure.

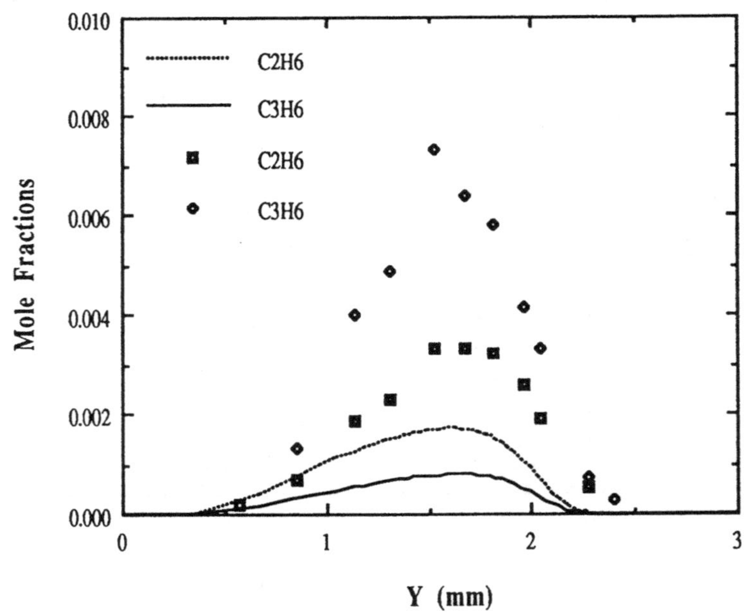

Fig. 15.4. Comparison of predictions of minor intermediate species with detailed mechanism (lines) and experimental results (points) for a propane-air diffusion flame with $a = 150/\mathrm{s}$ and 1 bar pressure.

[15.6]. For both flames there is reasonable agreement between predictions and measurements of the species profiles of H_2, CO, CO_2, as shown in Fig. 15.1 and Fig. 15.2. A comparison of the intermediate species for the case with a rate of strain of 150/s can be found in Fig. 15.3 and Fig. 15.4. The agreement for the concentration profiles of CH_4 and C_2H_2 is satisfactory. However, the concentrations of C_2H_4, C_2H_6 and C_3H_6 are generally underpredicted and significant discrepancies can be observed particularly for C_2H_4 and C_3H_6. Similar results have been found for the higher strain rate case. Since the two latter species are major pyrolysis products in propane diffusion flames, it is likely that some paths which are important for the formation of C_2H_4 and C_3H_6 in non-premixed propane flames are missing or have inappropriate rate constants.

Based on the analysis discussed above, it is possible to correct the underprediction of C_2H_4 and C_3H_6 profiles by the introduction of extra formation paths from the fuel to these two species. The focus of present study is, however, on the formulation of reduced kinetic mechanisms, and as any changes to a well formulated reaction mechanism can readily be implemented, the present mechanism is considered sufficiently accurate for the present purpose. However, further work is necessary to clarify the uncertainty in order to make the mechanism more comparable to experimental results.

15.4 A Reduced Nine Step Mechanism

Following the discussion in previous section, the C_3-chain together with C_3H_4 and C_3H_5 are eliminated from the original mechanism. This follows as a result of the slower rate of reaction (72) compared to reaction (71). A number of additional reactions, which are too slow to be important in propane diffusion flames, are also eliminated. This results in a skeleton mechanism which is made up of 56 reactions and can be represented by reactions 1–6, 8, 18–20, 24–25, 29–31, 33–36, 38, 40–48, 51, 52, 54–56, 58, 59, 61–63, 71, 73–75, 77–82 and 89–95. The rate constants for latter reaction steps are defined in Tab. 1 of Chap. 14. In comparison to similar work on premixed propane flame, see Chap. 8 of the present book, the use of this smaller set of reactions reduces the complexity of the resulting rate expressions for the reduced mechanisms.

By considering all the formation and consumption reactions for each species in this skeleton mechanism, the following system of truncated balance equations for non-steady state species can be formulated,

$$
\begin{aligned}
L([\mathrm{H}]) &= -w_1 + w_2 + w_3 - w_5 - w_6 + w_{18} + w_{24} - w_{25} \\
&\quad - w_{29} - w_{33} - w_{34} + w_{35} - w_{38} + w_{41} - w_{43} \\
&\quad + w_{44} + w_{46} + w_{51} - w_{52} - w_{56} + w_{58} \\
&\quad - w_{59} + w_{74} - w_{77} - w_{78} + w_{89} \\
&\quad + w_{90} + w_{91} - w_{93} + w_{94} \\
L([\mathrm{O}]) &= w_1 - w_2 + w_4 + w_{19} - w_{30} - w_{35} + w_{42} \\
&\quad - w_{44} - w_{45} - w_{46} - w_{63} - w_{79} - w_{80} \\
L([\mathrm{OH}]) &= w_1 + w_2 - w_3 - 2 \times w_4 + 2 \times w_6 - w_8 - w_{18} \\
&\quad + w_{30} - w_{31} - w_{40} - w_{47} - w_{54} - w_{61} + w_{79} \\
&\quad + w_{80} - w_{81} - w_{82} - w_{89} - w_{92} - w_{95} \\
L([\mathrm{H_2}]) &= -w_2 - w_3 + w_{25} + w_{29} + w_{33} + w_{38} - w_{41} \\
&\quad + w_{52} + w_{55} + w_{59} + w_{77} + w_{78} + w_{93} \qquad (15.1) \\
L([\mathrm{O_2}]) &= -w_1 - w_5 + w_8 - w_{19} - w_{42} - w_{62} - w_{94} \\
L([\mathrm{H_2O}]) &= w_3 + w_4 + w_8 + w_{31} + w_{40} + w_{47} + w_{54} \\
&\quad + w_{61} + w_{81} + w_{82} - w_{90} + w_{92} \\
L([\mathrm{CO}]) &= -w_{18} + w_{20} + w_{24} + w_{43} + 2 \times w_{44} + w_{45} \\
&\quad + w_{63} + w_{94} \\
L([\mathrm{CO_2}]) &= w_{18} - w_{20} \\
L([\mathrm{CH_4}]) &= w_{34} - w_{38} - w_{40} \\
L([\mathrm{C_2H_2}]) &= w_{41} - w_{45} - w_{46} - w_{47} - w_{48} + w_{51} + w_{55} \\
&\quad - w_{91} + w_{95} \\
L([\mathrm{C_2H_4}]) &= -w_{52} - w_{54} - w_{55} + w_{58} + w_{73} + w_{75} \\
L([\mathrm{C_3H_8}]) &= -w_{77} - w_{78} - w_{79} - w_{80} - w_{81} - w_{82}
\end{aligned}
$$

The corresponding system of balance equations for the steady state species are as follows,

$$
\begin{aligned}
0 = L([\mathrm{HO_2}]) &= w_5 - w_6 - w_8 \\
0 = L([\mathrm{CH}]) &= -w_{19} - w_{20} + w_{25} - w_{48} - w_{89} - w_{90} \\
0 = L([\mathrm{CHO}]) &= w_{19} + w_{20} - w_{24} + w_{29} + w_{30} \\
&\quad + w_{31} + w_{89} + w_{95} \\
0 = L([\mathrm{CH_2}]) &= -w_{25} + w_{33} + w_{43} + w_{45} - w_{91} \\
0 = L([\mathrm{CH_2O}]) &= -w_{29} - w_{30} - w_{31} + w_{35} + w_{62} + w_{90} \\
0 = L([\mathrm{CH_3}]) &= -w_{33} - w_{34} - w_{35} - 2 \times w_{36} + w_{38} + w_{40} \\
&\quad + 2 \times w_{56} + w_{71} + w_{73} + w_{75} \\
0 = L([\mathrm{C_2H}]) &= -w_{41} - w_{42} + w_{47}
\end{aligned}
$$

$$\begin{aligned}
0 = L([\mathrm{CHCO}]) &= w_{42} - w_{43} - w_{44} + w_{46} + w_{62} + w_{94} \\
0 = L([\mathrm{C_2H_3}]) &= -w_{51} + w_{52} + w_{54} + w_{63} + w_{71} \\
0 = L([\mathrm{C_2H_5}]) &= -w_{56} - w_{58} + w_{59} + w_{61} \\
0 = L([\mathrm{C_2H_6}]) &= w_{36} - w_{59} - w_{61} \\
0 = L([\mathrm{C_3H_2}]) &= w_{92} + w_{93} - w_{94} - w_{95} \qquad (15.2) \\
0 = L([\mathrm{C_3H_3}]) &= w_{48} - w_{62} - w_{63} + w_{91} - w_{92} - w_{93} \\
0 = L([\mathrm{C_3H_6}]) &= -w_{71} + w_{74} \\
0 = L([i\text{-}\mathrm{C_3H_7}]) &= -w_{75} + w_{78} + w_{80} + w_{82} \\
0 = L([n\text{-}\mathrm{C_3H_7}]) &= -w_{73} - w_{74} + w_{77} + w_{79} + w_{81}
\end{aligned}$$

The systematic reduction of complex kinetic mechanisms can be achieved by the use of the above steady state assumptions for intermediate species which leads to their elimination from the parent mechanism. The concentrations of the eliminated species are then estimated with appropriate assumptions in terms of the known species in the reduced scheme. Based on the computational results obtained with the detailed mechanism, the validity of the steady state assumptions for the intermediate species can be examined. This can be done by either calculating the convective-diffusive operator terms, which are represented by $L([X_i])$ in the above balance equations or, for species with similar transport properties, by simply comparing the rates of formation and consumption for each species.

The steady state assumptions are valid if the reactions by which intermediate species are formed are slower than those by which they are consumed. It has been found that most intermediate species can be assumed to be in steady state except C_3H_6, C_2H_6, C_2H_4, C_2H_2 and CH_4. However, the accuracy of these assumptions depends to some extent on combustion conditions and position in the flame. The approximations for some species (e.g. CH_2O and CHCO) will break down on the very rich side of the flame. The predicted concentrations of C_3H_6 and C_2H_6 are smaller than those of the other major intermediate species, so these two species are also eliminated by steady assumptions. On the other hand, although the steady state assumption for O and OH radicals are well justified, both species are retained in the reduced mechanism in order to reduce the stiffness of the kinetic equations. Moreover, as suggested in previous reduced mechanism for hydrocarbon flames, the H radical and the stable species O_2, CO, H_2, H_2O and CO_2 must be retained. As a result 12 non-steady state species remain in the reduced mechanism following the elimination of the 16 steady state species. Taking into account the atom-conservation conditions (for C, H and O) a total of nine global steps can be formulated to describe the chemistry.

The steady state approximations can be used to eliminate 16 of the remaining 56 reactions. The procedure of elimination is arbitrary. However, the resulting stiffness of the reduced reactions mechanism depends to some extent on how these approximations made. In the present study the steady state species are therefore used to eliminate the fastest consumption path of the corresponding species. By setting the balance equations (15.2) for the steady state species equal to zero, the following reactions 6, 24, 25, 31, 35, 42, 44, 51, 58, 59, 71, 77, 78, 90, 93, 94 are eliminated by the species HO_2, CHO, CH_2, CH_2O, CH_3, C_2H, CHCO, C_2H_3, C_2H_5, C_2H_6, C_3H_6, n-C_3H_7, i-C_3H_7, CH, C_3H_3 and C_3H_2 respectively. Following this procedure a system of balance equations for the remaining non-steady state species H, OH, O, H_2, H_2O, O_2, CO, CO_2, CH_4, C_2H_2, C_2H_4 and C_3H_8 can readily be formulated. For example the balance equation for the hydrogen radical is,

$$
\begin{aligned}
&L([\mathrm{H}]) + \{-L([\mathrm{HO_2}]) + L([\mathrm{CHO}]) + L([\mathrm{CH_2O}]) + 2\times L([\mathrm{CH_3}]) \\
&\qquad + L([\mathrm{CHCO}]) + L([\mathrm{C_2H}]) + L([\mathrm{C_2H_3}]) + L([\mathrm{C_2H_5}]) \\
&\qquad + 2\times L([\mathrm{CH}]) + L([\mathrm{CH_2}]) + 2\times L([\mathrm{C_3H_2}]) + L([\mathrm{C_3H_3}]) \\
&\qquad + 3\times L([\mathrm{C_3H_6}]) + 2\times L([n\text{-}\mathrm{C_3H_7}]) + 2\times L([i\text{-}\mathrm{C_3H_7}])\} \\
&= -w_1 + w_2 + w_3 - 2\times w_5 + w_8 + w_{18} - w_{19} - w_{20} - w_{29} - 2\times w_{33} \\
&\quad - 3\times w_{34} - 4\times w_{36} + w_{38} + 2\times w_{40} - w_{43} + w_{45} + 2\times w_{46} + w_{47} \\
&\quad - w_{48} + w_{54} + 2\times w_{56} + w_{61} + w_{62} + 2\times w_{74} + w_{77} + w_{78} + 2\times w_{79} \\
&\quad + 2\times w_{80} + 2\times w_{81} + 2\times w_{82} + w_{91} + w_{92} - w_{95}
\end{aligned}
$$

and for carbon monoxide,

$$
\begin{aligned}
&L([\mathrm{CO}]) + \{L([\mathrm{CHO}]) + L([\mathrm{CH_2O}]) + 2\times L([\mathrm{CHCO}]) + L([\mathrm{CH_3}]) \\
&\qquad + L([\mathrm{CH}]) + L([\mathrm{CH_2}]) + 2\times L([\mathrm{C_2H}]) + 3\times L([\mathrm{C_3H_2}]) \\
&\qquad + 3\times L([\mathrm{C_3H_3}]) + L([\mathrm{C_3H_6}]) + L([n\text{-}\mathrm{C_3H_7}]) \\
&\qquad + L([i\text{-}\mathrm{C_3H_7}])\} \\
&= -w_{18} + w_{20} - w_{34} - 2\times w_{36} + w_{38} + w_{40} - 2\times w_{41} + 2\times w_{45} \\
&\quad + 2\times w_{46} + 2\times w_{47} + 2\times w_{48} + 2\times w_{56} - 2\times w_{63} + w_{77} + w_{78} \\
&\quad + w_{79} + w_{80} + w_{81} + w_{82} + 2\times w_{91} - 2\times w_{95}
\end{aligned}
$$

and for propane,

$$
L([\mathrm{C_3H_8}]) = -w_{77} - w_{78} - w_{79} - w_{80} - w_{81} - w_{82} \tag{15.3}
$$

The convective-diffusive operators in the curl brackets correspond to those of the steady state species and are neglected. It can readily be shown that the above balance equations are consistent with the following nine step global mechanism.

(I)	$C_3H_8 + O + OH = C_2H_4 + CO + H_2O + H_2 + H$
(II)	$CH_4 + O + OH = CO + H_2O + H_2 + H$
(III)	$C_2H_4 + OH = C_2H_2 + H_2O + H$
(IV)	$C_2H_2 + O_2 = 2CO + H_2$
(V)	$CO + OH = CO_2 + H$
(VI)	$2H + M = H_2 + M$
(VII)	$O_2 + H = OH + O$
(VIII)	$H_2 + OH = H_2O + H$
(IX)	$H_2 + O = OH + H$

The corresponing global reactions rates are,

$$
\begin{aligned}
w_I &= w_{77} + w_{78} + w_{79} + w_{80} + w_{81} + w_{82} \\
w_{II} &= -w_{34} + w_{38} + w_{40} \\
w_{III} &= -w_{36} + w_{52} + w_{54} + w_{55} + w_{56} + w_{74} \\
w_{IV} &= -w_{36} - w_{41} + w_{45} + w_{46} + w_{47} + w_{48} \\
&\quad + w_{56} - w_{63} + w_{91} - w_{95} \\
w_V &= w_{18} - w_{20} \\
w_{VI} &= w_5 + w_{34} + w_{48} + w_{56} - w_{74} \\
w_{VII} &= w_1 + w_5 - w_8 + w_{19} + w_{36} - w_{45} - w_{46} - w_{56} \\
w_{VIII} &= w_3 + w_4 + w_8 - w_{29} - w_{30} - w_{33} - w_{36} \\
&\quad + w_{40} + w_{47} - w_{52} - w_{55} + w_{56} + w_{61} + w_{62} \\
&\quad - w_{74} + w_{81} + w_{82} + w_{92} \\
w_{IX} &= w_2 - w_4 + w_5 - w_8 + w_{30} - w_{33} - w_{36} - w_{43} \\
&\quad + w_{46} + w_{48} + w_{56} + w_{79} + w_{80} + w_{91} - w_{95}
\end{aligned}
\tag{15.4}
$$

The truncated steady state assumptions for all the species e.g. (15.2) can in principle be used to derive algebraic relations for the concentrations of these species as required by the reaction rate expressions in (15.4). However, the relative importance of the reactions changes under different combustion conditions and it is frequently better to consider a larger number of the reactions present in the original detailed mechanism. For the propyl radicals, the truncated balance expressions in (15.2) are used to determine their concentrations,

$$
[n\text{-}C_3H_7] = \frac{[C_3H_8] \times \{k_{77f}\,[H] + k_{79f}\,[O] + k_{81f}\,[OH]\} + k_{73b}\,[C_2H_4][CH_3]}{k_{77b}\,[H_2] + k_{79b}\,[OH] + k_{81b}\,[H_2O] + k_{73f}}
\tag{15.5}
$$

$$
[i\text{-}C_3H_7] = \frac{[C_3H_8] \times \{k_{78f}\,[H] + k_{80f}\,[O] + k_{82f}\,[OH]\}}{k_{75f} + k_{76f}\,[O_2] + k_{78b}\,[H_2] + k_{80b}\,[OH] + k_{82b}\,[H_2O]}
\tag{15.6}
$$

The steady state assumption for propene is not satisfactory in estimating propene concentration on the rich side of the propane flames. However, near the flame front the concentration can be approximated by following expression,

$$
[C_3H_6] = \frac{k_{71b}\,[C_2H_3][CH_3] + k_{74f}\,[n\text{-}C_3H_7] + k_{76f}\,[i\text{-}C_3H_7][O_2]}{k_{71f} + k_{74b}\,[H] + k_{76b}\,[HO_2]}
\tag{15.7}
$$

The steady state balance for the propargyl radical was approximated as,

$$
[C_3H_3] = \frac{k_{48f}\,[C_2H_2][CH] + k_{91f}\,[C_2H_2][CH_2] + W_1}{k_{48b} + k_{91b}\,[H] + k_{92f}\,[OH][1 - C_{92}] + k_{93f}\,[H][1 - C_{93}] + W_2}
\tag{15.8}
$$

where

$$
W_1 = k_{62b}\,[CHCO][CH_2O] + k_{63b}\,[C_2H_3][CO]
$$

$$W_2 = k_{62f}\,[O_2] + k_{63f}\,[O]$$

the factors C_{92} and C_{93} correspond to the fractions of C_3H_2 regenerating C_3H_3 via the reverse of reactions (92) and (93). They are written as,

$$C_{92} = \frac{k_{92b}\,[H_2O]}{k_{92b}\,[H_2O] + k_{93b}\,[H_2] + k_{94f}\,[O_2] + k_{95f}\,[OH]} \tag{15.9}$$

and

$$C_{93} = \frac{k_{93b}\,[H_2]}{k_{92b}\,[H_2O] + k_{93b}\,[H_2] + k_{94f}\,[O_2] + k_{95f}\,[OH]} \tag{15.10}$$

The final C_3 species is C_3H_2 which is approximated as,

$$[C_3H_2] = \frac{k_{92f}\,[C_3H_3][OH] + k_{93f}\,[C_3H_3][H] + k_{95b}\,[C_2H_2][CHO]}{k_{92b}\,[H_2O] + k_{93b}\,[H_2] + k_{94f}\,[O_2] + k_{95f}\,[OH]} \tag{15.11}$$

The C_2 species starts from ethane which is approximated as,

$$[C_2H_6] = \frac{k_{36f}\,[CH_3]^2 + k_{59b}\,[C_2H_5][H_2] + k_{61b}\,[C_2H_5][H_2O]}{k_{36b} + k_{59f}\,[H] + k_{61f}\,[OH]} \tag{15.12}$$

and the ethyl radical as,

$$[C_2H_5] = \frac{k_{56b}\,[CH_3]^2 + k_{58b}\,[C_2H_4][H] + k_{59f}\,[C_2H_6][H] + k_{61f}\,[C_2H_6][OH]}{k_{56f}\,[H] + k_{58f} + k_{59b}\,[H_2] + k_{61b}\,[H_2O]} \tag{15.13}$$

The vinyl radical,

$$[C_2H_3] = \frac{k_{49b}\,[C_2H_2][H_2] + k_{52f}\,[C_2H_4][H] + k_{54f}\,[C_2H_4][OH] + W_3}{k_{49f}\,[H] + k_{50f}\,[O_2] + k_{51f} + W_4} \tag{15.14}$$

where

$$W_3 = k_{51b}\,[C_2H_2][H] + k_{71f}\,[C_3H_6] + k_{63f}\,[C_3H_3][O] + k_{50b}\,[C_2H_2][HO_2]$$

$$W_4 = k_{52b}\,[H_2] + k_{54b}\,[H_2O] + k_{63b}\,[CO] + k_{71b}\,[CH_3]$$

The steady state balance applied for ethynyl was modified by a correction factor,

$$C_{41} = \frac{k_{41b}[H]}{k_{41b}[H] + k_{45f}\,[O] + k_{46f}\,[O] + k_{47f}\,[OH]} \tag{15.15}$$

required to reduce the influence of reaction 41 under very fuel rich conditions. The modified expression may be written as,

$$[C_2H] = \frac{k_{41b}\,[C_2H_2][H] \times \{1 - C_{41}\} + k_{42b}\,[CHCO][O] + k_{47f}\,[C_2H_2][OH]}{k_{41f}\,[H_2] + k_{42f}\,[O_2] + k_{47b}\,[H_2O]} \tag{15.16}$$

The final C_2 species is CHCO which is approximated as,

$$[CHCO] = \frac{k_{42f}\,[C_2H][O_2] + k_{43b}\,[CH_2][CO] + k_{46f}\,[C_2H_2][O] + W_5}{k_{42b}\,[O] + k_{43f}\,[H] + k_{44f}\,[O] + k_{46b}\,[H] + k_{62b}\,[CH_2O]} \tag{15.17}$$

where

$$W_5 = k_{62f}\,[C_3H_3][O_2] + k_{94f}\,[C_3H_2][O_2] + k_{44b}\,[CO]^2[H]$$

Starting with methyl radical, the C_1 species are approximated as,

$$[CH_3] = \frac{k_{73f}\,[n\text{-}C_3H_7] + k_{75f}\,[i\text{-}C_3H_7] + k_{38f}\,[CH_4][H] + W_6}{k_{33f}\,[H] + k_{34f}\,[H] + k_{35f}\,[O] + k_{37f}\,[O_2] + k_{38b}\,[H_2] + W_7} \quad (15.18)$$

where

$$W_6 = k_{33b}\,[CH_2][H_2] + k_{34b}\,[CH_4] + k_{35b}\,[CH_2O][H] + k_{53f}\,[C_2H_4][O] + k_{40f}\,[CH_4][OH] + k_{71f}\,[C_3H_6] + 2\times k_{36b}\,[C_2H_6] + 2\times k_{56f}\,[C_2H_5][H]$$

$$W_7 = k_{40b}\,[H_2O] + k_{71b}\,[C_2H_3] + k_{73b}\,[C_2H_4] + k_{75b}\,[C_2H_4] + 2\times\{k_{36f} + k_{56b}\}[CH_3]$$

The methylene radical as,

$$[CH_2] = \frac{k_{33f}\,[CH_3][H] + k_{43f}\,[CHCO][H] + k_{45f}\,[C_2H_2][O] + W_8}{k_{25f}\,[H] + k_{26f}\,[O] + k_{27f}\,[O_2] + k_{28f}\,[O_2] + k_{91f}\,[C_2H_2] + W_9} \quad (15.19)$$

where

$$W_8 = k_{25b}\,[CH][H_2] + k_{91b}\,[C_3H_3][H]$$

$$W_9 = k_{33b}\,[H_2] + k_{43b}\,[CO] + k_{45b}\,[CO]$$

The formaldehyde,

$$[CH_2O] = \frac{[CHO]\times\{k_{29b}\,[H_2O] + k_{30b}\,[OH] + k_{31b}\,[H_2O]\} + W_{10}}{k_{29f}\,[H] + k_{30f}\,[O] + k_{31f}\,[OH] + k_{32f}\,[M] + W_{11}} \quad (15.20)$$

where

$$W_{10} = k_{35f}\,[CH_3][O] + k_{37f}\,[CH_3][O_2] + k_{62f}\,[C_3H_3][O_2] + k_{90f}\,[CH][H_2O] + k_{32b}\,[CHO][H][M]$$

$$W_{11} = k_{35b}\,[H] + k_{37b}\,[OH] + k_{62b}\,[CHCO] + k_{90b}\,[H]$$

The methenyl and formyl radicals are approximated as,

$$[CH] = \frac{k_{25f}\,[CH_2][H] + k_{20b}\,[CHO][CO] + k_{19b}\,[CHO][O] + W_{12}}{k_{19f}\,[O_2] + k_{20f}\,[CO_2] + k_{25b}\,[H_2] + k_{48f}\,[C_2H_2] + W_{13}} \quad (15.21)$$

where

$$W_{12} = k_{48b}\,[C_3H_3] + k_{88b}\,[CO][H] + k_{89b}\,[CHO][H] + k_{90b}\,[CH_2O][H]$$

$$W_{13} = k_{88f}\,[O] + k_{89f}\,[OH] + k_{90f}\,[H_2O]$$

and

$$[\mathrm{CHO}] = \frac{\mathrm{W}_{14} + \mathrm{W}_{15} + k_{95f}\,[\mathrm{C_3H_2}][\mathrm{OH}] + k_{24b}\,[\mathrm{CO}][\mathrm{H}][\mathrm{M}]}{k_{21f}\,[\mathrm{H}] + k_{22f}\,[\mathrm{OH}] + k_{23f}\,[\mathrm{O_2}] + k_{24f}\,[\mathrm{M}] + k_{89b}\,[\mathrm{H}] + k_{95b}\,[\mathrm{C_2H_2}]} \tag{15.22}$$

where

$$\mathrm{W}_{14} = [\mathrm{CH}] \times \{k_{19f}\,[\mathrm{O_2}] + k_{20f}\,[\mathrm{CO_2}] + k_{89f}\,[\mathrm{OH}]\}$$

$$\mathrm{W}_{15} = [\mathrm{CH_2O}] \times \{k_{29f}\,[\mathrm{H}] + k_{30f}\,[\mathrm{OH}] + k_{31f}\,[\mathrm{OH}] + k_{32f}\,[\mathrm{M}]\}$$

The approximation used for hydrogen-peroxide was,

$$[\mathrm{H_2O_2}] = \frac{k_{12f}\,[\mathrm{OH}]^2[M] + k_{11f}\,[\mathrm{HO_2}]^2 \times \{1 - \mathrm{C}_{11}\} + \mathrm{W}_{16}}{k_{13f}\,[\mathrm{H}] + k_{14f}\,[\mathrm{OH}] + k_{12b}\,[\mathrm{M}] + k_{11b}\,[\mathrm{O_2}]} \tag{15.23}$$

where C_{11} is a correction factor,

$$\mathrm{C}_{11} = \frac{k_{11f}\,[\mathrm{HO_2}]^2}{k_{11f}\,[\mathrm{HO_2}]^2 + k_{12f}\,[\mathrm{OH}]^2[\mathrm{M}] + \{k_{13b}\,[\mathrm{OH}] + k_{14b}\,[\mathrm{HO_2}]\} \times [\mathrm{H_2O}]} \tag{15.24}$$

required to reduce the influence of reaction (11) on the lean side of the flame and

$$\mathrm{W}_{16} = \{k_{13b}\,[\mathrm{OH}] + k_{14b}\,[\mathrm{HO_2}]\} \times [\mathrm{H_2O}]$$

Finally, the HO_2 radical is approximated as,

$$[\mathrm{HO_2}] = \frac{k_{5f}\,[\mathrm{H}][\mathrm{O_2}][\mathrm{M}] + k_{6b}\,[\mathrm{OH}]^2 + k_{7b}\,[\mathrm{H_2}][\mathrm{O_2}] + k_{8b}\,[\mathrm{H_2O}][\mathrm{O_2}] + \mathrm{W}_{17}}{k_{5b}\,[\mathrm{M}] + k_{8f}\,[\mathrm{OH}] + k_{10f}\,[\mathrm{O}] + k_{14b}\,[\mathrm{H_2O}] + \mathrm{W}_{18}} \tag{15.25}$$

where

$$\mathrm{W}_{17} = k_{9b}\,[\mathrm{H_2O}][\mathrm{O}] + k_{14f}\,[\mathrm{H_2O_2}][\mathrm{OH}]$$

$$\mathrm{W}_{18} = [k_{6f} + k_{7f} + k_{9f}]\,[\mathrm{H}] + 2 \times k_{11f}\,[\mathrm{HO_2}]$$

All the steady state approximation derived above are explicit but remain coupled and must be solved iteratively. The above reduced mechanism assumes that as customary the steady state species are neglected in the elemental balance so that the mass fractions of the non-steady state species add up to unity. This assumption therefore implicitly ignores the elemental mass stored in the steady state species.

15.5 A Reduced Seven Step Mechanism

Further reductions of the derived 9-step mechanism can be achieved by the elimination of more species from the reduced mechanism. However, results of computations with the 9-step mechanism show that steady state assumptions for CH_4, C_2H_2 and C_2H_4 are not valid for propane diffusion flames. Therefore it is better to keep these three species in the reduced mechanism to avoid erroneous results and numerical stiffness. Introducing steady state assumptions for O and OH radicals leads to the following 7-step mechanism,

$$\begin{array}{ll}
(I) & C_3H_8 + H_2O + 2H = C_2H_4 + CO + 4H_2 \\
(II) & CH_4 + H_2O + 2H = CO + 4H_2 \\
(III) & C_2H_4 + M = C_2H_2 + H_2 + M \\
(IV) & C_2H_2 + O_2 = 2CO + H_2 \\
(V) & CO + H_2O = CO_2 + H_2 \\
(VI) & 2H + M = H_2 + M \\
(VII) & O_2 + 3H_2 = 2H_2O + 2H
\end{array}$$

with the corresponding reaction rates,

$$\begin{aligned}
w_I &= w_{77} + w_{78} + w_{79} + w_{80} + w_{81} + w_{82} \\
w_{II} &= -w_{34} + w_{38} + w_{40} \\
w_{III} &= -w_{36} + w_{52} + w_{54} + w_{55} + w_{56} + w_{74} \\
w_{IV} &= -w_{36} - w_{41} + w_{45} + w_{46} + w_{47} + w_{48} \\
&\quad + w_{56} - w_{63} + w_{91} - w_{95} \\
w_V &= w_{18} - w_{20} \\
w_{VI} &= w_5 + w_{34} + w_{48} + w_{56} - w_{74} \\
w_{VII} &= w_1 + w_5 - w_8 + w_{19} + w_{36} - w_{45} - w_{46} - w_{56}
\end{aligned} \tag{15.26}$$

The approximations for the steady state species are the same as discussed in the 9-step mechanism with an additional approximation for the O atom,

$$[O] = \frac{k_{1f}\,[O_2][H] + k_{2b}\,[OH][H] + k_{4f}\,[OH]^2 + W_{19}}{k_{1b}\,[OH] + k_{2f}\,[H_2] + k_{4b}\,[H_2O] + W_{20}} \tag{15.27}$$

where

$$\begin{aligned}
W_{19} = {} & k_{19f}\,[CH][CO_2] + k_{30b}\,[CHO][OH] + k_{35b}\,[CH_2O][H] + k_{42f}\,[C_2H][O_2] \\
& + k_{45b}\,[CH_2][CO] + k_{46b}\,[CHCO][H] + k_{63b}\,[C_2H_3][CO]
\end{aligned}$$

$$\begin{aligned}
W_{20} = {} & k_{9b}\,[H_2O] + k_{10f}\,[HO_2] + 2\times k_{17f}\,[O][M] + k_{19b}\,[CHO] + k_{26f}\,[CH_2] \\
& + k_{30f}\,[CH_2O] + k_{35f}\,[CH_3] + k_{42b}\,[CHCO] + k_{44f}\,[CHCO] \\
& + \{k_{45f} + k_{46f}\}\times[C_2H_2] + k_{63f}\,[C_3H_3] + \{k_{79f} + k_{80f}\}\times[C_3H_8]
\end{aligned}$$

and for the hydroxyl radical,

$$[OH] = \frac{k_{1f}\,[O_2][H] + k_{2f}\,[H_2][O] + k_{3b}\,[H_2O][H] + W_{21}}{k_{1b}\,[O] + k_{2b}\,[H] + k_{3f}\,[H_2] + W_{22}} \tag{15.28}$$

where

$$W_{21} = 2\times k_{4b}\,[H_2O][O] + k_{18b}\,[CO_2][H] + k_{8b}\,[H_2O][O_2] + k_{16b}\,[H_2O][M]$$

$$\begin{aligned} W_{22} = {} & 2 \times k_{4f}\,[OH] + 2 \times k_{6b}\,[OH] + k_{8f}\,[HO_2] + 2 \times k_{12f}\,[OH][M] \\ & + k_{13b}\,[H_2O] + k_{14f}\,[H_2O_2] + k_{16f}\,[H][M] + k_{18f}\,[CO] \\ & + k_{22f}\,[CHO] + k_{31f}\,[CH_2O] + k_{40f}\,[CH_4] + k_{47f}\,[C_2H_2] \\ & + k_{54f}\,[C_2H_4] + \{k_{81f} + k_{82f}\} \times [C_3H_8] + k_{84f}\,[CH] \\ & + k_{92f}\,[C_3H_3] + k_{95f}\,[C_3H_2] \end{aligned}$$

The above approximations for the O and OH radicals include most of the reactions which are important for the formation and consumption of these two species in propane diffusion flames. However, reactions involving HO_2 and H_2O_2 are removed from the formation paths for O and OH radicals. This is necessary to prevent errors in the approximations for HO_2 and H_2O_2 imparting on the approximations for O and OH radicals while they are solved iteratively.

15.6 Comparison of Detailed and Reduced Mechanisms

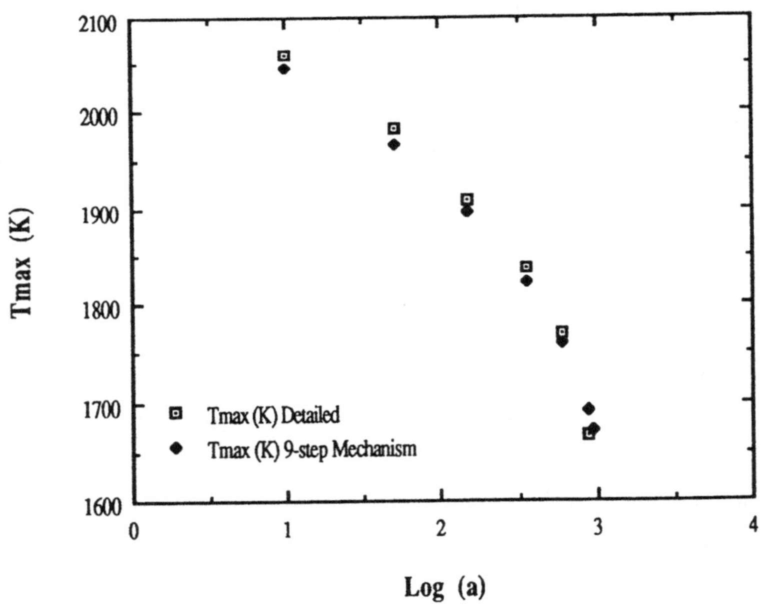

Fig. 15.5. Comparison of maximum flame temperature obtained as a function of rate of strain with detailed and 9-step mechanisms at 1 bar pressure.

The first parameter to be studied with the reduced mechanism was the variation of the peak temperature in the flame as a function of strain rate. Flames at atmospheric pressure were computed with the detailed mechanism at $a = 10, 50, 150, 350, 600, 880$/s with the last value very close to the extinction point. The experimental extinction point measured by Tsuji and Yamaoka

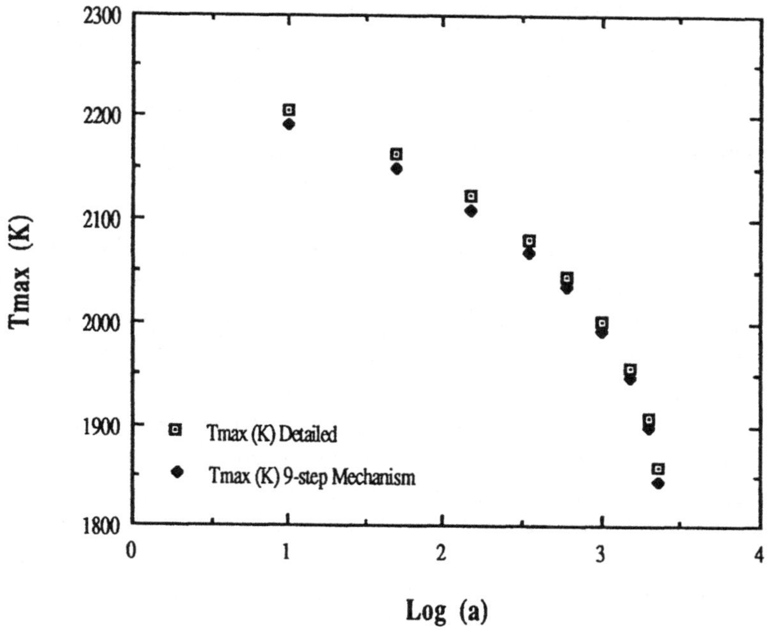

Fig. 15.6. Comparison of maximum flame temperature obtained as a function of rate of strain with detailed and 9-step mechanisms at 10 bar pressure.

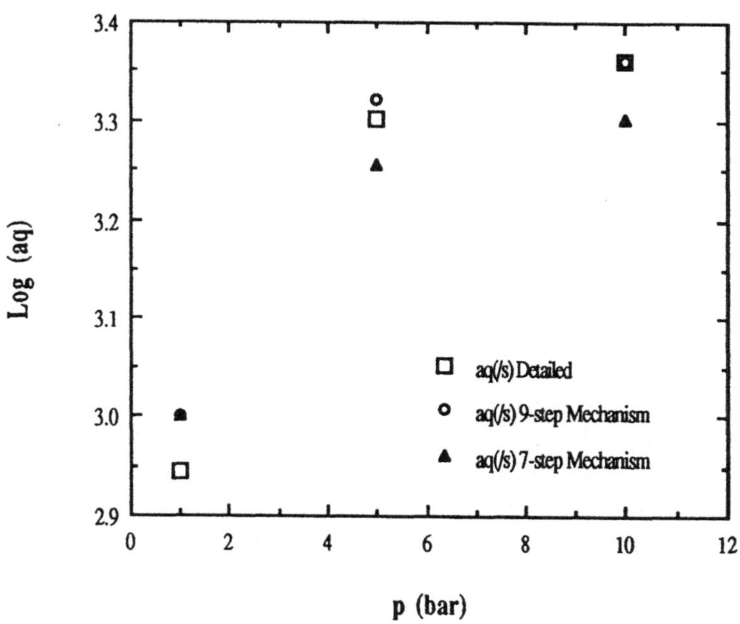

Fig. 15.7. The approximate extinction points as a function of pressure obtained with detailed and reduced chemistry.

[15.6], for a small cylinder, was 675/s which is in acceptable agreement with the present results. For detailed discussion see [15.4]. A comparison of the results obtained with the 9-step mechanism is shown in Fig. 15.5. The agreement is satisfactory with the extinction point for the reduced mechanism at around 1000/s. The same procedure was repeated at pressures of 5 and 10 bar. The results obtained for a pressure of 10 bar can be found in Fig. 15.6. The results are again satisfactory and the agreement between the extinction points is equally satisfactory in the elevated pressure cases as can be seen in Fig. 15.7 where the extinction points have been plotted as a function of pressure. Similar extinction points are predicted by the 7-step and 9-step mechanisms at a pressure of 1 bar while the extinction point is underpredicted by up to 15% with the 7-step mechanism at a pressure of 10 bar.

The reduced mechanisms were validated further by comparing the computed species concentration profiles with those predicted by the detailed mechanism. A comparison of species profiles between the detailed and the reduced mechanisms for a flame with a rate of strain of $a = 150/s$ and 1 bar pressure is shown in Fig. 15.8–15.14. Profiles are plotted against the mixture fraction Z assuming unity Lewis numbers as outlined in Chap. 1. There is very little difference for the stable species C_3H_8, O_2, H_2O, H_2, CO and CO_2 predicted by the 9-step mechanism as shown in Fig. 15.8 and Fig. 15.9. The CO and H_2 profiles are overpredicted with the 9-step mechanism which is due to a larger proportion of the elemental mass fraction stored in these species following the introduction of steady state assumptions for many of the hydrocarbon species eliminated from the original detailed mechanism. Similar results are also obtained for the predictions of major species with the 7-step mechanism.

Regarding the prediction of the key radical concentration, e.g. H, O, OH and CH_3, comparisons are shown in Fig. 15.10 and Fig. 15.11 for the predictions with the 9-step and 7-step mechanisms respectively. Although some degree of overprediction of these radicals can be observed, the agreement is remarkably good with the differences for these radicals less than 15% in the case of the 7-step mechanism, where steady state approximations have been used for O and OH radicals. Similar agreement is also attained at both low and high strain rate conditions at elevated pressures. The good agreement for O, OH and CH_3 predictions shows the validity of the steady state assumption for these species in propane diffusion flames. This is very important for the proper behaviour of reduced mechanisms for non-premixed propane flames as CH_3 is a major pyrolysis product of propane and as it is also a parent species for a number of C_1- and C_2-species e.g. CH_2 and C_2H_6. Acceptable approximations for these species depend to a large extent on the accurate prediction for CH_3.

Comparisons for some of the major intermediate species, CH_4, C_2H_2 and C_2H_4 predicted by the 9-step and 7-step mechanisms can be found in Fig. 15.12 and Fig. 15.13 respectively. The agreement is satisfactory for both reduced mechanisms. This is important as these non-steady state species are the major pyrolysis products in propane diffusion flames. The flame structure of propane diffusion flames is also determined by the relative concentrations of these

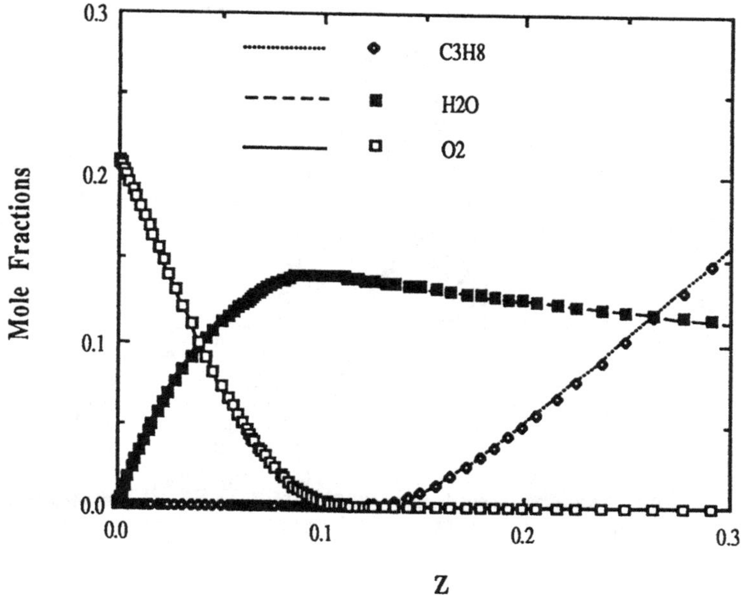

Fig. 15.8. Comparison of predictions of major species obtained with detailed (points) and 9-step (lines) mechanisms for a propane-air diffusion flame with $a = 150/\mathrm{s}$ and 1 bar pressure.

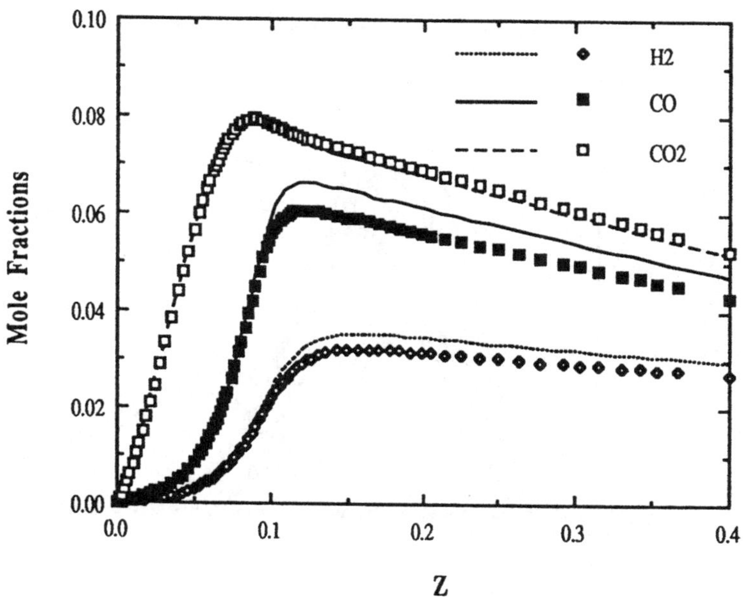

Fig. 15.9. Comparison of predictions of major species obtained with detailed (points) and 9-step (lines) mechanisms for a propane-air diffusion flame with $a = 150/\mathrm{s}$ and 1 bar pressure.

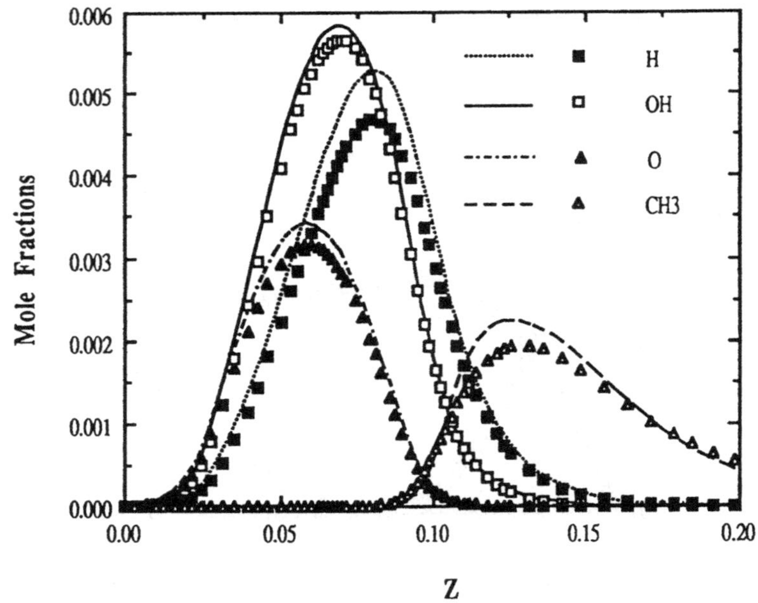

Fig. 15.10. Comparison of predictions of major radical species obtained with detailed (points) and 9-step (lines) mechanisms for a propane-air diffusion flame with $a = 150/s$ and 1 bar pressure.

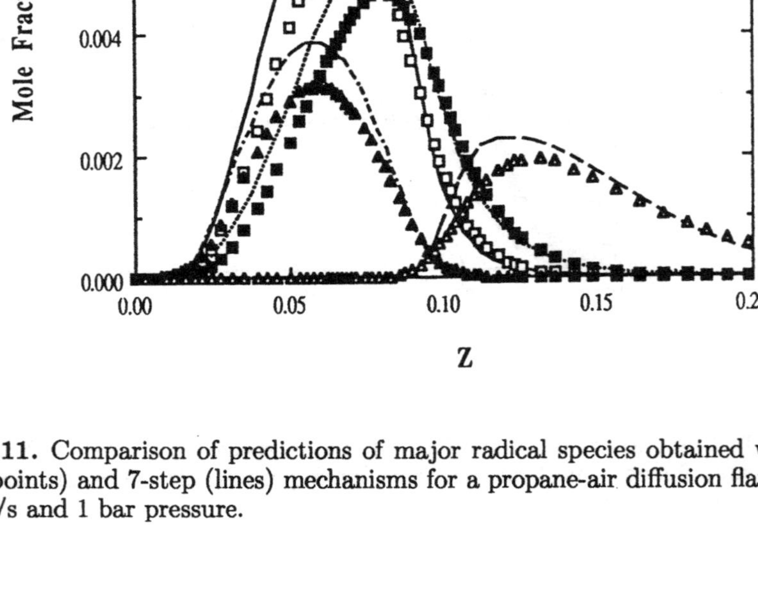

Fig. 15.11. Comparison of predictions of major radical species obtained with detailed (points) and 7-step (lines) mechanisms for a propane-air diffusion flame with $a = 150/s$ and 1 bar pressure.

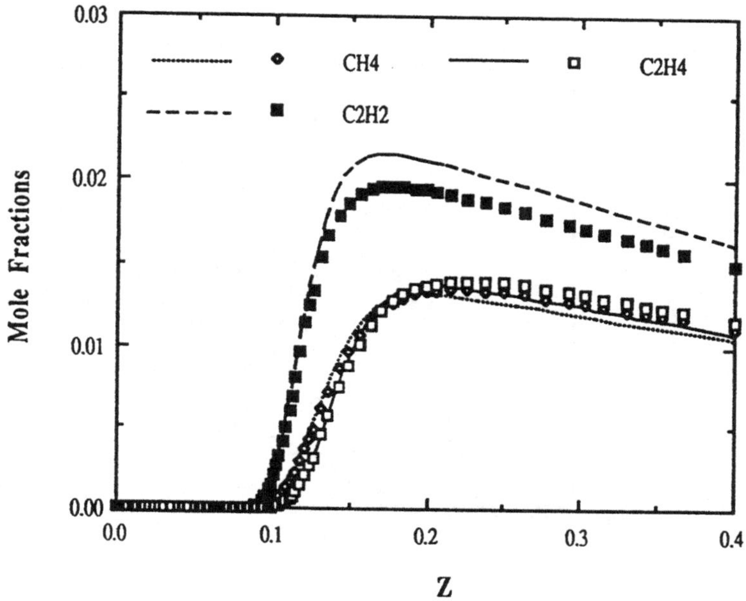

Fig. 15.12. Comparison of predictions of major intermediate species obtained with detailed (points) and 9-step (lines) mechanisms for a propane-air diffusion flame with $a = 150/\mathrm{s}$ and 1 bar pressure.

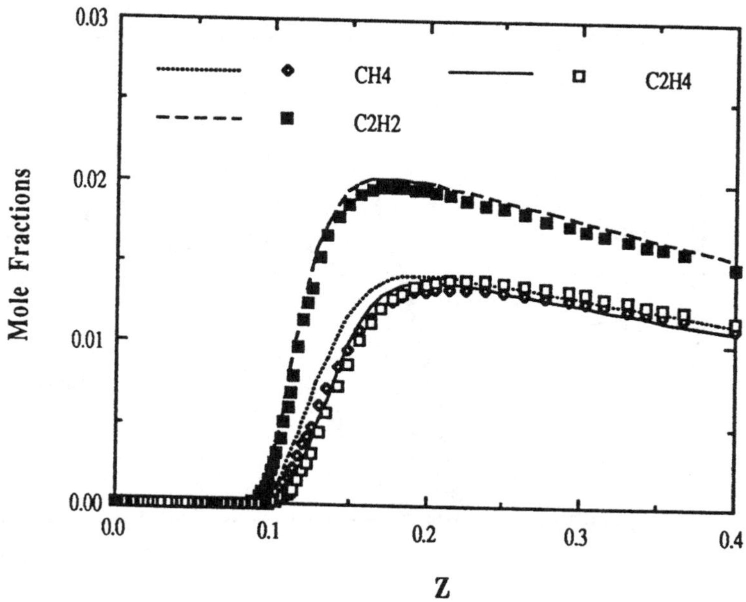

Fig. 15.13. Comparison of predictions of major intermediate species obtained with detailed (points) and 7-step (lines) mechanisms for a propane-air diffusion flame with $a = 150/\mathrm{s}$ and 1 bar pressure.

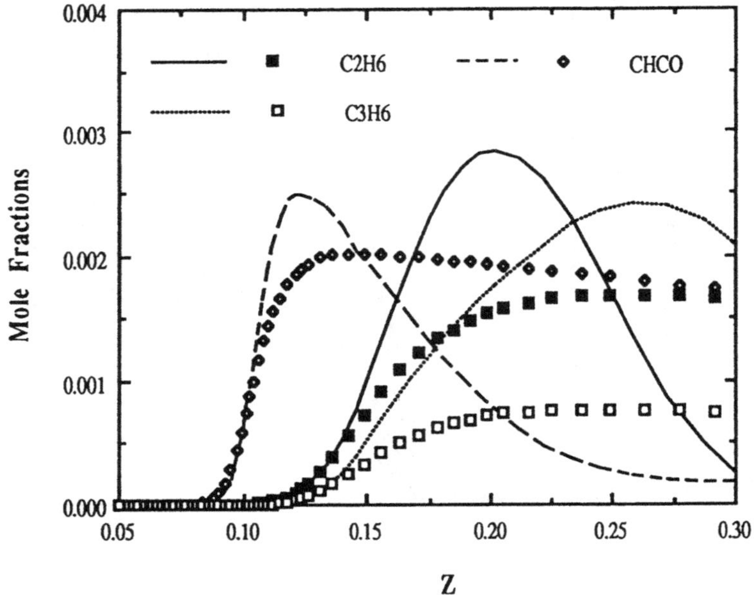

Fig. 15.14. Comparison of predictions of minor intermediate species obtained with detailed (points) and 9-step (lines) mechanisms for a propane-air diffusion flame with $a = 150/\mathrm{s}$ and 1 bar pressure.

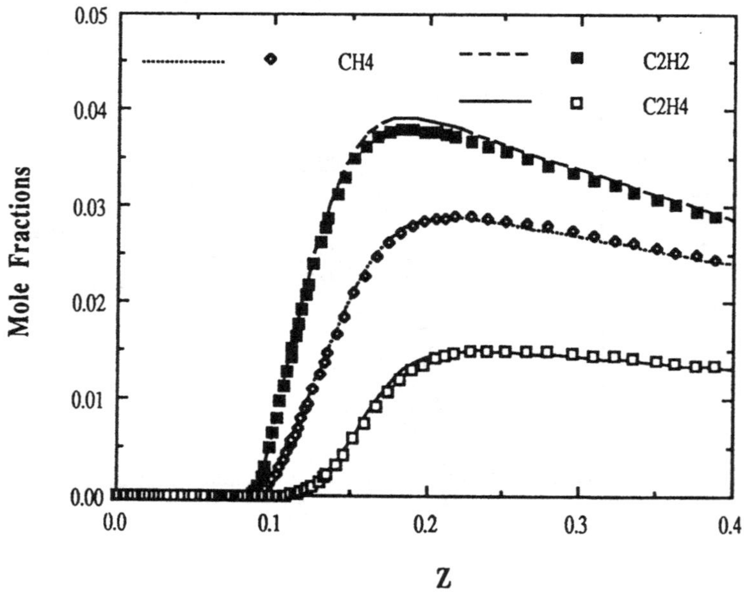

Fig. 15.15. Comparison of predictions of major intermediate species obtained with detailed (points) and 9-step (lines) mechanisms for a propane-air diffusion flame with $a = 10/\mathrm{s}$ and 1 bar pressure.

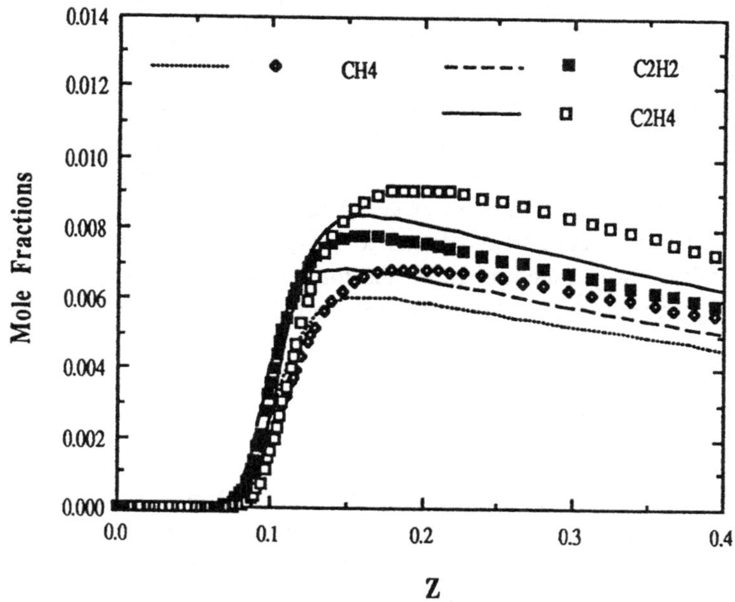

Fig. 15.16. Comparison of predictions of major intermediate species obtained with detailed (points) and 9-step (lines) mechanisms for a propane-air diffusion flame with $a = 2000/\mathrm{s}$ and 10 bar pressure.

species. It has been found that the use of steady state approximations for these species introduce significant errors on the rich side of propane flames which also increases the numerical stiffness and leads to unacceptable changes in the flame structure.

A comparison of the minor intermediate species, CHCO, C_2H_6 and C_3H_6 predicted by the 9-step mechanism can be found in Fig. 15.14. The predicted profiles of these steady state species with the reduced mechanism match very well with those predicted by the detailed mechanism near the flame front. However, significant overpredictions of these species with the reduced mechanism are observed in the rich part of the flame at mixture fractions above 0.15 for C_2H_6 and C_3H_6. For CHCO overpredictions start at a mixture fraction of 0.11. Furthermore the concentration profiles drop too quickly on the rich side of the flame for CHCO and C_2H_6. This is a typical feature for species for which the steady state assumptions break down in this part of the flame. Similar behaviour has been found under other conditions resulting in varying degree of overpredictions. The effect of this problem on the overall quality of predictions is less than might be anticipated as reaction rates are low in this part of the flame.

Comparisons for the major intermediate species can be found for a flame at low rate of strain of $a = 10/\mathrm{s}$ and 1 bar pressure in Fig. 15.15, and a flame at high rate of strain of $a = 2000/\mathrm{s}$ and 10 bar pressure in Fig. 15.16. These correspond to the extreme conditions in the range of computations of the present study. The agreement is very good in the low strain rate case while modest under-predictions for CH_4, C_2H_4 and C_2H_4 can be observed at the high strain rate used in the computation of the elevated pressure case.

Finally, the computational efficiency of the reduced mechanisms has to be addressed. It has been shown that both the 9-step and the 7-step mechanisms can in general reproduce the features of the original detailed mechanism very well. However, a significant increase in numerical stiffness has been found for the 7-step mechanism which leads to longer computational times than those required by 9-step mechanism. Thus for studies where the number of species is not of paramount importance the 9-step mechanism may be a better choice.

15.7 Conclusions

Computations of non-premixed propane flames have been performed for a large range of strain and pressure conditions using a planar counterflow geometry and a detailed mechanism for the combustion of propane. The results of these computations show that with the current mechanism CH_4, C_2H_2 and C_2H_4 are the major intermediate species and that steady state assumptions for these species are not well justified for propane diffusion flames.

Two reduced mechanisms have been formulated by the systematic reduction of the detailed mechanism. The twelve remaining species in a 9-step mechanism are C_3H_8, C_2H_4, C_2H_2, CH_4, H, O, OH, O_2, H_2, H_2O, CO and CO_2. Further elimination of the O and OH radicals leads to a 7-step mechanism. Steady state approximations have been used to calculate the concentrations of the eliminated species from the above non-steady state species. Computations using both reduced mechanisms show that the extinction points predicted by the detailed mechanism can be reproduced very well from 1 to 10 bar pressure. Major species and radical concentrations can also be well reproduced by both reduced mechanisms. However, the steady state approximations for CHCO, C_2H_6, C_3H_6 have been found not to be well justified on the rich side of propane diffusion flames and improved descriptions for these species are highly desirable.

Acknowledgement

The authors wish to thank the MoD at RAE Pyestock for in part supporting this work.

References

[15.1] Warnatz, J., Comb. Sci. and Tech. **34**, 177–200, 1983
[15.2] Westbrook, C.K. and Pitz, W.J., Comb. Sci. and Tech. **37**, 117–152, 1984
[15.3] Paczko, G., Lefdal, P.M., Peters, N., Twenty-First Symposium (International) on Combustion, The Combustion Institute, pp. 739–748, 1986
[15.4] Jones, W.P. and Lindstedt, R.P., Comb. Sci. and Tech. **61** 31–49, 1988
[15.5] Evens, S. and Simmons, R.F., Twenty-Second Symposium (International) on Combustion, The Combustion Institute, pp. 1433–1439, 1988
[15.6] Tsuji, H. and Yamaoka, I., Twelfth Symposium (International) on Combustion, The Combustion Institute, pp. 997–1005, 1969
[15.7] Rao, V.S. and Skinner, G.B., J. Phys. Chem. **93**, 1869–1876, 1989

16. Reduced Kinetic Mechanisms for Counterflow Methanol Diffusion Flames

C. M. Müller
Institut für Technische Mechanik, RWTH Aachen, W-5100 Aachen, Germany

K. Seshadri
Department of Applied Mechanics and Engineering Sciences, University of California, San Diego, La Jolla, California 92093, U.S.A.

J. Y. Chen
Combustion Research Facility, Sandia National Laboratories, Livermore, California 94551–0969, U.S.A.

16.1 Introduction

Studies on the oxidation of methanol in laminar flames are of practical importance, because methanol is considered for use as an alternate fuel for propulsion of internal combustion engines. Numerous experimental and theoretical studies on the combustion of methanol in premixed flames are available and they have been briefly reviewed in Chap. 9. However, such an extensive study is not available for diffusion flames. The present chapter concerns the structure and extinction characteristics of methanol-air diffusion flames.

Numerical calculations have been performed previously using detailed sets of elementary chemical kinetic mechanisms [16.1], [16.2], and suitably reduced chemical kinetic mechanisms [16.2] to determine the structure and critical conditions of extinction of counterflow methanol-air diffusion flames. The results of these numerical calculations show that the principal path for the oxidation of methanol is CH_3OH–CH_2OH–CH_2O–HCO–CO, H_2–CO_2, H_2O [16.1], [16.2], which is similar to that in premixed flames. The results also show that as in hydrocarbon-air flames [16.3], [16.4], [16.5] the structure of methanol-air diffusion flames can be subdivided into two layers, an inner layer where methanol reacts with radicals to form CO and H_2, and an oxidation layer where H_2 and CO oxidize to form H_2O and CO_2. The results of these numerical calculations also show that [16.1], [16.2] steady-state approximations are valid for the species CH_2OH and HCO, but not for CH_2O. A 5-step mechanism was deduced previously [16.2] by introducing steady-state approximations for CH_2OH, HCO, O, OH, HO_2 and H_2O_2. To explore the impact of making a steady-state approximation for CH_2O, this 5-step mechanism was further reduced to four steps by eliminating CH_2O [16.2]. Comparison of calculated results for the stagnation point diffusion flame showed reasonable agreement among the different mechanisms [16.2]. However, the 4-step mechanism was found to yield considerably higher values for the peak concentrations of CH_2O when compared with those predicted by the detailed mechanism or by the 5-step mechanism [16.2]. The results of the numerical calculations reported here confirm these observations.

To test the predictions of the chemical kinetic mechanism shown in Table 1 of Chap. 1, the results of the numerical calculations employing this mechanism are compared with previous experimental measurements [16.6]. The results of these calculations are also used to determine those species for which steady-state approximations are valid. Suitable reduced chemical kinetic mechanisms are then deduced and results of the calculations employing these reduced mechanism are compared with those obtained using the detailed elementary chemical kinetic mechanism.

16.2 Formulation

16.2.1 Description of the Experimental Measurements

Experiments have been performed previously [16.6] to determine the structure of diffusion flames stabilized in the stagnation point boundary layer by directing an oxidizing gas stream vertically downward onto the vaporizing surface of pools of methanol. A schematic illustration of the experimental configuration employed in [16.6] is shown in Fig. 16.1. In these experiments, the oxidizer duct was located 1 cm above the fuel surface [16.6]. A diffusion flame can be indefinitely stabilized in this configuration. Further details of the burner are given elsewhere [16.6]. Critical conditions of extinction of the flame were measured over a wide parameteric range. The velocities of the oxidizer stream at the exit of the duct were recorded as a function of its composition, at flame extinction [16.6]. Concentration profiles of stable species in the flame were measured by removing gas samples from the flame using a quartz microprobe, and the samples were analyzed using a gas chromatograph [16.6]. Temperature profiles were measured by thermocouples. These measurements were made for $Y_{O_2e} = 0.148$ and $-V_e = 0.377\,\mathrm{m/s}$, where Y_{O_2e} and $-V_e$ are respectively the mass fraction of oxygen and axial velocity of the oxidizer stream at the exit of the duct. All measurements were made at a value of the total pressure $p = 1\,\mathrm{atm}$, and with the temperature of the oxidizer stream at the exit of the duct $T_e = 298\,\mathrm{K}$. Further details of the procedure for making these measurements are outlined in [16.6]. These measurements are compared with the results of the numerical calculations described here.

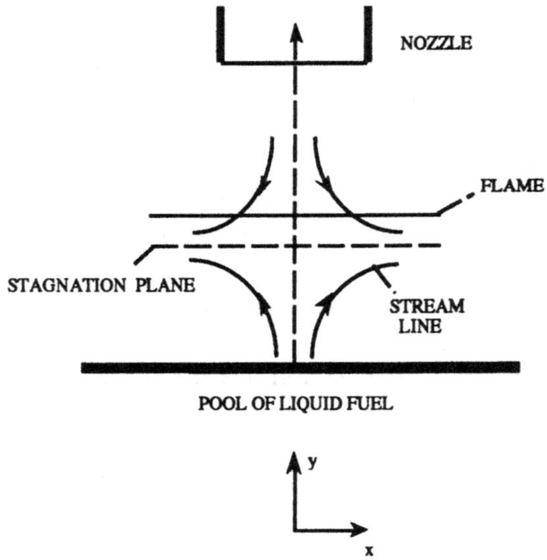

Fig. 16.1. A schematic illustration of the experimental configuration used in the measurements [16.6].

16.2.2 Formulation of the Numerical Problem

The governing equations for the numerical problem are described in detail in Chap. 1, and the notation introduced in Chap. 1 is used here. All calculations were performed in the axisymmetric configuration. Two different sets of boundary conditions were employed to solve these equations. The first set was used to compare the results of the numerical calculations with the experimental measurements, in an attempt to test the predictions of the proposed starting mechanism. The second set of boundary conditions was used to compare the results of the calculations using the starting mechanism with those calculated using the reduced chemical kinetic mechanisms, in an attempt to test the predictions of these reduced mechanisms.

To facilitate comparisons of numerical calculations with experimental measurements, the primary oxidizer stream was presumed to be located at $y = L$ and the surface of the liquid fuel at $y = 0$, where y is the axial coordinate measured from the surface of the liquid pool. The calculations were performed for prescribed values of the composition and velocity of the oxidizer stream at the exit of the duct, and were identical to those used in the experiment. The flow field in the experimental configuration, as shown in Fig. 16.1, consists of an inner viscous flow region near the fuel surface, which extends slightly beyond the stagnation plane and into the oxidizer side and an outer, inviscid and rotational region extending from the oxidizer side of the stagnation plane to the exit of the oxidizer duct [16.7], where the tangential component of the flow velocity vanishes. Hence, in the outer flow the strain rate is a function of the axial coordinate. Introducing as in Chap. 1 the notation $u = G\,x$ where u is the radial component of the flow velocity, x is the radial coordinate, and G

is a function of the axial coordinate, the boundary conditions at the exit of the oxidizer duct located at $y = L$ can be written as

$$\begin{aligned} &\rho v = \rho_e V_e ; \quad G = 0 , \quad Y_{O_2} = Y_{O_2 e} ; \\ &Y_k = 0 , \quad k \neq O_2 , N_2 ; \quad Z = 0 ; \quad T = T_e \end{aligned} \tag{16.1}$$

where ρ denotes the gas density, v the axial component of the flow velocity, T the gas temperature, and Y_i the mass fraction of species i. Subscript e denotes conditions at the exit of the oxidizer duct. At the surface of the liquid pool located at $y = 0$, the value of the tangential component of the flow velocity is presumed to be zero (no slip) and the appropriate interface balance conditions can be written as

$$\begin{aligned} G &= 0 \\ \rho_w v_w Y_{kw} + j_{kw} &= 0 \qquad k \neq F \\ \rho_w v_w (1 - Y_{Fw}) - j_{Fw} &= 0 \\ \rho_w v_w (1 - Z_w) + \left[\frac{\lambda}{c_p} \frac{dZ}{dy} \right]_w &= 0 \\ \left[\lambda \frac{dT}{dy} \right]_w - \rho_w v_w h_L &= 0 \\ T &= T_w \end{aligned} \tag{16.2}$$

where j_i is the diffusive flux of species i as defined in Chap. 1, λ is the thermal conductivity of the gas mixture and h_L is the latent heat of vaporization of the fuel, which is presumed to be known. Subscript F and w refer to the fuel and conditions on the gas side of the liquid-gas interface respectively. For simplicity the surface temperature T_w is presumed to be equal to the boiling point of methanol. The mass burning rate of the liquid fuel $\rho_w v_w$ is an unknown and will be determined as a part of the solution.

For the second set of calculations, the external flow field is presumed to be irrotational [16.7]. The fuel stream is presumed to flow from $y = -\infty$ toward the stagnation plane located at $y = 0$, and the oxidizer stream flows toward the stagnation plane from $y = \infty$. Computations were performed for fixed values of the strain rate a, which is defined as the derivative of the component of the flow velocity parallel to the flame sheet with respect to the coordinate parallel to the flame sheet, and evaluated in the external oxidizer stream. Boundary conditions applied in the fuel stream are

$$\begin{aligned} &G = a\sqrt{\rho_\infty / \rho_{-\infty}} ; \\ &Y_F = Y_{F-\infty} = 1.0 ; \qquad Y_i = 0 \quad i \neq F \\ &Z = 1 ; \qquad T = T_{-\infty} \end{aligned} \tag{16.3}$$

where Z is the mixture fraction defined in Chap. 1 and $u_\infty = ax$ and a is the characteristic strain rate. Subscripts $-\infty$ and ∞ refer to conditions in the ambient fuel stream and the ambient oxidizer stream respectively. In the calculations the value of $T_{-\infty}$ was set equal to the boiling point of methanol at the prevailing value of p [16.8]. Boundary conditions applied in the oxidizer stream are

$$\begin{aligned} &G = a\,; && u/u_\infty = 1.0\,; \qquad T = T_\infty \\ &Y_{O_2} = Y_{O_2\infty}\,; && Y_{N_2} = Y_{N_2\infty}\,; \\ &Y_i = 0 \quad i \neq O_2\,, N_2\,; && Z = 0 \end{aligned} \tag{16.4}$$

In addition to (16.3) and (16.4) the condition $v = 0$ at $y = 0$ is also applied.

Results of those numerical calculations, which were performed for fixed values of a are plotted using the mixture fraction Z as the independent variable. The quantity Z was calculated using the differential equation shown in Chap. 1. The transport model [16.9] and the numerical procedure [16.10] are similar to those described in the literature.

16.2.3 Chemical Kinetic Mechanism

The chemical kinetic mechanism employed in the numerical calculations is essentially the C_1 mechanism shown in Table 1 of Chap. 1 excluding reactions involving CH, CH_2 and CH_3. Thus, the mechanism includes reactions 1–18, 21–24, 29–32 and 83–87. This elementary chemical kinetic mechanism will be referred to as the starting mechanism and was used to deduce the reduced mechanisms.

16.3 Comparison between Numerical Calculations and Experimental Measurements

Numerical calculations were performed using the starting mechanism at conditions identical to those used in the experiment. First the measured and calculated structure of the flame are compared, and later the predicted critical conditions of flame extinction are compared with the measurements. Boundary conditions (16.1) and (16.2) were employed in the calculations with $T_w = 338\,\mathrm{K}$.

In Fig. 16.2 the measured and calculated values of temperature are shown. The profiles in Fig. 16.2 are aligned reasonable well, and the peak values and the shape of the profiles agree fairly well. In Fig. 16.3 the measured and calculated profiles of CH_3OH, O_2, H_2O and CO_2 are compared. At the position where the value of the temperature is a maximum the measured values of CH_3OH and O_2 are higher than the corresponding calculated values. This

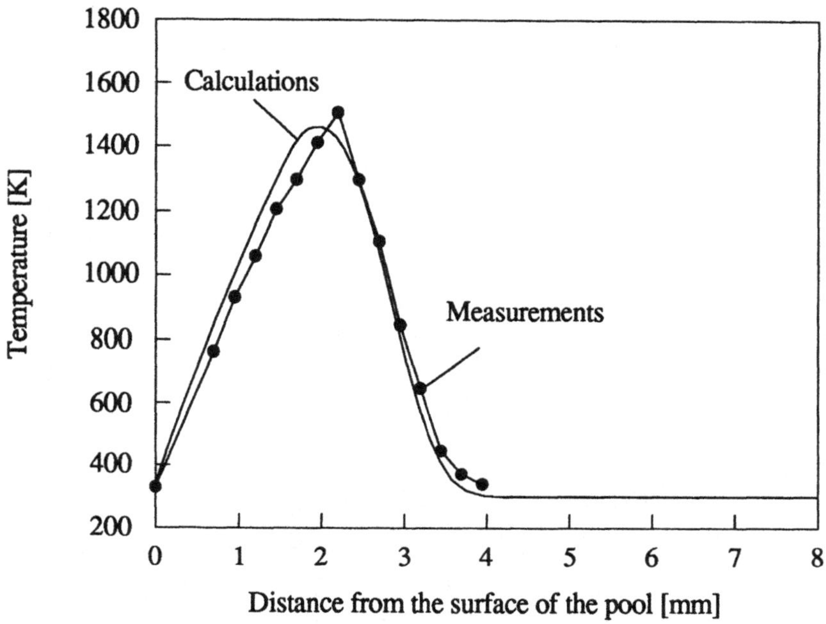

Fig. 16.2. Comparison between the measured (line through points) and calculated temperature profile using the starting mechanism with $Y_{O_2e} = 0.148$, $-V_e = 0.377\,\mathrm{m/s}$, $T_e = 298\,\mathrm{K}$ and $T_W = 338\,\mathrm{K}$.

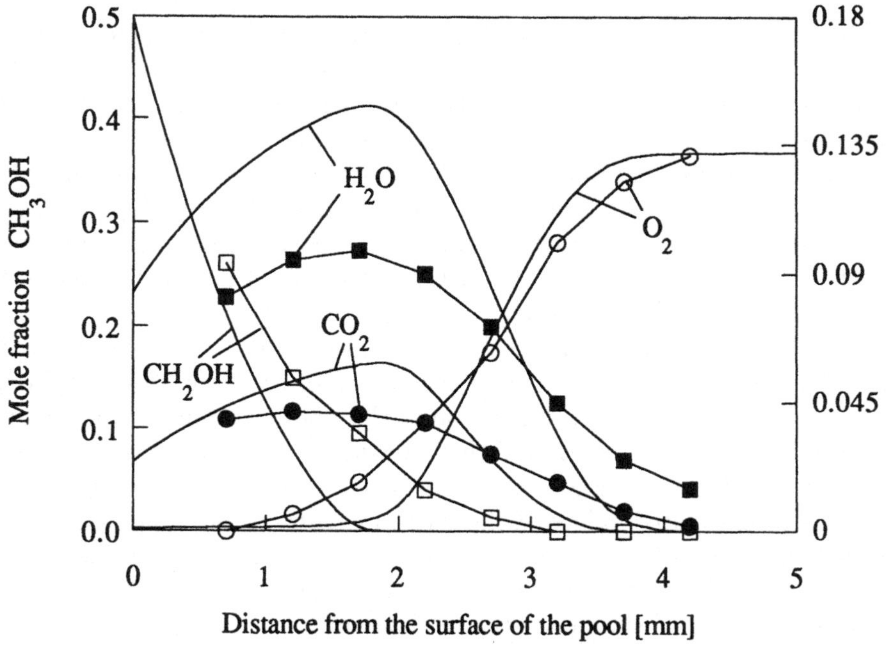

Fig. 16.3. Comparison between the measured (line through points) and calculated mole fractions of CH_3OH, O_2, H_2O and CO_2 using the starting mechanism with $Y_{O_2e} = 0.148$, $-V_e = 0.377\,\mathrm{m/s}$, $T_e = 298\,\mathrm{K}$ and $T_W = 338\,\mathrm{K}$.

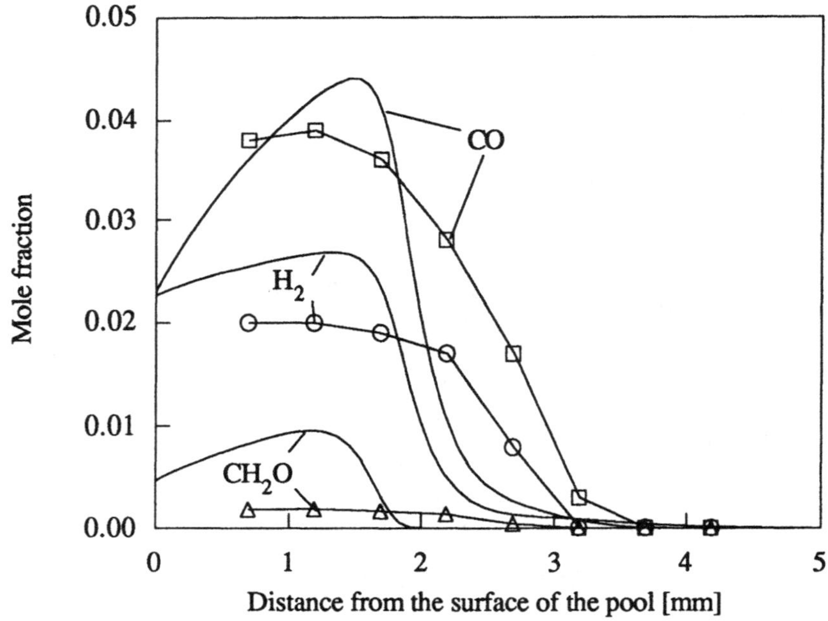

Fig. 16.4. Comparison between the measured (line through points) and calculated mole fractions of CO, H_2 and CH_2O using the starting mechanism with $Y_{O_2e} = 0.148$, $-V_e = 0.377\,\mathrm{m/s}$, $T_e = 298\,\mathrm{K}$ and $T_w = 338\,\mathrm{K}$.

could be attributed to local quenching associated with introducing a microprobe into the flame. Figure 16.3 also shows the measured values of H_2O to be considerably lower than the corresponding calculated values, and is attributed to condensation of water vapor in the sampling lines [16.6]. In Fig. 16.4 the measured and calculated values of the major stable intermediate species H_2, CO and CH_2O are compared. The measured values of H_2 and CH_2O are considerably lower than the calculated values and is attributed to uncertainities associated with measuring these quantities, particularly due to the high diffusivity of H_2 and the high reactivity of CH_2O [16.6].

In Fig. 16.5 the measured and calculated values of the velocity of the oxidizer stream at the exit of the duct as a function of its composition at flame extinction are compared. The line through the points in Fig. 16.5 represent the experimental measurements and the line without the points represents results of the calculations. Figure 16.5 shows fairly good agreement between the measured and calculated values of the critical conditions of flame extinction. Since extinction conditions are very sensitive to the choice of the kinetics, it can be concluded that the starting mechanism is capable of predicting fairly accurately a number of characteristics of the flame.

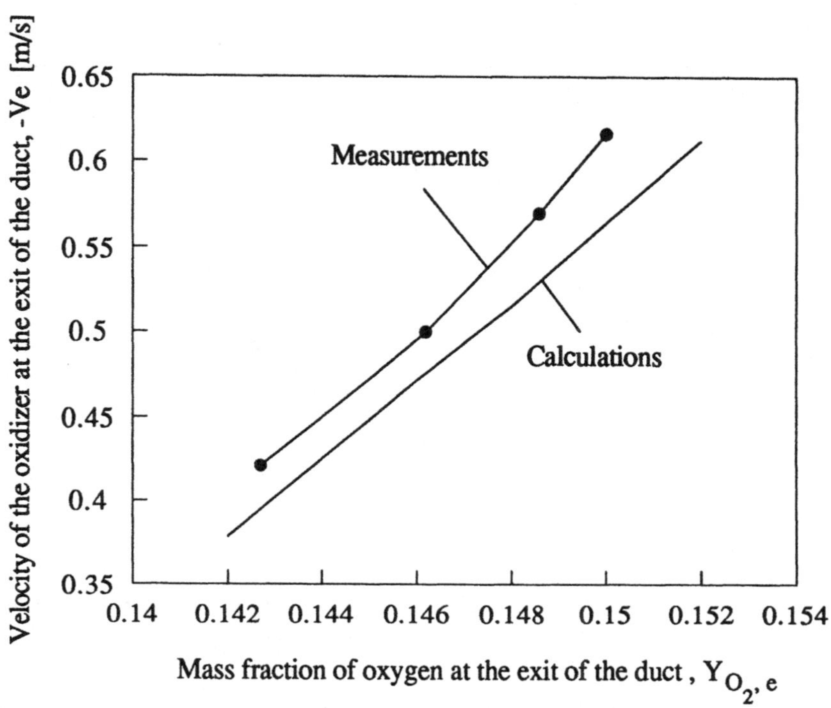

Fig. 16.5. Comparison between the measured (line through points) and calculated values of the velocity of the oxidizer stream as a function of its composition at flame extinction. The calculations were performed with the starting mechanism with $T_e = 298\,\mathrm{K}$ and $T_w = 338\,\mathrm{K}$.

16.4 Reduced Kinetic Mechanism

To deduce a reduced chemical kinetic mechanism, which is capable of describing fairly accurately the structure and critical conditions of extinction of the flame, it is necessary to identify those species for which steady-state approximations are valid. In Figs. 16.6–16.9 the terms accounting for convection, diffusion, chemical production and consumption are shown as a function of the mixture fraction Z for the species CH_2OH, CH_2O, H and OH. The calculations were performed using the starting mechanism together with the boundary conditions (16.3) and (16.4), for $p = 1\,\mathrm{bar}$, $a = 100\,\mathrm{s}^{-1}$, $1-Y_{N_2\infty} = Y_{O_2\infty} = 0.232$, $T_\infty = 298\,\mathrm{K}$ and $T_{-\infty} = 338\,\mathrm{K}$. Figure 16.9 shows that for the radical OH, the convective and diffusive terms are negligibly small, and the chemical production rates are nearly equal to the consumption rates everywhere. Hence, steady-state approximation is valid for OH to a high degree of accuracy. Similarly, Figs. 16.6 and 16.8 show that it is reasonable to introduce steady-state approximations for CH_2OH and H. However, it is evident from Fig. 16.7 that a steady-state approximation is not valid for CH_2O. In Figs. 16.10–16.12 data similar to Figs. 16.6–16.9 are shown for $a = 700\,\mathrm{s}^{-1}$, again with $p = 1\,\mathrm{bar}$, $1 - Y_{N_2\infty} = Y_{O_2\infty} = 0.232$, $T_\infty = 298\,\mathrm{K}$ and $T_{-\infty} = 338\,\mathrm{K}$, for the species CH_2OH, H and OH. Figures 16.10–16.12 show that steady-state approximations are again reasonably accurate for CH_2OH and OH, but not for H. Plots

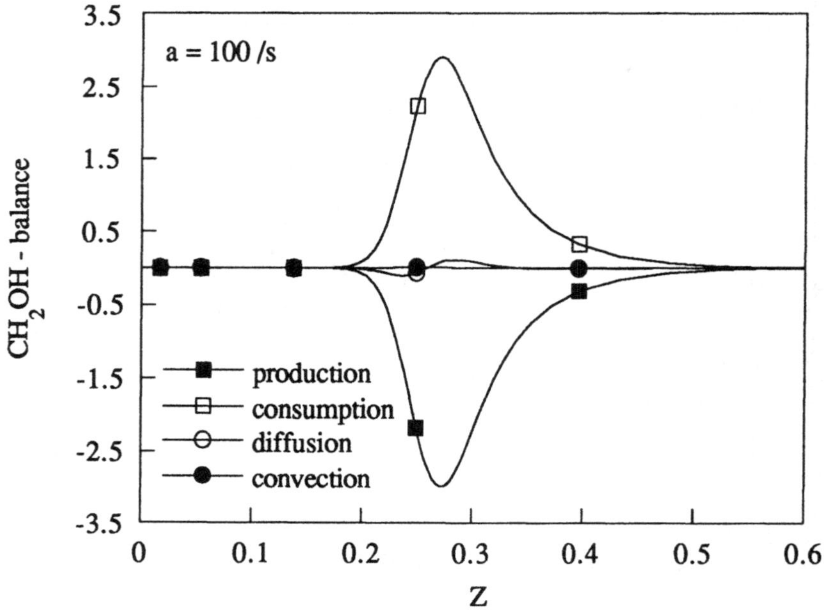

Fig. 16.6. Balance of convection, diffusion, production and consumption of CH_2OH, calculated using the starting mechanism with $a = 100\,s^{-1}$, $p = 1\,bar$, $T_{-\infty} = 338\,K$, $Y_{O_2\infty} = 0.232$ and $T_\infty = 298\,K$.

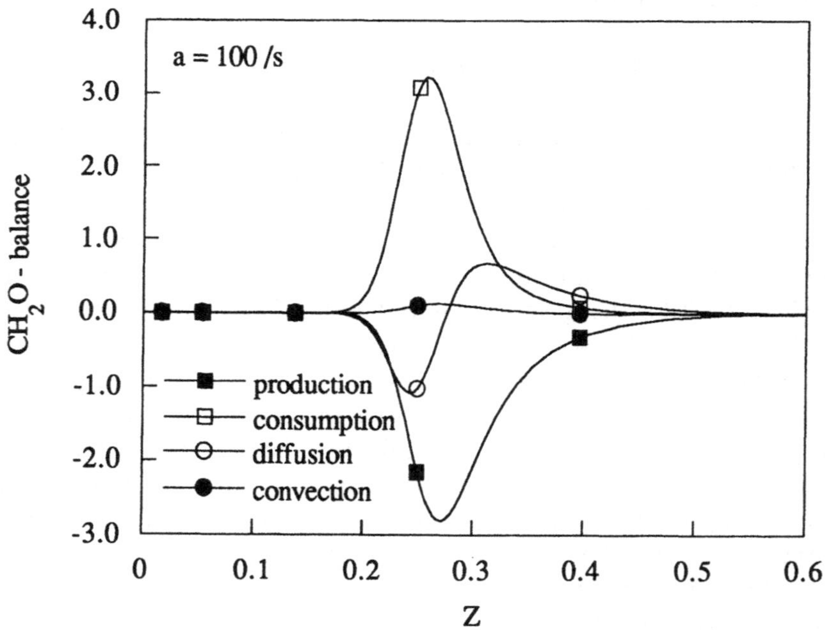

Fig. 16.7. Balance of convection, diffusion, prodution and consumption of CH_2O, calculated using the starting mechanism with $a = 100\,s^{-1}$, $p = 1\,bar$, $T_{-\infty} = 338\,K$, $Y_{O_2\infty} = 0.232$ and $T_\infty = 298\,K$.

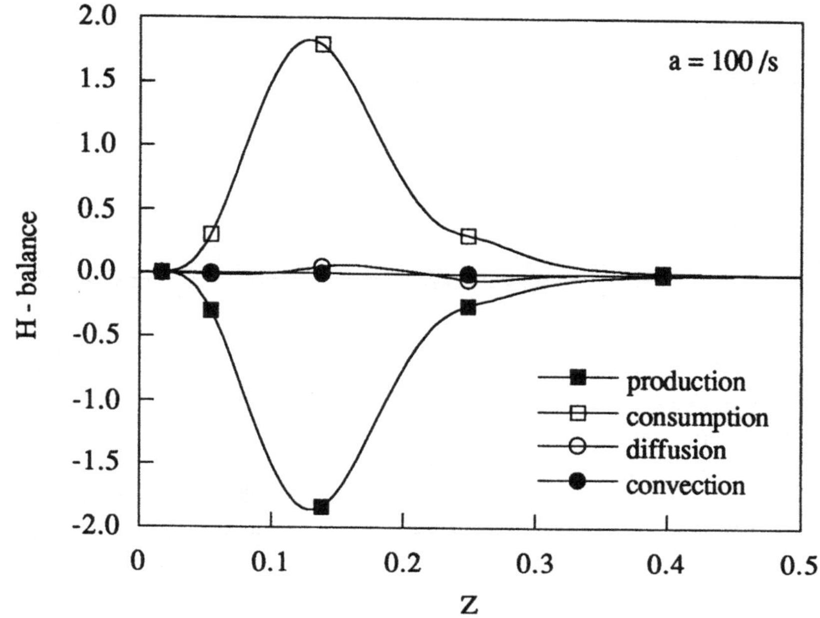

Fig. 16.8. Balance of convection, diffusion, production and consumption of H, calculated using the starting mechanism with $a = 100\,\mathrm{s}^{-1}$, $p = 1\,\mathrm{bar}$, $T_{-\infty} = 338\,\mathrm{K}$, $Y_{O_2\infty} = 0.232$, and $T_\infty = 298\,\mathrm{K}$.

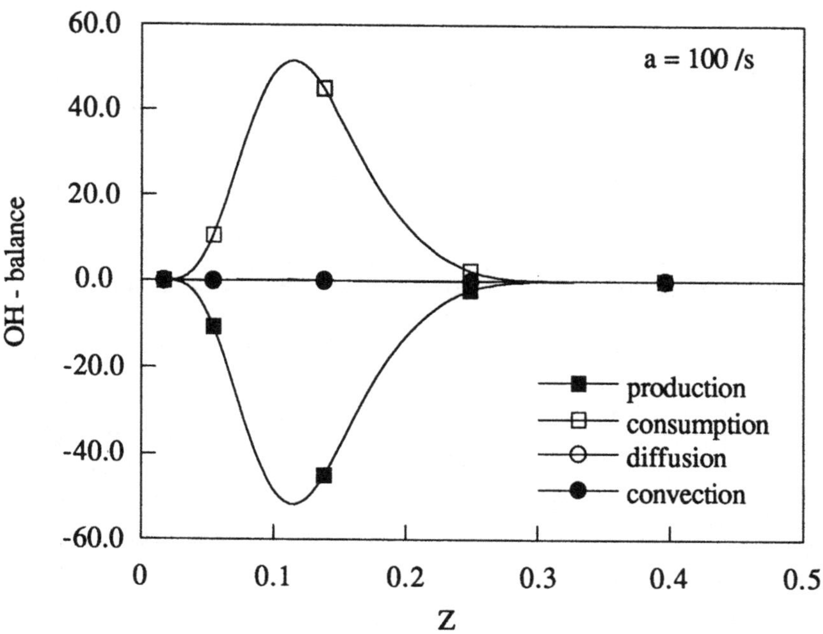

Fig. 16.9. Balance of convection, diffusion, production and consumption of OH, calculated using the starting mechanism with $a = 100\,\mathrm{s}^{-1}$, $p = 1\,\mathrm{bar}$, $T_{-\infty} = 338\,\mathrm{K}$, $Y_{O_2\infty} = 0.232$, and $T_\infty = 298\,\mathrm{K}$.

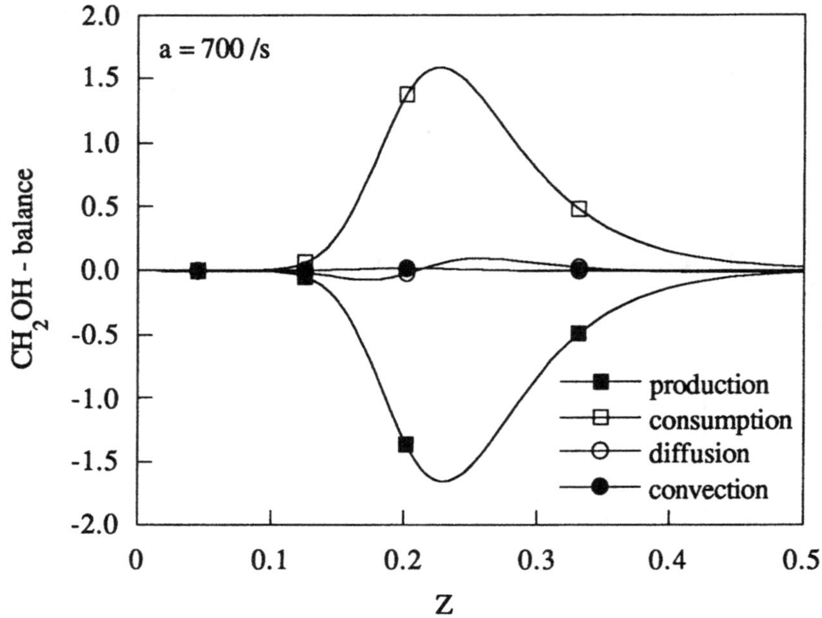

Fig. 16.10. Balance of convection, diffusion, production and consumption of CH_2OH, calculated using the starting mechanism with $a = 700\,s^{-1}$, $p = 1$ bar, $T_{-\infty} = 338$ K, $Y_{O_2\infty} = 0.232$, and $T_\infty = 298$ K.

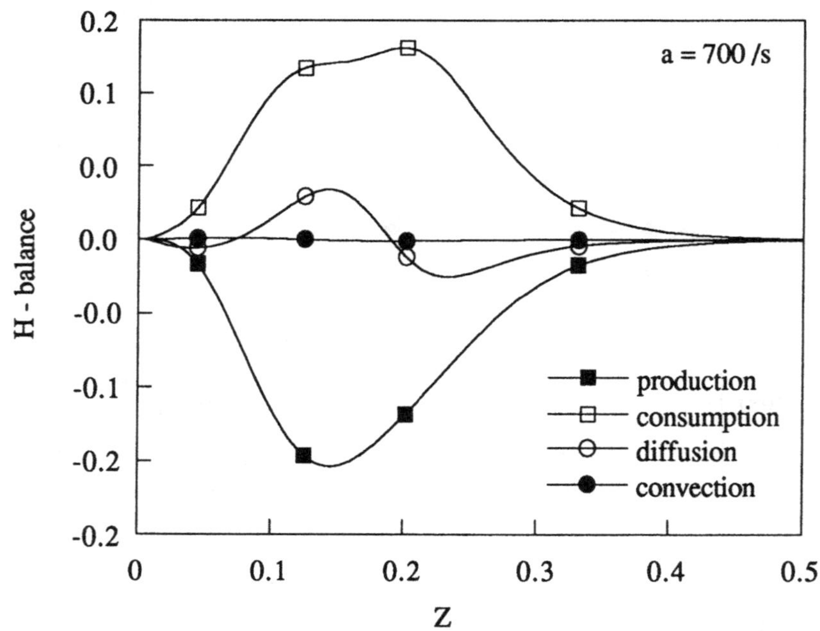

Fig. 16.11. Balance of convection, diffusion, production and consumption of H, calculated using the starting mechanism with $a = 700\,s^{-1}$, $p = 1$ bar, $T_{-\infty} = 338$ K, $Y_{O_2\infty} = 0.232$, and $T_\infty = 298$ K

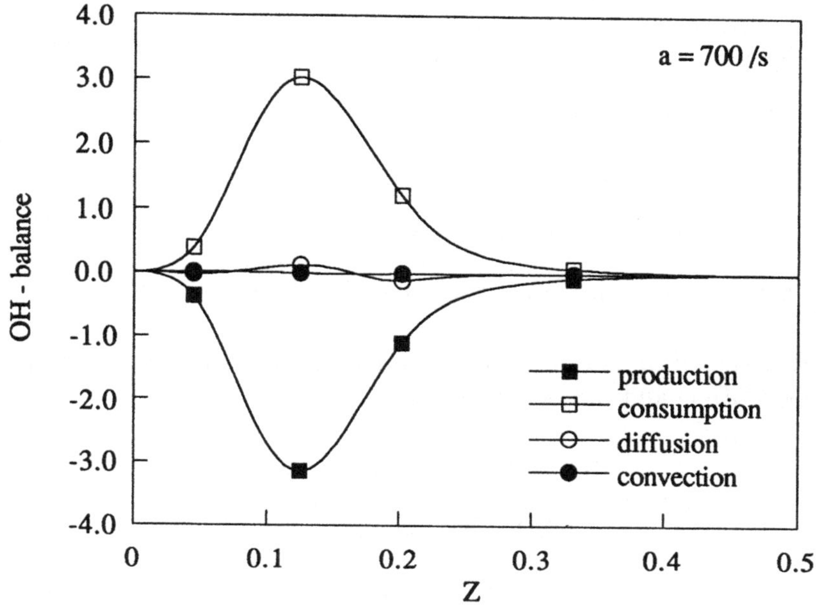

Fig. 16.12. Balance of convection, diffusion, production and consumption of OH, calculated using the starting mechanism with $a = 700\,\mathrm{s}^{-1}$, $p = 1\,\mathrm{bar}$, $T_{-\infty} = 338\,\mathrm{K}$, $Y_{O_2\infty} = 0.232$, and $T_{\infty} = 298\,\mathrm{K}$.

similar to Figs. 16.6–16.12 were also prepared for HCO, O, HO_2 and H_2O_2 and they show that steady-state approximations are valid for these species.

A 5-step mechanism can be deduced from the starting mechanism by introducing steady-state approximations for the species CH_2OH, HCO, OH, O, HO_2 and H_2O_2 and eliminating these species from the starting mechanism with steps 83, 24, 3, 2, 8 and 14 respectively [16.11]. The resulting mechanism can be written as

$$CH_3OH + 2H = 2H_2 + CH_2O \qquad \text{(Ia)}$$

$$CH_2O = CO + H_2 \qquad \text{(Ib)}$$

$$CO + H_2O = CO_2 + H_2 \qquad \text{(II)}$$

$$H + H + M = H_2 + M \qquad \text{(III)}$$

$$3H_2 + O_2 = 2H + 2H_2O \qquad \text{(IV)}$$

The reaction rates for the overall steps Ia–IV can be expressed in terms of elementary reaction rates as

$$
\begin{aligned}
w_{\mathrm{Ia}} &= w_{86} + w_{87} \\
w_{\mathrm{Ib}} &= w_{29} + w_{30} + w_{31} + w_{32} \\
w_{\mathrm{II}} &= w_{18f} - w_{18b} \\
w_{\mathrm{III}} &= w_{5f} - w_{5b} + w_{12f} - w_{12b} + w_{15} + w_{16} \\
&\quad + w_{17} + w_{21} + w_{22} + w_{23} - w_{32} - w_{85} \\
w_{\mathrm{IV}} &= w_{1f} - w_{1b} + w_{6} + w_{9} - w_{12f} + w_{12b} + w_{13} - w_{17}
\end{aligned}
\tag{16.5}
$$

A 4-step mechanism can be deduced by introducing a steady-state approximation for CH_2O and can be written as

$$CH_3OH + 2H = 3H_2 + CO \tag{I'}$$

$$CO + H_2O = CO_2 + H_2 \tag{II'}$$

$$H + H + M = H_2 + M \tag{III'}$$

$$3H_2 + O_2 = 2H + 2H_2O \tag{IV'}$$

The reaction rates for the overall steps I′–IV′ can be expressed in terms of the elementary rates as

$$
\begin{aligned}
w_{\mathrm{I'}} &= w_{86} + w_{87} \\
w_{\mathrm{II'}} &= w_{18f} - w_{18b} \\
w_{\mathrm{III'}} &= w_{5f} - w_{5b} + w_{12f} - w_{12b} + w_{15} + w_{16} \\
&\quad + w_{17} + w_{21} + w_{22} + w_{23} - w_{32} - w_{85} \\
w_{\mathrm{IV'}} &= w_{1f} - w_{1b} + w_{6} + w_{9} - w_{12f} + w_{12b} + w_{13} - w_{17}
\end{aligned}
\tag{16.6}
$$

Prior to performing numerical calculations using the reduced mechanisms, it is necessary to deduce algebraic expressions for the steady-state concentrations of the eliminated species.

16.5 Steady-State Assumptions

The overall reaction rates for the reduced chemical kinetic mechanism contain the eliminated species CH_2OH, HCO, HO_2, H_2O_2, OH and O therefore it is necessary to obtain explicit algebraic expressions for the steady-state concentrations of these species, which can be written as

$$[\mathrm{CH_2OH}] = \frac{(k_{86}\,[\mathrm{H}] + k_{87}\,[\mathrm{OH}])\,[\mathrm{CH_3OH}]}{k_{83}\,[\mathrm{H}] + k_{84}\,[\mathrm{O_2}] + k_{85}\,[\mathrm{M}]}$$

$$[\mathrm{HCO}] = \frac{(k_{29}[\mathrm{H}] + k_{30}[\mathrm{O}] + k_{31}[\mathrm{OH}] + k_{32}[\mathrm{M}])[\mathrm{CH_2O}]}{k_{21}\,[\mathrm{H}] + k_{22}\,[\mathrm{OH}] + k_{23}\,[\mathrm{O_2}] + k_{24f}\,[\mathrm{M}]} + \frac{k_{24b}[\mathrm{CO}][\mathrm{H}][\mathrm{M}]}{k_{21}\,[\mathrm{H}] + k_{22}\,[\mathrm{OH}] + k_{23}\,[\mathrm{O_2}] + k_{24f}\,[\mathrm{M}]} \tag{16.7}$$

$$[\mathrm{H_2O_2}] = \frac{k_{11}\,[\mathrm{HO_2}]^2 + k_{12f}\,[\mathrm{OH}]^2\,[\mathrm{M}] + k_{14b}\,[\mathrm{H_2O}]\,[\mathrm{HO_2}]}{k_{12b}\,[\mathrm{M}] + k_{13}\,[\mathrm{H}] + k_{14f}\,[\mathrm{OH}]}$$

$$[\mathrm{HO_2}] = \frac{-b + \sqrt{b^2 + 4ac}}{2a}$$

where

$$a = \frac{(2k_{12b}\,[\mathrm{M}] + 2k_{13}\,[\mathrm{H}] + k_{14f}\,[\mathrm{OH}])\,k_{11}}{k_{12b}\,[\mathrm{M}] + k_{13}\,[\mathrm{H}] + k_{14f}\,[\mathrm{OH}]}$$

$$b = k_{5b}\,[\mathrm{M}] + (k_6 + k_7 + k_9)\,[\mathrm{H}] + k_8\,[\mathrm{OH}] + k_{10}\,[\mathrm{O}] + \frac{(k_{12b}\,[\mathrm{M}] + k_{13}\,[\mathrm{H}])\,k_{14b}\,[\mathrm{H_2O}]}{k_{12b}\,[\mathrm{M}] + k_{13}\,[\mathrm{H}] + k_{14f}\,[\mathrm{OH}]}$$

$$c = (k_{5f}\,[\mathrm{H}]\,[\mathrm{M}] + k_{84}\,[\mathrm{CH_2OH}] + k_{23}\,[\mathrm{HCO}])\,[\mathrm{O_2}] + \frac{k_{12f}\,k_{14f}\,[\mathrm{OH}]^3\,[\mathrm{M}]}{k_{12b}\,[\mathrm{M}] + k_{13}\,[\mathrm{H}] + k_{14f}\,[\mathrm{OH}]}$$

here $[x_i]$ represents the molar concentration of species i and [M] is the molar concentration of the third body. The steady-state concentrations of CH_2OH, HCO, H_2O_2 and HO_2 contain the quantities [O] and [OH], for which additional algebraic relations must be deducted. The steady-state concentration of O atoms in terms of the steady-state concentration of OH can be written as

$$[\mathrm{O}] = \frac{k_{1f}\,[\mathrm{H}]\,[\mathrm{O_2}] + k_{2b}\,[\mathrm{H}]\,[\mathrm{OH}] + k_{4f}\,[\mathrm{OH}]^2}{k_{1b}\,[\mathrm{OH}] + k_{2f}\,[\mathrm{H_2}] + k_{4b}\,[\mathrm{H_2O}] + k_{30}\,[\mathrm{CH_2O}]} \tag{16.8}$$

where the influence the elementary reactions 9, 10 and 17 are neglected. The steady-state concentration of the radical OH neglecting the elementary reactions 6, 8, 10, 12, 13, 14 and 22 can be determined from the equation

$$\alpha\,[\mathrm{OH}]^2 + \beta\,[\mathrm{OH}] - \gamma = 0 \tag{16.9}$$

where

$$\alpha = 3k_{1b}k_{4f}[\mathrm{OH}] + (2k_{2b}[\mathrm{H}] + k_{3f}[\mathrm{H_2}] + k_{16}[\mathrm{H}][\mathrm{M}]$$

$$+ k_{18f}[CO] + k_{31}[CH_2O] + k_{87}[CH_3OH])k_{1b}$$

$$+ (k_{2f}[H_2] + k_{30}[CH_2O])k_{4f}$$

$$\begin{aligned}\beta = {} & (k_{2f}[H_2] + k_{4b}[H_2O] + k_{30}[CH_2O]) \\ & (k_{3f}[H_2] + k_{16}[H][M] + k_{18f}[CO] \\ & + k_{31}[CH_2O] + k_{87}[CH_3OH]) \\ & - (k_{2b}k_{4b}[H_2O] + k_{1b}k_{3b}[H_2O] + k_{1b}k_{18b}[CO_2])[H]\end{aligned} \tag{16.10}$$

$$\begin{aligned}\gamma = {} & 3k_{1f}k_{4b}[O_2][H_2O][H] \\ & + (k_{2f}[H_2] + k_{30}[CH_2O])\, 2k_{1f}[O_2][H] \\ & + (k_{2f}[H_2] + k_{4b}[H_2O] + k_{30}[CH_2O]) \\ & (k_{3b}[H_2O][H] + k_{18b}[CO_2][H])\end{aligned}$$

Since the quantity α contains the steady-state concentration of OH, (16.9) is a cubic equation in [OH]. For simplicity, to the leading order the quantity [OH] appearing in the expression for α is replaced by $[OH]^*$, where $[OH]^*$ is the concentration of OH calculated assuming that the elementary reaction 3 is in partial equilibrium. Thus,

$$[OH]^* = \frac{[H_2O]\,[H]}{K_3\,[H_2]} \tag{16.11}$$

where K_3 is the equilibrium constant of reaction 3. This approximation reduces (16.9) from a cubic equation to a quadratic equation in [OH], which can be readily solved. This approximation can be justified for reasons similar to those introduced previously for premixed methane-air flames [16.12]. In the expression for [OH], to the leading order it is necessary to include the elementary reaction $1f$, since its rate is large close to the inner layer [16.3]–[16.5], where most of the fuel specific chemistry occurs. To satisfy the matching conditions to the oxidizer stream, the backward reaction $1b$ would have to be considered, although its rate is small near the inner layer. Since $[OH]^*$ appears as a product with the rate constant of reaction $1b$ in (16.10), it may be viewed as a second order term that is retained only to allow proper matching with the oxidizer stream. Equation 16.9 allows [OH] to be calculated explicity in terms of the species appearing in the 5-step mechanism Ia–IV. Hence, the quantities [O], $[HO_2]$, $[H_2O_2]$, [HCO] and $[CH_2OH]$ can be calculated from (16.8) and (16.7).

Numerical calculations using the 4-step mechanism I′–IV′ requires an algebraic expression for the steady state concentration of CH_2O, which can be written as

$$[CH_2O] = \frac{(k_{83}\,[H] + k_{84}\,[O_2] + k_{85}\,[M])\,[CH_2OH]}{k_{29}\,[H] + k_{30}\,[O] + k_{31}\,[OH] + k_{32}\,[M]} \tag{16.12}$$

The approximation used here for calculating the quantity α defined in (16.9) introduced some minor numerical difficulties in resolving the structure of the flame close to the ambient oxidizer stream. This difficulty was found to occur only at conditions far from extinction, and did not prevent the program from converging to a stable solution. This problem was primarily attributed to the calculation of $[OH]^*$ from (16.11) in regions on the oxidizer side of the flame where the value of $[H_2]$ is small. To overcome this problem the quantity $[H_2]$ appearing in the denominator of (16.11) was replaced by $[H_2]^*$, where $[H_2]^* = [H_2]$ for $0.01\,([H_2])_m < [H_2]$ and $[H_2]^* = 0.01\,([H_2])_m$ for $[H_2] < 0.01\,([H_2])_m$, where $([H_2])_m$ is the peak value of $[H_2]$. By use of this approximation it was possible to resolve the structure of the flame close to the ambient oxidizer stream, and for all values of the strain rate.

In principle a three-step mechanism can be derived from the 4-step mechanism I′–IV′ by introducing a steady state approximation for the radical H, which from Figs. 16.8 and 16.11 appears to be marginally valid. However, the resulting steady-state relations for CH_2OH, CH_2O, HCO, H_2O_2, HO_2, OH, O and H would be algebraically complicated, hence it is not attempted here.

16.6 Comparison between Starting Mechanism, 5-Step Mechanism and 4-Step Mechanism

16.6.1 Calculations at Fixed Values of the Strain Rate

Numerical calculations were performed using the starting mechanism, 5-step mechanism, and 4-step mechanism at a fixed value of the strain rate. Boundary conditions (16.3) and (16.4) were employed with $a = 100\,\mathrm{s}^{-1}$, with $p = 1\,\mathrm{bar}$, $T_{-\infty} = 338\,\mathrm{K}$, $Y_{O_2\infty} = 0.232$, and $T_\infty = 298\,\mathrm{K}$.
The results are plotted in Figs. 16.13–16.17. Figure 16.13 shows that the profiles of temperature and fuel and oxygen predicted by the various mechanisms to agree well, although the reduced mechanisms predict slightly higher values of temperature than the starting mechanism. Figure 16.14 shows fairly good agreement between the predictions of the various mechanisms for the profiles of CO_2 and H_2O. Figure 16.15 shows that the predictions of the profiles of H_2, CO and CH_2O obtained using the 5-step mechanism to agree well with those obtained using the starting mechanism. However, in comparison, the 4-step mechanism predicts slightly higher values of CO and H_2, especially on the fuel side of the flame and considerably higher values of CH_2O. The relatively high values of CH_2O predicted by the 4-step mechanism is clearly due to the steady-state approximation introduced for this species, which according to the results shown in Fig. 16.7 is inaccurate. Figure 16.16 shows that the reduced mechanisms predict lower values of CH_2OH on the fuel side of the flame, and considerably higher values of HO_2 everywhere in the flame, in comparisons to those predicted by the starting mechanism. Figure 16.16 also shows that the profile of HCO predicted by the 5-step mechanism to be nearly similar to that

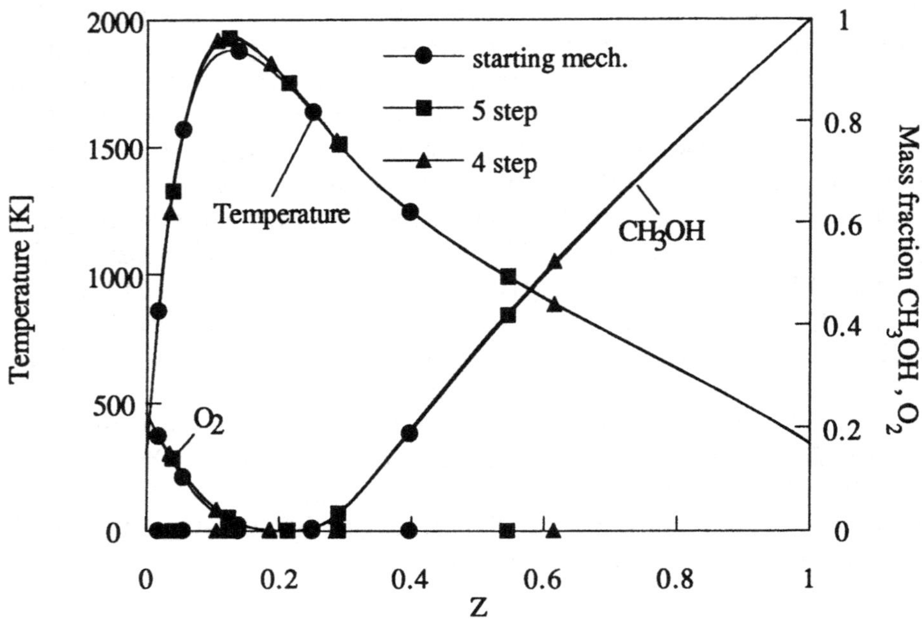

Fig. 16.13. Temperature profiles and profiles for the mass fraction of fuel and oxygen calculated for $a = 100\,s^{-1}$, $p = 1$ bar, $T_{-\infty} = 338$ K, $Y_{O_2\infty} = 0.232$, and $T_\infty = 298$ K.

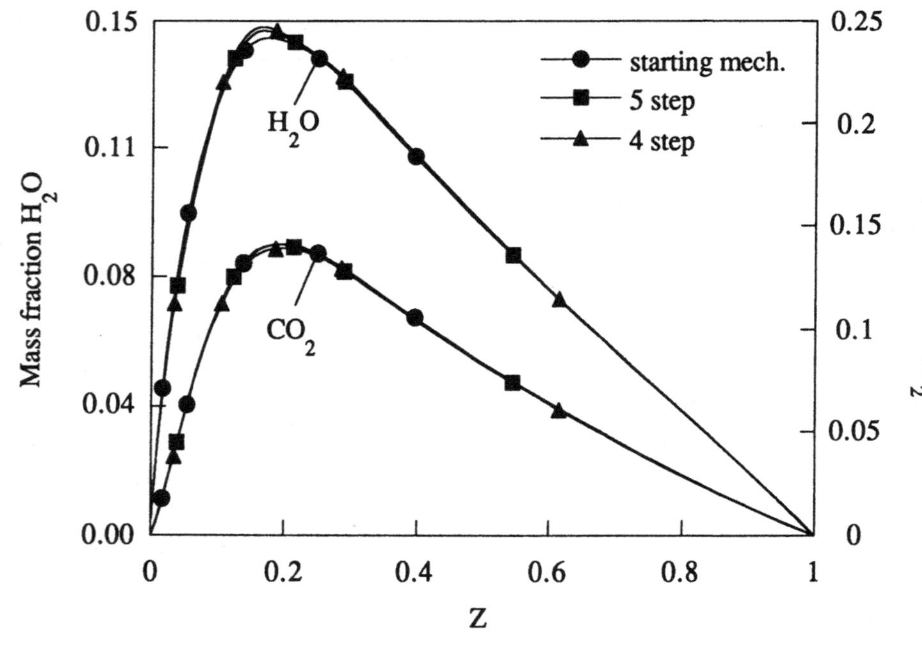

Fig. 16.14. Profiles for the mass fraction of H_2O and CO_2 calculated for $a = 100\,s^{-1}$, $p = 1$ bar, $T_{-\infty} = 338$ K, $Y_{O_2\infty} = 0.232$, and $T_\infty = 298$ K.

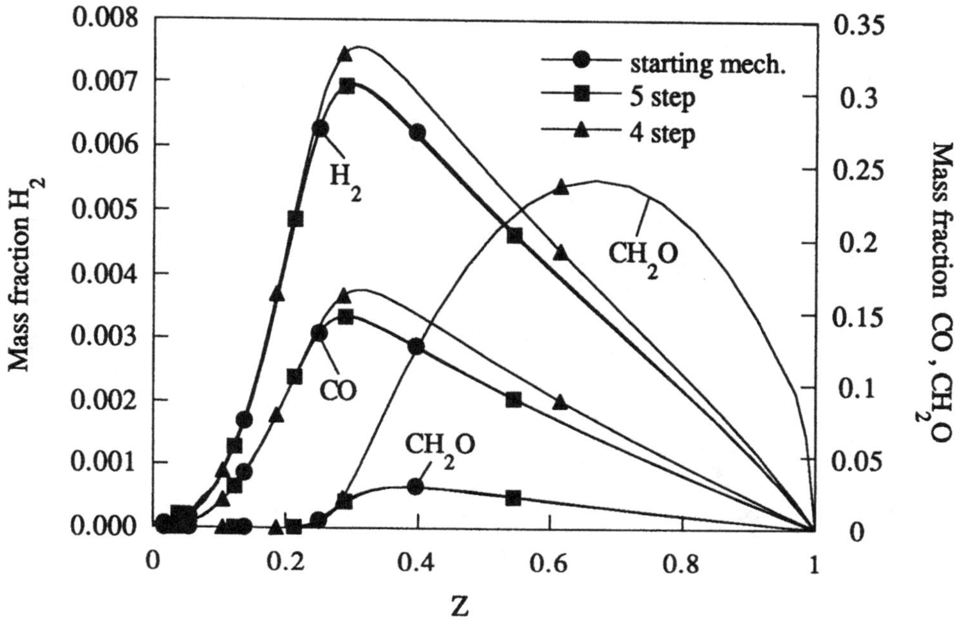

Fig. 16.15. Profiles for the mass fraction of H_2, CO, and CH_2O calculated for $a = 100\,s^{-1}$, $p = 1$ bar, $T_{-\infty} = 338$ K, $Y_{O_2\infty} = 0.232$, and $T_\infty = 298$ K.

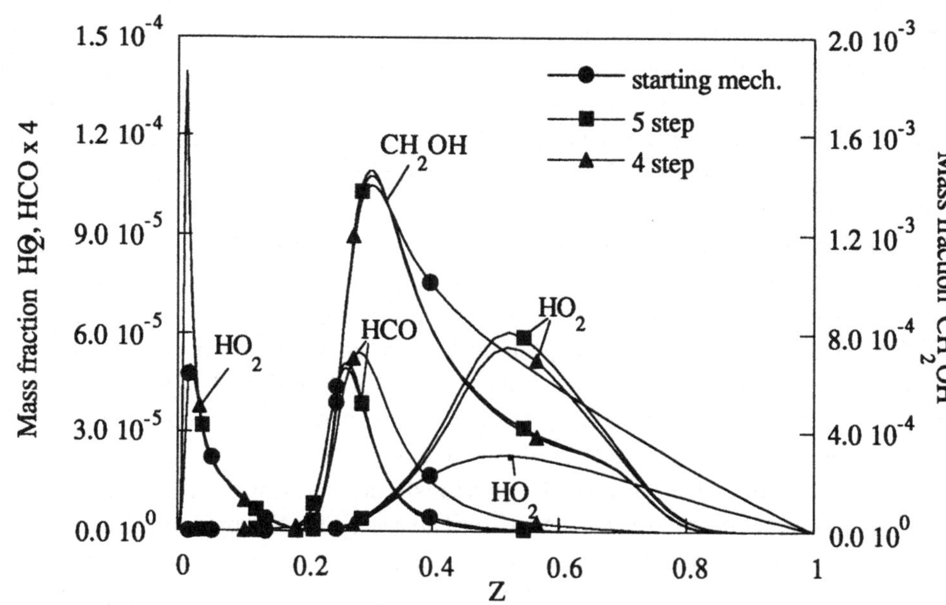

Fig. 16.16. Profiles for the mass fraction of CH_2OH, HCO and HO_2 calculated for $a = 100\,s^{-1}$, $p = 1$ bar, $T_{-\infty} = 338$ K, $Y_{O_2\infty} = 0.232$, and $T_\infty = 298$ K.

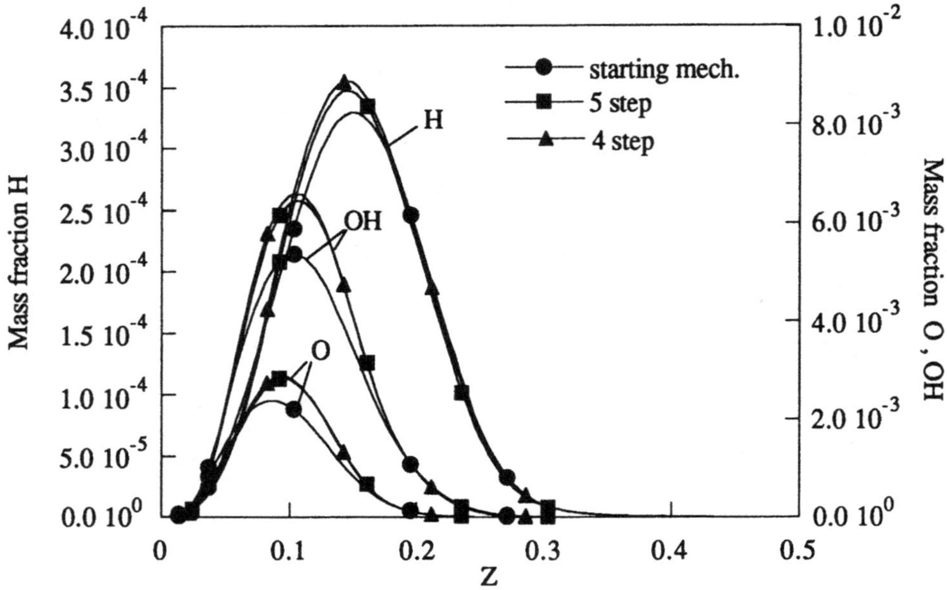

Fig. 16.17. Profiles for the mass fraction of H, OH, and O calculated for $a = 100\,\text{s}^{-1}$, $p = 1\,\text{bar}$, $T_{-\infty} = 338\,\text{K}$, $Y_{O_2\infty} = 0.232$, and $T_\infty = 298\,\text{K}$.

obtained using the starting mechanism. In comparison the profile for HCO predicted by the 4-step mechanism is shifted toward the fuel side, although the peak value of this quantity agrees reasonbly well with that calculated using the starting mechanism. Figure 16.17 shows that the profiles for H predicted by the various mechanisms are nearly in agreement. However, the reduced mechanisms predict much higher values of OH and O in comparison to those predicted by the starting mechanism.

In Figs. 16.18 and 16.19 the global reaction rates w_{Ia}, w_{Ib}, w_{II}, w_{III}, and w_{IV} for the 5-step mechanism, and $w_{\text{I}'}$, $w_{\text{II}'}$, $w_{\text{III}'}$, and $w_{\text{VI}'}$ for the 4-step mechanism are plotted for $a = 700\,\text{s}^{-1}$, with $p = 1\,\text{bar}$, $T_{-\infty} = 338\,\text{K}$, $Y_{O_2\infty} = 0.232$, and $T_\infty = 298\,\text{K}$. The global rates were normalized by division by $\rho_\infty a/W_F = 2.5471 \times 10^4\,\text{mol/(m}^3\,\text{s)}$. Figures 16.18 and 16.19 show that as in methane-air flames [16.3], [16.4], [16.5], in the region where chemical reactions occur, two distinct layers can be identified.

In one layer, which is similar to the previously defined inner layer for methane-air flames [16.4] the global reaction II for the 5-step mechanism, or II′ for the 4-step mechanism is unimportant. Also in this inner layer the major contribution to the global reaction III or III′ is from the elementary reaction 85, with all other elementary reactions contributing negligibly to this global step in this layer. Therefore, in the inner layer methanol reacts with radicals to form primarily CO and H_2 and some CO_2 and H_2O. In the other layer, which is similar to the previously defined oxidation layer for methane-air flames [16.3],

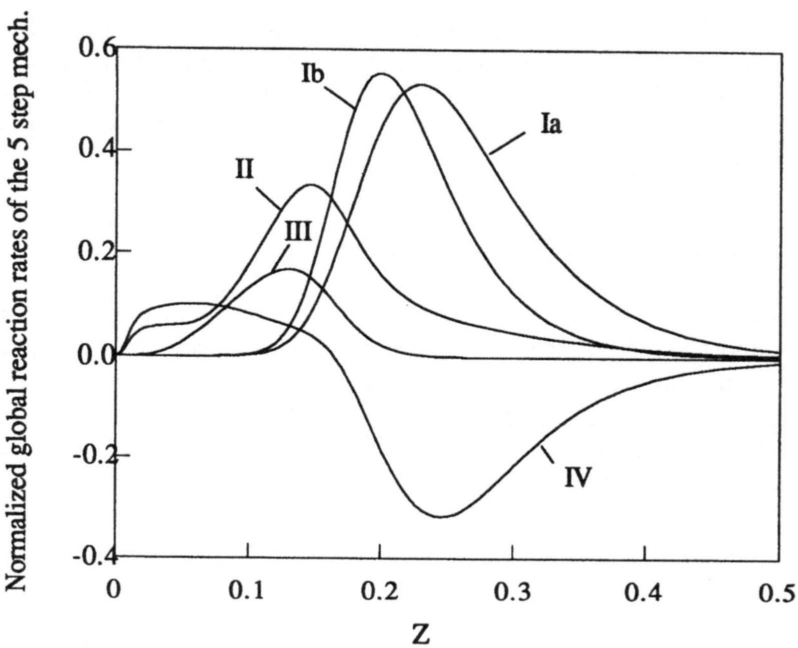

Fig. 16.18. Global reaction rates w_{Ia}, w_{Ib}, w_{II}, w_{III}, and w_{IV} for the 5-step mechanism calculated for $a = 700\,\mathrm{s}^{-1}$, $p = 1\,\mathrm{bar}$, $T_{-\infty} = 338\,\mathrm{K}$, $Y_{O_2\infty} = 0.232$, and $T_{\infty} = 298\,\mathrm{K}$.

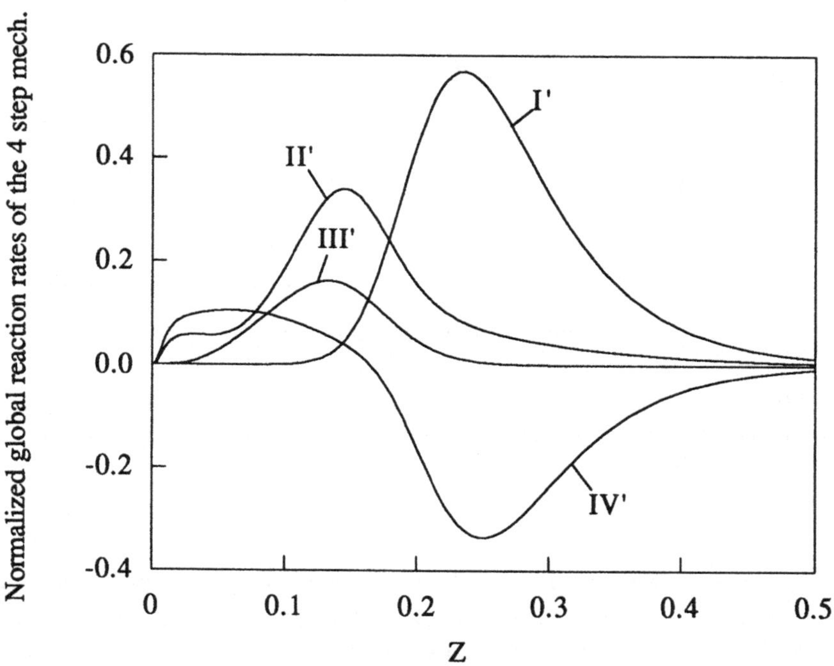

Fig. 16.19. Global reaction rates w_{I}, w_{II}, w_{III}, and w_{IV} for the 4-step mechanism calculated for $a = 700\,\mathrm{s}^{-1}$, $p = 1\,\mathrm{bar}$, $T_{-\infty} = 338\,\mathrm{K}$, $Y_{O_2\infty} = 0.232$, and $T_{\infty} = 298\,\mathrm{K}$.

[16.4], [16.5], the global reactions Ia and Ib for the 5-step mechanism, or the global reaction I′ for the 4-step mechanism is unimportant. In addition in this oxidation layer the elementary reactions 21, 22, 23, 32, and 85 contribute negligibly to the global reaction III or III′. Therefore in the oxidation layer the CO and H_2 are oxidized to CO_2 and H_2O.

16.6.2 Critical Conditions of Extinction

Calculations were performed using the starting mechanism, the 5-step mechanism, and the 4-step mechanism to determine the critical conditions of flame extinction. Boundary conditions (16.3), and (16.4) were employed in these calculations and the results are shown in Figs. 16.20, 16.21 and 16.22. In Fig. 16.20, the strain rate at extinction is plotted as a function of the mole fraction of oxygen in the ambient oxidizer stream.

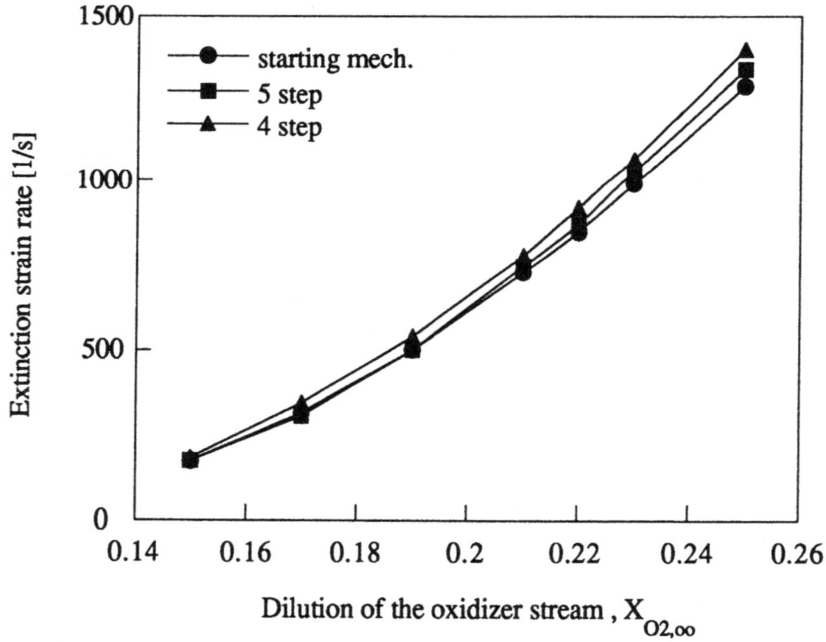

Fig. 16.20. Values of the strain rate a, as a function of the mole fraction of oxygen in the ambient oxidizer stream at flame extinction calculated for $p = 1$ bar, $T_{-\infty} = 338$ K, and $T_\infty = 298$ K.

These calculations were performed for $p = 1$ bar, $T_{-\infty} = 338$ K and $T_\infty = 298$ K. In Fig. 16.21, the strain rate at extinction is plotted as a function of the total pressure. These calculations were performed for $Y_{O_2\infty} = 0.232$, and $T_\infty = 298$ K. The value of $T_{-\infty}$ was chosen to be equal to the boiling point of methanol, which at a given value of p, and was calculated from the relation [16.8] $T_{-\infty} = 1862/(5.536 - \log_{10} p)$, where the units of $T_{-\infty}$ and p are

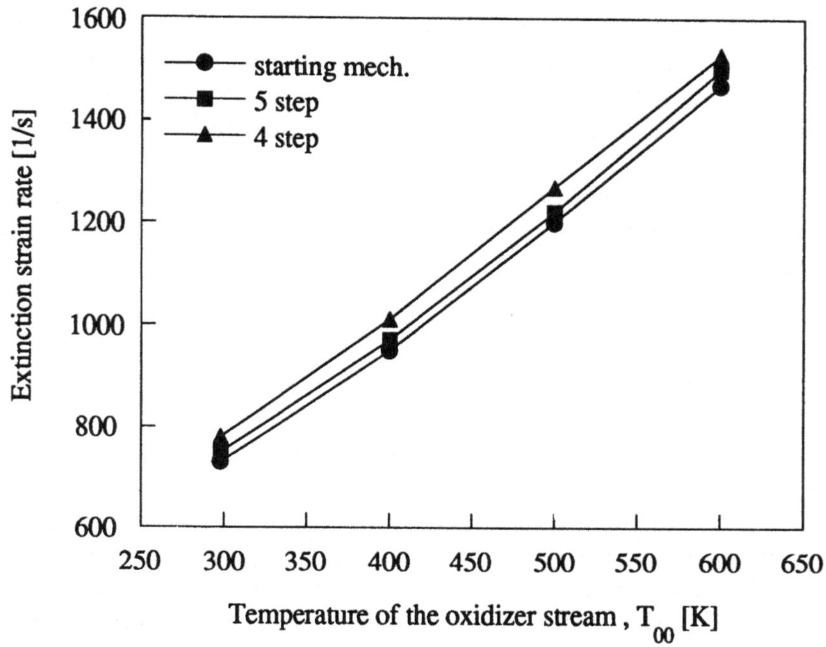

Fig. 16.21. Values of the strain rate a, at flame extinction as a function of the temperature of the oxidizer stream, T_∞ at $p = 1$ bar.

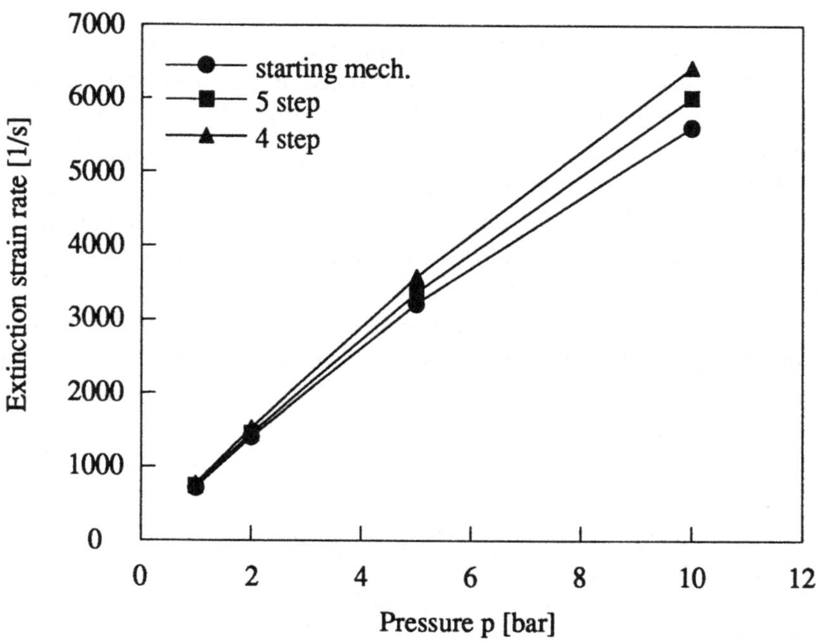

Fig. 16.22. Values of the strain rate a, at flame extinction as a function of the total pressure p, for $Y_{O_2\infty} = 0.232$ and $T_\infty = 298\,\mathrm{K}$.

in Kelvin and bar respectively. In Fig. 16.22, the strain rate at extinction is plotted as a function of the temperature of the ambient oxidizer stream. These calculations were performed for $p = 1\,\mathrm{bar}$, $T_{-\infty} = 338\,\mathrm{K}$ and $Y_{O_2\infty} = 0.232$. Figures 16.20, 16.21, and 16.22 show that the critical conditions of extinction predicted by the various mechanisms agree reasonably well. At all conditions tested here the results obtained using the 5-step mechanism appears to agree better with the results of the starting mechanism, than the results obtained using the 4-step mechanism.

16.7 Conclusions

A 5-step mechanism and a 4-step mechanism have been proposed for analyzing the structure of counterflow, methanol-air diffusion flames. These reduced mechanisms were deduced from a starting mechanism, which is essentially the C_1 mechanism shown in Table 1 of Chap. 1, excluding reactions involving CH, CH_2 and CH_3. The predictions of the starting mechanism were tested, by comparing them with the results of previous experimental measurements [16.6], and were found to agree reasonably well. Hence, it can be concluded that the starting mechanism is capable of predicting a number of characteristics of methanol-air diffusion flames.

The predictions of the reduced mechanisms were tested by comparing them with the predictions of the starting mechanism over a wide parametric range of pressure (1 bar to 10 bar), dilution and preheat temperatures of the oxidizer stream. It was found that the predicted profiles for those species, for which steady-state approximation was not introduced in the reduced mechanism, to agree well, with the predictions of the starting mechanism. The predicted profiles of those species, for which steady state approximation was introduced in the reduced mechanism, agreed reasonably well with the predictions of the starting mechanism. The only exception was the predicted profile of CH_2O by the 4-step mechanism, which was considerably higher than that predicted by the starting mechanism. The critical conditions of extinction, calculated using the reduced mechanisms, were in reasonably good agreement with those calculated using the starting mechanism. In general the results obtained using the 5-step mechanism agreed better with the results obtained using the starting mechanism than to those obtained using the 4-step mechanism.

Plots of the global reaction rates of the reduced mechanisms as a function of the mixture fraction showed that as in hydrocarbon-air flames, the structure of methanol-air flames can be subdivided into two layers, an inner layer where methanol reacts with radicals to form primarily CO and H_2, and an oxidation layer where H_2 and CO are oxidized to H_2O and CO_2. Therefore, it appears that previously developed asymptotic analysis for methane-air flames [16.3], [16.4], [16.5] can be extended to methanol-air flames.

Acknowledgement

This work was partly supported by the Stiftung Volkswagenwerk. The authors are indepted to N. Peters and F. Mauss for many helpful discussions.

References

[16.1] Seshadri, K., Trevino, C. and Smooke, M. D., Combustion Flame 76, III, 1989.
[16.2] Chen, J. Y., Combustion Flame, **78**, 127,1991.
[16.3] Peters, N. and Williams, F. A., Combustion Flame **68**, 185, 1987.
[16.4] Seshadri, K. and Peters, N., Combustion Flame, **81**, No. 2, **96**, 1990.
[16.5] Seshadri, K. and Peters, N., Combustion Flame **73**, 23, 1988.
[16.6] Seshadri, K., Combustion Flame **33**, 197, 1978.
[16.7] Seshadri, K. and Williams, F. A., Int. J. Heat Mass Transfer **21**, 251,1978.
[16.8] Landolt-Börnstein, Zahlenwerte und Funktionen II/2a, Springer, 1960.
[16.9] Kee, R. J., Warnatz, J. and Miller, J. A. Sandia Report SAND 83-8209, UC-32, Livermore, 1983.
[16.10] Smooke, M. D., J. Computational Physics, No. 1, 48, 72,1982.
[16.11] Peters, N., Numerical Solution of Combustion Phenomena. Lecture Notes in Physics 241, 90, 1985.
[16.12] Peters, N., in: Reduced Kinetic Mechanisms an Asymptotic Approximations for Methane-Air Flames (M. D. Smooke, Ed.), Chapter 3, Springer, 1991.

17. Reduced Kinetic Mechanisms and NO_x Formation in Diffusion Flames of CH_4/C_2H_6 Mixtures

S. H. Chung, S. R. Lee
Department of Mechanical Engineering, Seoul National University, Seoul 151-742, Korea

F. Mauss, N. Peters
Institut für Technische Mechanik, RWTH Aachen, W-5100 Aachen, Germany

17.1 Introduction

Technical fuels very often consist of a mixture of several components with different physical and chemical properties. Since reduced mechanisms are typically derived for one-component fuels, the question arises whether the same reduction procedure can be used for multicomponent-fuels and whether these mechanisms are then valid for variable compositions. A prominent example for a gaseous multicomponent fuel is natural gas. Table 17.1 shows the composition of natural gas from various sources in Europe.

While the dominating component is always methane (CH_4), the concentrations of additional components vary largely for different sources. The second largest concentration of a reacting species for these sources is ethane (C_2H_6), and there are large variations of the essentially inert component N_2. To simplify the analysis, we here consider mixtures of methane and ethane, in particular a 85% CH_4, 15% C_2H_6 mixture as a model for natural gas. This corresponds the closest to natural gas from the North Sea, assuming that propane and all the higher hydrocarbons can be represented by ethane. We analyse the structure of a diffusion flame because natural gas is used mainly in house hold and industrial appliances under essentially non-premixed and partially premixed conditions. Since much is known about reduced kinetic mechanisms and the flame structure of methane diffusion flames (Chap. 13), this chapter will focus on complementary aspects only. Besides questions related to the addition of the second component C_2H_6, we will also calculate the production of thermal NO_x in counterflow diffusion flames. In particular we will test the potential of reduced mechanisms to accurately predict the formation of thermal NO_x, which is very sensitive to inaccuracies of the calculated temperature and radical concentrations.

Table 17.1. Composition of three types of natural gas in Europe.

Origin	Netherlands	North Sea	URSS
methane	82.4	85.7	97.0
ethane	3.3	9.0	1.1
propane	0.57	2.4	0.4
i-butane	0.09	0.25	0.1
n-butane	0.1	0.48	0.1
i-pentane	0.023	0.05	0.03
n-pentane	0.027	0.05	0.03
hexane	0.072	0.02	0.02
helium	0.047	—	0.02
N_2	12.324	0.6	1.0
CO_2	1.031	1.4	0.2

17.2 Reaction Paths in the Oxidation Mechanism of CH_4/C_2H_6 Mixtures and for NO_x-Formation

With CH_4 and C_2H_6 as fuels the full mechanism shown in Table 1 of Chap. 1 allows the formation of C_3-hydrocarbons up to n-C_3H_7 via the reactions 71b and 74b. Therefore all reation steps up to 74b were used as a subset of this full mechanism and will be called the starting mechanism. The reaction mechanism of CH_4 starts from reactions 38f, 39 and 40f and proceeds through 35 or 37 either to the C_1-chain, or through reaction 36 towards C_2H_6. This starts the C_2-chain which proceeds through C_2H_5, C_2H_4, C_2H_3 to C_2H_2 and from thereon by reactions 45 and 46 either back to the C_1-chain or by reaction 47f to C_2H. Another possibility, which is important for soot formation, is reaction 48 which forms C_3H_3, which may combine with another C_3H_3 to form benzene C_6H_6 and thereby the first aromatic ring. This sequence will not be considered here. However, we have included reactions 71b and 74b, which opens a path to C_3-hydrocarbons. These will be reduced to C_2- and C_1-hydrocarbons again by the many possibilities provided by the reactions 62–73.

To calculate the formation of NO_x in these counterflow diffusion flames, we have addded the following thermal NO mechanism

Table 17.2. Nitrogen chemistry

No.	Reaction	A mole, cm^3, sec	n	E KJ/mole
N1f	$O + N_2 = N + NO$	1.9E14	0.0	319.03
N1b	$N + NO = O + N_2$	4.220E13	0.0	4.25
N2f	$N + O_2 = NO + O$	1.129E10	1.0	27.83
N2b	$NO + O = N + O_2$	2.4E09	1.0	161.67
N3f	$N + OH = NO + H$	4.795E13	0.0	5.23
N3b	$NO + H = N + OH$	1.3E14	0.0	205.85
N4f	$NO + HO_2 = NO_2 + OH$	3.0E12	0.5	10.04
N4b	$NO_2 + OH = NO + HO_2$	1.196E13	0.5	47.23
N5f	$NO + OH = NO_2 + H$	5.9E12	0.0	129.49
N5b	$NO_2 + H = NO + OH$	2.350E14	0.0	0.04

17.3 Boundary Conditions

The governing equations for the numerical problem are described in detail in Chap. 1, and the notation introduced in Chap. 1 is used here. All calculations were performed in the planar ($j = 0$) configuration using the potential-flow boundary conditions. As discussed in Chap. 1, the mixture fraction Z is calculated as a function of the independent spatial coordinate, y, using its balance equation (1.10). Boundary conditions are applied at $y = y_{-\infty}$ and $y = y_{\infty}$, where the subscripts $-\infty$ and $+\infty$ identify conditions in the ambient oxidizer stream and fuel stream, respectively. In the vicinity of $y_{-\infty}$ the continuity equation implies that $G = -\mathrm{d}v_{-\infty}/\mathrm{d}y = a$, where a is the strain rate. The boundary conditions at the ambient oxidizer stream are therefore

$$\begin{aligned} &G = a\,;\ T = 298\mathrm{K}\,;\ \rho = \rho_{-\infty}\,;\ Y_{O_2} = 0.232\,; \\ &Y_{N_2} = 0.768\,;\ Y_i = 0\,, i \neq O_2\,, N_2\,;\ Z = 0 \quad \text{at } y = y_{-\infty}\,, \end{aligned} \tag{17.1}$$

and at the ambient fuel stream

$$\begin{aligned} &G = \sqrt{(\rho_{\infty}/\rho_{-\infty})}a\,;\ T = 298\mathrm{K}\,;\ \rho = \rho_{\infty}\,; \\ &Y_{CH_4} = Y_{CH_4,\infty}\,;\ Y_{C_2H_6} = Y_{C_2H_6,\infty}\,; \\ &Y_i = 0\,,\ i \neq CH_4\,, C_2H_6\,;\ Z = 1 \text{ at } y = y_{\infty}\,, \end{aligned} \tag{17.2}$$

The normal velocity v at the boundaries is calculated by integrating the continuity equation.

17.4 Reduced Kinetic Mechanisms

In order to test different levels of approximation in particular with respect to NO_x formation, we have developed a nine step and a five step reduced mechanism. In addition to these we have reduced the nitrogen chemistry to two global steps

17.4.1 Nine-Step Reduced Mechanism

In the nine-step reduced mechanism, HO_2, H_2O_2, CH, CHO, CH_2, CH_2O, CH_3, C_2H, CHCO, C_2H_3, C_2H_5, C_3H_3, C_3H_4, C_3H_5, C_3H_6 and n-C_3H_7 are assumed in steady state. This provides 16 equations that can be used to eliminate 16 reaction rates from remaining balance equations. From these relations, the reaction steps 7, 12, 19, 24, 25, 31, 38, 42, 44, 51, 59, 62, 68, 70, 71 and 73 are eliminated in the balance equations of the remaining species by suitable linear combinations with the steady state relations. Then, the following nine-step mechanism results

$$\begin{array}{ll}
\mathrm{I} & C_2H_6 + 2H = C_2H_4 + 2H_2 \\
\mathrm{II} & CH_4 + H + OH = CO + 3H_2 \\
\mathrm{III} & C_2H_4 + 2H = C_2H_2 + 2H_2 \\
\mathrm{IV} & C_2H_2 + 2OH = 2CO + 2H_2 \\
\mathrm{V} & CO + OH = CO_2 + H \\
\mathrm{VI} & O_2 + H = OH + O \\
\mathrm{VII} & H_2 + OH = H_2O + H \\
\mathrm{VIII} & H + H + M = H_2 + M \\
\mathrm{IX} & H_2 + O = OH + H
\end{array} \tag{17.3}$$

Also, we derive a reduced mechanism for nitrogen chemistry. In general, the concentration of N is much lower than that of NO and NO_2, which suggests the steady-state approximation for N. At high strain rates, the NO_2 production rate on the fuel side is so large that NO_2 may not be assumed to be in steady state. Thus a two-step reduced mechanism of nitrogen chemistry, considering NO and NO_2, may be written as

$$\begin{array}{ll}
\mathrm{X} & N_2 + O_2 = 2NO \\
\mathrm{XI} & NO + OH = NO_2 + H
\end{array} \tag{17.4}$$

The reaction rates are

$$
\begin{aligned}
w_{\mathrm{I}} &= -w_{36} + w_{56} + w_{57} + w_{58} \\
w_{\mathrm{II}} &= w_{33} + w_{35} + 2w_{36} + w_{37} - 2w_{56} + w_{72} \\
w_{\mathrm{III}} &= -w_{36} + w_{52} + w_{53} + w_{54} + w_{55} + w_{56} \\
&\quad + w_{64} + w_{65} + w_{66} + w_{67} - w_{72} + w_{74} \\
w_{\mathrm{IV}} &= -w_{36} - w_{41} + w_{45} + w_{46} + w_{47} + w_{48} + w_{53} \\
&\quad + w_{56} - w_{63} + w_{64} \\
w_{\mathrm{V}} &= w_{18} - w_{20} + w_{28} \\
w_{\mathrm{VI}} &= w_{1} + w_{6} + w_{9} + w_{11} - w_{14} - w_{17} - w_{20} - w_{26} \\
&\quad + w_{33} + w_{37} - w_{41} + w_{43} + w_{45} + w_{47} - w_{63} \\
&\quad + w_{64} \quad (-w_{\mathrm{N1}} + w_{\mathrm{N2}} + w_{\mathrm{N4}}) \\
w_{\mathrm{VII}} &= w_{3} + w_{4} + w_{8} + w_{9} + w_{13} + w_{14} + w_{16} + w_{22} \\
&\quad - w_{29} - w_{30} - w_{32} + w_{35} + w_{37} + w_{40} + w_{47} + w_{48} \\
&\quad + w_{54} + w_{61} - w_{63} + w_{64} + w_{65} + w_{67} \\
w_{\mathrm{VIII}} &= w_{5} - w_{11} + w_{13} + w_{14} + w_{15} + w_{16} + w_{17} \\
&\quad + w_{21} + w_{22} + w_{23} - w_{32} + w_{34} + 2w_{36} + w_{48} + w_{49} \\
&\quad + w_{50} - w_{52} - w_{53} - w_{54} - w_{55} - w_{56} - w_{58} - w_{64} \\
&\quad - w_{66} - w_{67} - w_{69} + 2w_{72} - 2w_{74} \\
w_{\mathrm{IX}} &= w_{2} - w_{4} + w_{6} + w_{10} + w_{11} - w_{14} + w_{17} + w_{26} \\
&\quad + w_{27} + w_{28} + w_{30} + w_{35} + w_{37} + w_{39} - w_{41} - w_{43} \\
&\quad + w_{45} + 2w_{46} + w_{47} + 2w_{48} + w_{53} + w_{60} - w_{63} \\
&\quad + 2w_{64} + w_{65} + w_{66} \quad (+w_{\mathrm{N4}}) \\
w_{\mathrm{X}} &= w_{\mathrm{N1}} \\
w_{\mathrm{XI}} &= w_{\mathrm{N4}} + w_{\mathrm{N5}}
\end{aligned}
\tag{17.5}
$$

The reactions in parenthesis are used only when the reduced mechanism including the nitrogen chemistry is considered. The concentrations of the steady state species are obtained from the algebraic steady state relations.
For $[CH_3]$ this leads to

$$
\begin{aligned}
[\mathrm{CH_3}] &= (b_1^2/4 + a_1)^{1/2} - b_1/2 \\
a_1 &= \{w_{38f} + w_{39} + w_{40f} + w_{53} \\
&\quad + 2Z_{56f}(w_{58b} + w_{59} + w_{60} + w_{61})\}/N_{\mathrm{CH_3}} \\
b_1 &= \{(k_{33f} + k_{34})[\mathrm{H}] + k_{35}[\mathrm{O}] + k_{37}[\mathrm{O_2}] \\
&\quad + k_{38b}[\mathrm{H_2}] + k_{40b}[\mathrm{H_2O}] + Z_{72}k_{71b}[\mathrm{C_2H_3}]\}/N_{\mathrm{CH_3}} \\
N_{\mathrm{CH_3}} &= 2\{k_{36} + (1 - Z_{56f})k_{56b}\}
\end{aligned}
\tag{17.6}
$$

The other species are as follows.

$$
\begin{aligned}
[n\text{-}C_3H_7] &= \omega_{74b}/(k_{73} + k_{74f}) \\
[C_3H_6] &= \omega_{71b}/\{k_{71f} + (k_{72} + Z_{73}k_{74b})[H]\} \\
[C_3H_5] &= (Z_{72}\omega_{71b} + \omega_{69b})/(k_{69f} + k_{70}[H]) \\
[C_3H_4] &= (\omega_{72} + Z_{64b}\omega_{48})/\{(1 - Z_{64b})k_{64f} + (k_{65} + k_{66})[O] \\
&\quad + (k_{67} + k_{68})[OH]\} \\
[C_3H_3] &= (\omega_{48} + \omega_{64f})/\{k_{64f}[O_2] + k_{63}[O] + k_{64b}[H]\} \\
[C_2H_5] &= (\omega_{56b} + \omega_{58b} + \omega_{59} + \omega_{60} + \omega_{61}) \\
&\quad /\{k_{56f}[H] + k_{57}[O_2] + k_{58f}\} \\
[C_2H_3] &= (\omega_{51b} + \omega_{52f} + \omega_{54f})/\{k_{49}[H] + k_{50}[O_2] + k_{51f} \\
&\quad + k_{52}[H_2] + k_{54b}[H_2O]\} \\
[C_2H] &= (\omega_{41b} + \omega_{47f})/\{k_{41f}[H_2] + k_{42}[O_2] + k_{47b}[H_2O]\} \\
[CH_2] &= [\omega_{33f} + Z_{43f}\{Z_{42}(\omega_{41b} + \omega_{47f}) + \omega_{46} \\
&\quad + Z_{62}Z_{64f}Z_{72}\omega_{71b}\}]/N_{CH_2} \\
N_{CH_2} &= (1 - Z_{25b} - Z_{43f}Z_{62}Z_{48})k_{25f}[H] + k_{26}[O] \\
&\quad + (k_{27} + k_{28})[O_2] + k_{33b}[H_2] + (1 - Z_{43f})k_{43b}[CO] \\
[CHCO] &= (\omega_{43b} + \omega_{42} + \omega_{46} + \omega_{62})/(k_{43f}[H] + k_{44}[O]) \\
[CH] &= \omega_{25f}/(k_{19}[O_2] + k_{20}[CO_2] + k_{25b}[H_2] + k_{48}[C_2H_2]) \\
[CH_2O] &= (\omega_{35} + \omega_{37} + \omega_{62} + \omega_{65} + \omega_{67})/(k_{29}[H] + k_{30}[O] \\
&\quad + k_{31}[OH] + k_{32}[M']) \\
[CHO] &= (\omega_{19} + \omega_{20} + \omega_{24b} + \omega_{29} + \omega_{30} + \omega_{31} + \omega_{32} \\
&\quad + \omega_{66} + \omega_{68})/(k_{21}[H] + k_{22}[OH] + k_{23}[O_2] + k_{24f}[M']) \\
[HO_2] &= (b_2^2/4 + a_2)^{1/2} - b_2/2 \\
a_2 &= (\omega_{5f} + Z_{14f}\omega_{12f} + \omega_{23} + \omega_{50} + \omega_{57})/k_{11}(2 - Z_{14f}) \\
b_2 &= \{k_{5b}[M'] + (k_6 + k_7 + k_9)[H] + k_8[OH] + k_{10}[O] \\
&\quad + (1 - Z_{14f})k_{14b}[H_2O]\}/k_{11}(2 - Z_{14f}) \\
[H_2O_2] &= (\omega_{11} + \omega_{12f} + \omega_{14b})/(k_{12b}[M'] + k_{13}[M] + k_{14f}[OH]) \\
[N] &= (\omega_{N1f} + \omega_{N2b} + \omega_{N3b})/(k_{N1b}[NO] \\
&\quad + k_{N2f}[O_2] + k_{N3f}[OH])
\end{aligned}
\tag{17.7}
$$

In this equation we have defined

$$
\begin{aligned}
Z_{14f} &= k_{14f}[OH]/(k_{12b}[M'] + k_{13}[H] + k_{14f}[OH]) \\
Z_{25b} &= k_{25b}[H_2]/(k_{19}[O_2] + k_{20}[CO_2] + k_{25b}[H_2] + k_{48}[C_2H_2]) \\
Z_{42} &= k_{42}[O_2]/(k_{41f}[H_2] + k_{42}[O_2] + k_{47b}[H_2O]) \\
Z_{43f} &= k_{43f}[H]/(k_{43f}[H] + k_{44}[O])
\end{aligned}
$$

$$\begin{aligned}
Z_{48} &= k_{48}[C_2H_2]/(k_{19}[O_2] + k_{20}[CO_2] + k_{25b}[H_2] + k_{48}[C_2H_2]) \\
Z_{62} &= k_{62}[O_2]/(k_{62}[O_2] + k_{63}[O] + (1 - Z_{64f})k_{64b}[H]) \\
Z_{64f} &= k_{64f}/\{k_{64f} + (k_{65} + k_{66})[O] + (k_{67} + k_{68})[OH]\} \\
Z_{64b} &= k_{64b}[H]/(k_{62}[O_2] + k_{63} + k_{64}[H]) \\
Z_{56f} &= k_{56f}[H]/(k_{56f}[H] + k_{57}[O_2] + k_{58f}) \\
Z_{72} &= k_{72}[H]/\{k_{71f} + (k_{72} + Z_{73}k_{74b})[H]\} \\
Z_{73} &= k_{73}/(k_{73} + k_{74f})
\end{aligned} \tag{17.8}$$

17.4.2 Five-Step Reduced Mechanism

We obtain the following reduced mechanism by steady state assumption of O, OH, C_2H_2 and C_2H_4 as well as the species of the nine-step mechanism. In addition we have eliminated reactions 45 and 54.

$$\begin{array}{ll}
\text{I} & C_2H_6 + 2H_2O + 6H = 2CO + 8H_2 \\
\text{II} & CH_4 + H_2O + 2H = CO + 4H_2 \\
\text{III} & CO + H_2O = CO_2 + H_2 \\
\text{IV} & H + H + M = H_2 + M \\
\text{V} & 3H_2 + O_2 = 2H_2O + 2H
\end{array} \tag{17.9}$$

Two-step reduced mechanism of nitrogen chemistry is now

$$\begin{array}{ll}
\text{VI} & N_2 + O_2 = 2NO \\
\text{VII} & NO + O_2 + 2H_2 = NO_2 + 2H + H_2O
\end{array}$$

The rates of these global steps are

$$\begin{aligned}
w_{\rm I} &= -w_{36} + w_{56} + w_{57} + w_{58} \\
w_{\rm II} &= w_{33} + w_{35} + 2w_{36} + w_{37} - w_{53} - 2w_{56} + w_{72} \\
w_{\rm III} &= w_{18} - w_{20} + w_{28} \\
w_{\rm IV} &= w_5 - w_{11} + w_{13} + w_{14} + w_{15} + w_{16} + w_{17} \\
&\quad + w_{21} + w_{22} + w_{23} - w_{32} + w_{34} + 2w_{36} + w_{48} + w_{49} \\
&\quad + w_{50} - w_{56} - w_{57} - 2w_{58} + w_{65} - w_{69} + w_{72} - w_{74} \\
w_{\rm V} &= w_1 + w_6 + w_9 + w_{11} - w_{14} - w_{17} - w_{20} - w_{26} \\
&\quad + w_{33} + w_{37} + w_{43} - w_{46} - w_{48} - w_{53} + w_{57} + w_{58} \\
&\quad (-w_{\rm N1} + w_{\rm N2} + w_{\rm N5}) \\
w_{\rm VI} &= w_{\rm N1} \\
w_{\rm VII} &= w_{\rm N4} + w_{\rm N5}
\end{aligned} \tag{17.10}$$

The reactions in parenthesis are used only when the reduced mechanism of the nitrogen chemistry is considered. In order to obtain mathematically tractable expressions for OH, we determine the concentrations of OH from partial equilibrium of w_3. Also we determine the concentration of O from a simplified form of its steady state relation involving only w_1, w_2 and w_4.

$$\begin{aligned} [\mathrm{OH}] &= [\mathrm{H}]\,[\mathrm{H_2O}]/(K_3[\mathrm{H_2}]) \\ [\mathrm{O}] &= (k_{1f}[\mathrm{H}]\,[\mathrm{O_2}] + k_{2b}[\mathrm{OH}]\,[\mathrm{H}] + k_{4f}[\mathrm{OH}]\,[\mathrm{OH}])/(k_{1b}[\mathrm{OH}] \\ &\quad + k_{2f}[\mathrm{H_2}] + k_{4b}[\mathrm{H_2O}]) \end{aligned} \tag{17.11}$$

As we assume steady-state for C_2H_2 and C_2H_4, the equations for calculating C_2H_3 and C_2H for the five-step mechanism are different from those for the nine-step mechanism. The concentrations of C_2H_3, CH_2, C_2H_2 and C_2H_4 are calculated as follows.

$$\begin{aligned} [\mathrm{C_2H_3}] &= (Z_{51b} + Z_{52f} + Z_{54f})(Z_{57} + Z_{58f})(\omega_{58b} + \omega_{59} \\ &\quad + \omega_{60} + \omega_{61})/\mathrm{N_{C_2H_3}} \\ \mathrm{N_{C_2H_3}} &= (1 - Z_{51b})(k_{49}[\mathrm{H}] + k_{50}[\mathrm{O_2}] + k_{51f}) \\ &\quad + (1 - Z_{51b}Z_{55} - Z_{52f} - Z_{54f})(k_{52b}[\mathrm{H_2}] + k_{54b}[\mathrm{H_2O}]) \\ [\mathrm{C_2H_4}] &= \{\omega_{52b} + \omega_{54b} + (Z_{57} + Z_{58f})(\omega_{58b} + \omega_{59} \\ &\quad + \omega_{60} + \omega_{61})\}/\mathrm{N_{C_2H_4}} \\ \mathrm{N_{C_2H_4}} &= \{k_{52f} + (1 - Z_{57} - Z_{58f})k_{58b}\}[\mathrm{H}] + k_{53}[\mathrm{O}] \\ &\quad + k_{54f}[\mathrm{OH}] + k_{55}[\mathrm{M'}] \\ [\mathrm{C_2H_2}] &= \{\omega_{49} + \omega_{50} + \omega_{51f} + Z_{55}(\omega_{52b} + \omega_{54b}) \\ &\quad + (Z_{57} + Z_{58f})(\omega_{58b} + \omega_{59} + \omega_{60} + \omega_{61})\}/\mathrm{N_{C_2H_2}} \\ \mathrm{N_{C_2H_2}} &= (k_{45} + k_{46})[\mathrm{O}] + (1 - Z_{41f} - Z_{47b})k_{47f}[\mathrm{OH}] \\ &\quad + \{(1 - Z_{41f} - Z_{47b})k_{41b} + k_{51b}\}[\mathrm{H}] \\ [\mathrm{CH_2}] &= [\omega_{33f} + \omega_{45} + Z_{43f}\{Z_{42}(\omega_{41b} + \omega_{47f}) + \omega_{46}\}]/\mathrm{N_{CH_2}} \\ \mathrm{N_{CH_2}} &= (1 - Z_{25b})k_{25f}[\mathrm{H}] + \{k_{26} + (1 - Z_{43f})k_{43b}\}[\mathrm{O}] \\ &\quad + (k_{27} + k_{28})[\mathrm{O_2}] + k_{33b}[\mathrm{H_2}] \\ Z_{41f} &= k_{41f}[\mathrm{H_2}]/(k_{41f}[\mathrm{H_2}] + k_{42}[\mathrm{O_2}] + k_{47b}[\mathrm{H_2O}]) \\ Z_{47b} &= k_{47b}[\mathrm{H_2O}]/(k_{41f}[\mathrm{H_2}] + k_{42}[\mathrm{O_2}] + k_{47b}[\mathrm{H_2O}]) \\ Z_{51b} &= k_{51b}[\mathrm{H}]/\mathrm{N_{C_2H_2}} \\ Z_{52f} &= k_{52f}[\mathrm{H}]/\{k_{52f}[\mathrm{H}] + k_{53}[\mathrm{O}] + k_{54f}[\mathrm{OH}] + k_{55}[\mathrm{M'}] \\ &\quad + (1 - Z_{57} - Z_{58f})k_{58b}[\mathrm{H}]\} \\ Z_{54f} &= k_{54f}[\mathrm{OH}]/\{k_{52f}[\mathrm{H}] + k_{53}[\mathrm{O}] + k_{54f}[\mathrm{OH}] + k_{55}[\mathrm{M'}] \\ &\quad + (1 - Z_{57} - Z_{58f})k_{58b}[\mathrm{H}]\} \\ Z_{55} &= k_{55}[\mathrm{M'}]/\{k_{52f}[\mathrm{H}] + k_{53}[\mathrm{O}] + k_{54f}[\mathrm{OH}] + k_{55}[\mathrm{M'}] \\ &\quad + (1 - Z_{57} - Z_{58f})k_{58b}[\mathrm{H}]\} \\ Z_{57} &= k_{57}[\mathrm{O_2}]/(k_{56f}[\mathrm{H}] + k_{57}[\mathrm{O_2}] + k_{58f}) \\ Z_{58f} &= k_{58f}/(k_{56f}[\mathrm{H}] + k_{57}[\mathrm{O_2}] + k_{58f}) \end{aligned} \tag{17.12}$$

17.5 Extinction Strain Rates for Mixtures of CH_4 and C_2H_6

In order to investigate the influence of additions of C_2H_6 to CH_4 as a fuel, diffusion flames mixtures of both fuels were calculated based on the starting mechanism. The maximum temperature is plotted as a function of the inverse of the strain rate in Fig. 17.1. The percentage of CH_4 in the mixture varies from 0 to 100 in terms of mole fraction $X_{CH_4,\infty}$. This is related to the mass fraction $Y_{CH_4,\infty}$ by

$$Y_{CH_4,\infty} = X_{CH_4,\infty} W_{CH_4}/\overline{W} \tag{17.13}$$

where the mean molecular weight is

$$\overline{W} = X_{CH_4,\infty} W_{CH_4} + (1 - X_{CH_4,\infty}) W_{C_2H_6} \tag{17.14}$$

The mass fraction of C_2H_6 in the fuel stream is, of course, $Y_{C_2H_6,\infty} = 1 - Y_{CH_4,\infty}$.

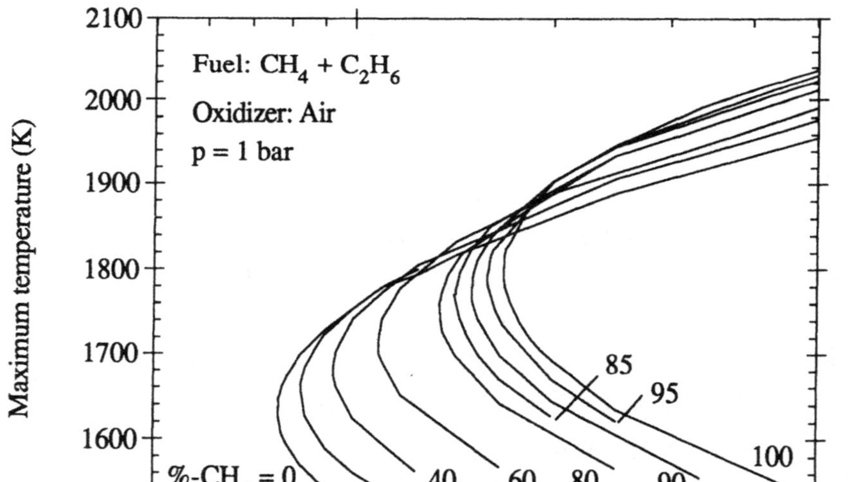

Fig. 17.1. Maximum temperatures as a function of strain rate for mixtures of methane and ethane exhibiting turning point behavior

It is seen from Fig. 17.1 that the temperature is a double valued function of the strain rate. These results were obtained by using an adaptive continuation algorithm [17.1]. The method used here deviates somewhat from the one used in [17.1]. Starting from a solution of high strain rate near extinction, the strain rate is reduced to lower values, causing the solution to converge to the lower branch solution. Subsequently, the strain rate was increased giving the upper branch solutions. Extinction is defined by the vertical tangent to this curve where $\mathrm{d}a/\mathrm{d}T = 0$. For pure CH_4 as a fuel the extinction strain rate is the smallest and for pure C_2H_6 the largest. This indicates that the addition of ethane to a methane diffusion flame makes the flame more resistent to flame stretch because the chemistry becomes faster.

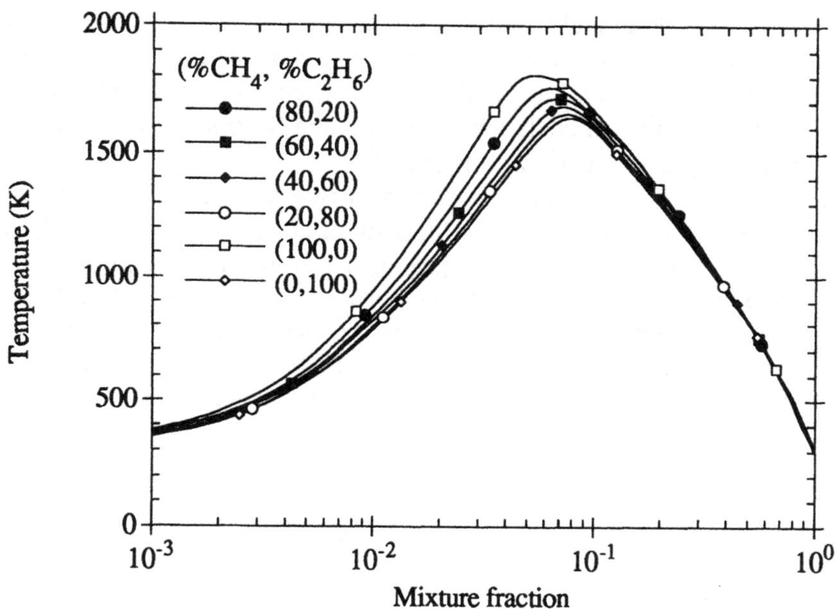

Fig. 17.2. Temperature profiles at extinction as a function of mixture fraction for mixtures of methane and ethane.

Temperature profiles at extinction are shown in Fig. 17.2 as a function of the mixture fraction. It is seen that the maximum of the temperature profile shifts to larger values of the mixture fraction when C_2H_6 is added to CH_4. This shift is larger than the shift in the stoichiometric mixture fraction which may be calculated from

$$Z_{st} = \left[1 + \frac{W_{O_2}}{Y_{O_2,-\infty}}\left(\frac{2Y_{CH_4,\infty}}{W_{CH_4}} + \frac{7}{2}\frac{Y_{C_2H_6,\infty}}{W_{C_2H_6}}\right)\right]^{-1} \tag{17.15}$$

yielding $Z_{st} = 0.0585$ for pure C_2H_6 as compared to $Z_{st} = 0.0548$ for pure CH_4. The shift is associated with a decrease of the maximum temperature when C_2H_6 is added. This leads to the following argument: At extinction heat generation by chemical reactions just balances heat loss out of the reaction zone by convection and diffusion. Therefore, if one assumes that the heat release increases with temperature due to Arrhenius kinetics of the reaction rates, a decrease of the maximum temperature at extinction indicates that chemistry becomes faster when C_2H_6 is added, because it effectively balances the losses even at lower temperatures. This is in agreement with the higher strain rates at extinction.

It is interesting to compare the profiles of the radicals H, OH and O at extinction which are shown in Figs. 17.3, 17.4 and 17.5, respectively. While the mole fraction of H increases by nearly a factor of two for pure C_2H_6 as compared to pure CH_4, O increases very little and OH even decreases. The increase of H indicates that the C_2-chain, pobably due to the important reaction $58f$, generates more H radicals and contains less chain breaking steps than the C_1-chain (cont. Chap. 5 of this book). Since H radicals initiate the chain breaking mechanism via reaction 1f, the overall effect is a faster chemistry. The larger amount of OH radicals for pure methane flames can probably be explained by the higher extinction temperature.

To summarize these observations extinction strain rates and the maximum extinction temperatures are shown as a function of the mole fraction of C_2H_6 in the fuel mixture in Fig. 17.6. This figure also shows that these quantities change approximately linearly with the percentage of C_2H_6 as indicated by the straight lines. The effect of pressure on the extinction strain rate is shown in Fig. 17.7 for pure methane and a 85% CH_4, 15% C_2H_6 mixture. In both cases the extinction strain rate increases up to pressures of 10 bar and then levels off.

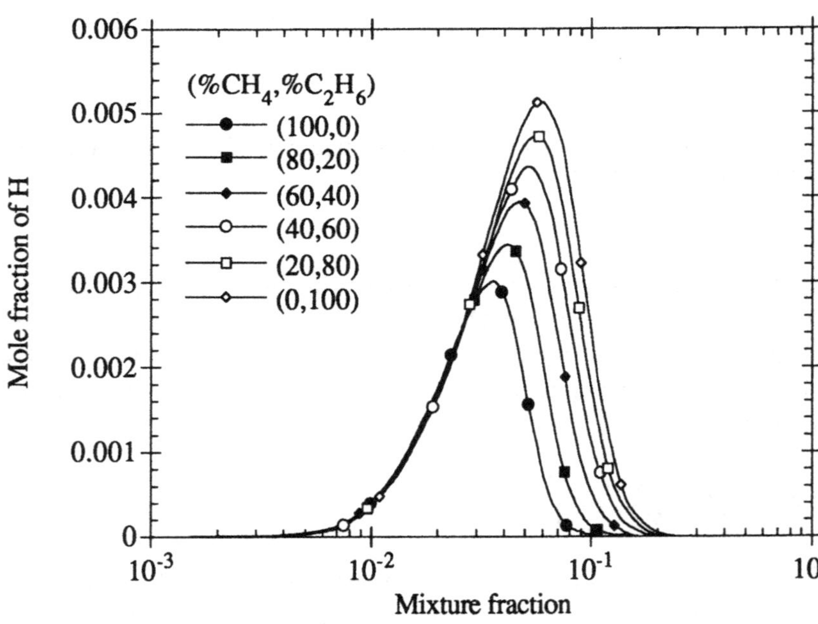

Fig. 17.3. Profiles of H radical at extinction as a function of mixture fraction for mixtures of methane and ethane.

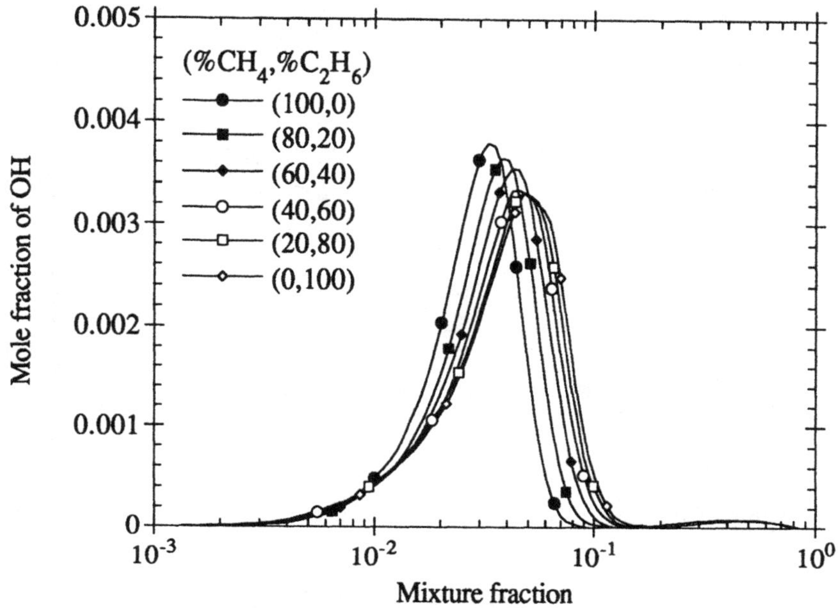

Fig. 17.4. Profiles of OH radical at extinction as a function of mixture fraction for mixtures of methane and ethane.

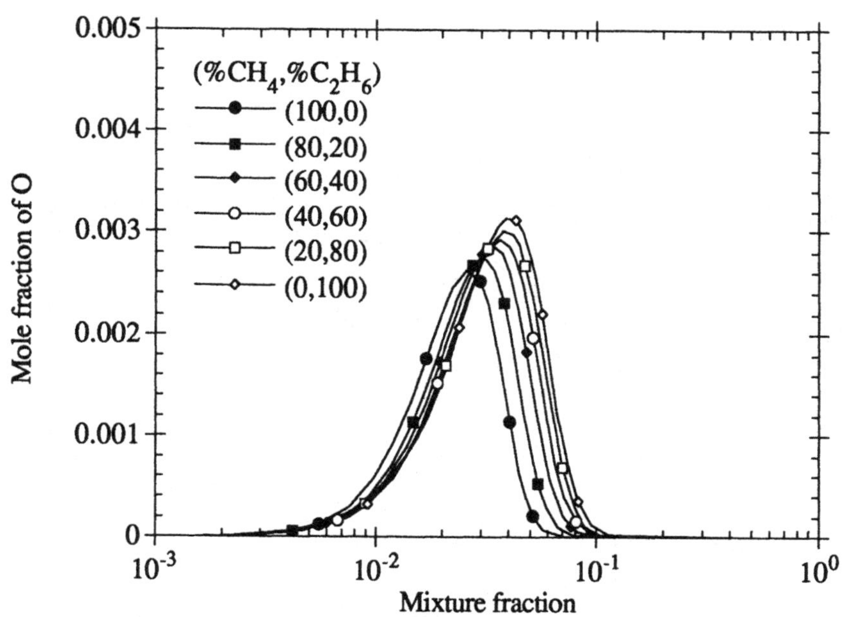

Fig. 17.5. Profiles of O radical at extinction as a function of mixture fraction for mixtures of methane and ethane.

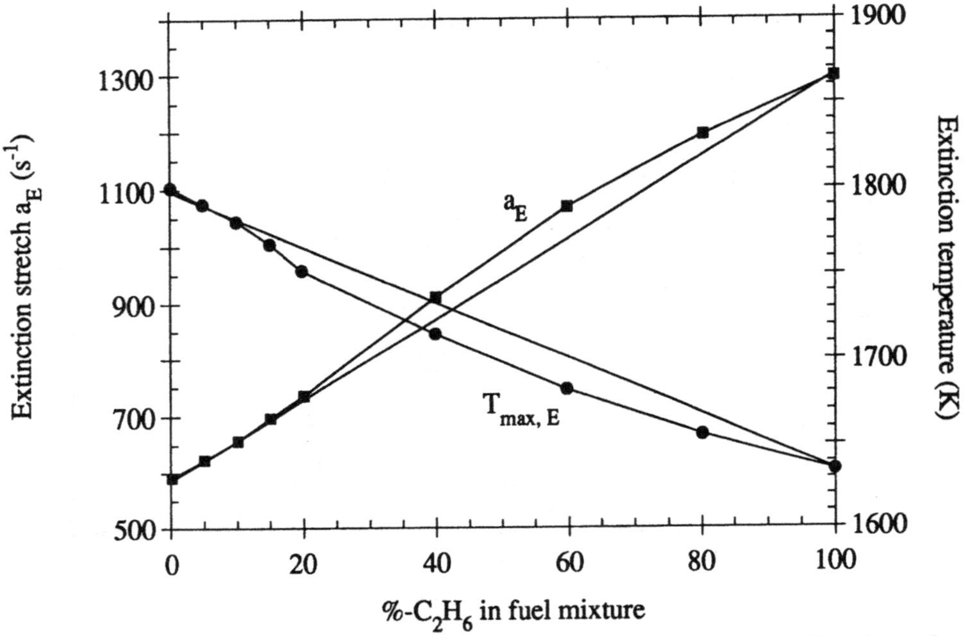

Fig. 17.6. Extinction strain rates and maximum temperatures as a function of ethane mole fraction in the fuel mixture.

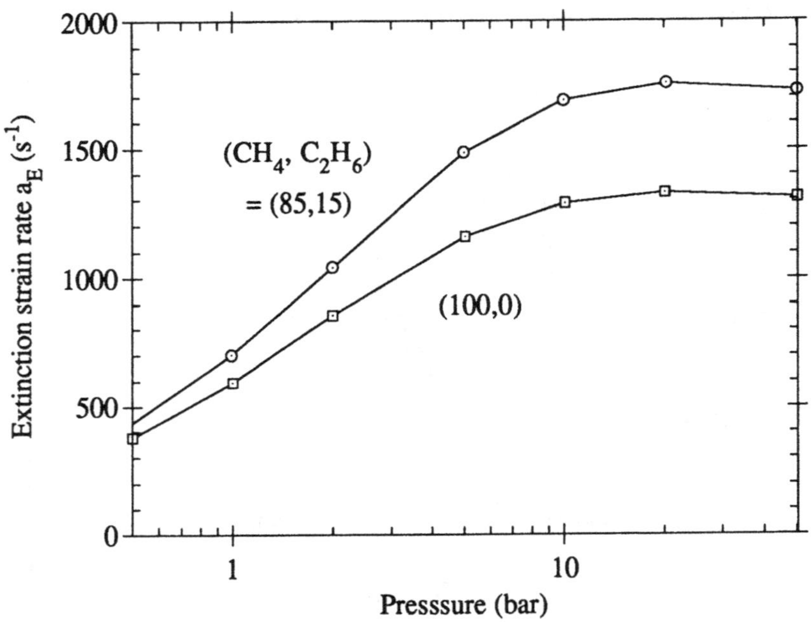

Fig. 17.7. Extinction strain rates as a function of pressure.

17.6 Comparison between Starting Mechanism, the 9-Step Mechanism and the 5-Step Mechanism

For the case of a 85% CH_4, 15% C_2H_6 mixture, which was introduced above as a model fuel representing natural gas, results obtained with the 9-step and 5-step mechanism were compared with those obtained from the starting mechanism. The maximum temperature is plotted in Fig. 17.8 as a function of the strain rate. For low strain rates up to 400/sec the maximum temperature is larger about 50 K for the 9-step mechanism and by about 100 K for the 5-step mechanism. The nine-step mechanism performs quite well even at extinction. This statement is, however, only valid for a narrow range of pressures. When extinction strain rates from both reduced mechnaisms are compared with those from the starting mechanism for increasing pressure in Fig. 17.9, they are seen to be considerably smaller and to level off and then decrease at around 3 bars, while that from the starting mechanism still increases. In Fig. 17.10 extinction strain rates and extinction temperatures are plotted as a function of the percentage of C_2H_6 in the fuel. The 9-step mechanism in its present formulation performed well until 60% C_2H_6 but the 5-step mechanism only up to 20% C_2H_6. These differences are probably due to the steady assumptions for C_2H_2 and C_2H_4 and to the additional truncations, in particular those that led to the expressions for O and OH in (17.11). The results in Figs. 17.9 and 17.10 show that the reduced mechanisms derived here which were based on steady state approximations valid essentially for methane combustion at 1 atm are not suited for pure C_2H_6.

17.7 NO_x Formation for a Fuel Mixture of 85% CH_4, 15% C_2H_6

The formation of NO_x in natural gas flames is a problem of major practical concern. In considering NO_x as the sum of NO and NO_2 concentrations, one recognizes the fact that the chemical conversion of NO to NO_2 through reaction N4 and N5 during the combustion process does not really change the pollution problem. In principle there are two major contributions to NO formation: the thermal NO formed through the extended Zeldovich mechanism namely reactions N1–N3, and the "prompt NO" mechanism wherein hydrocarbon fragments attack bimolecular nitrogen, producing atomic nitrogen, cyanides, and amines, which subsequently oxidize to nitric oxide. Which one of these mechanisms predominates depends mainly on the flame temperature. Prompt NO is not very sensitive to temperature, it contributes between ten to thirty ppm to the total NO production in hydrocarbon flames, while thermal NO may contribute up to several hundred ppm. It is clear that the high NO levels that occur in practical systems can only substantially be reduced by reducing thermal NO formation.

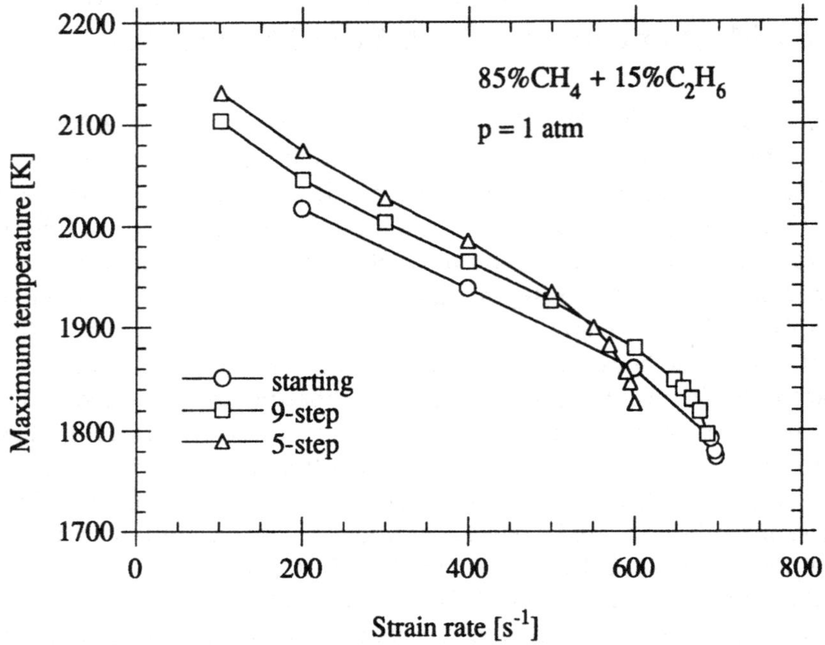

Fig. 17.8. Comparison of maximum temperatures as a function of the strain rate for the starting and reduced mechanisms.

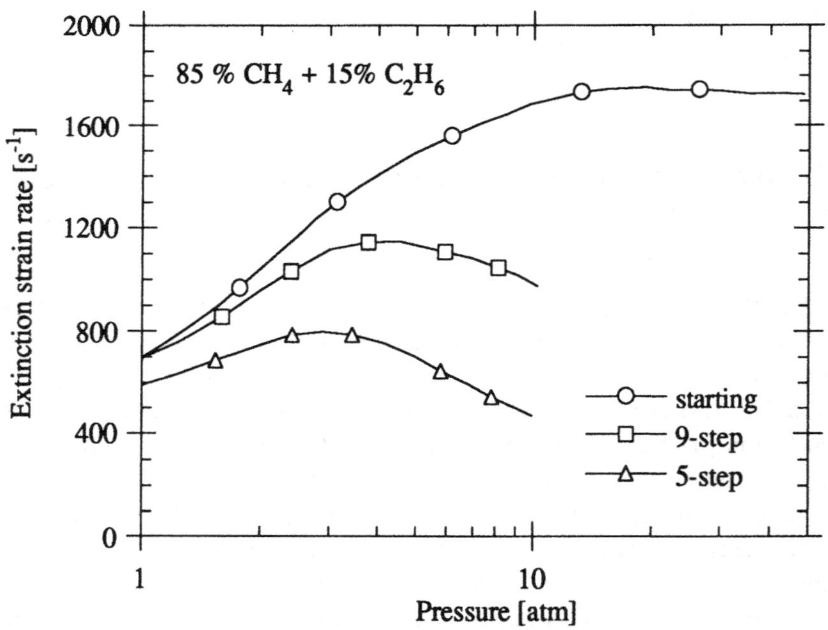

Fig. 17.9. Comparison of extinction strain rates as a function of pressure for the starting and reduced mechanisms.

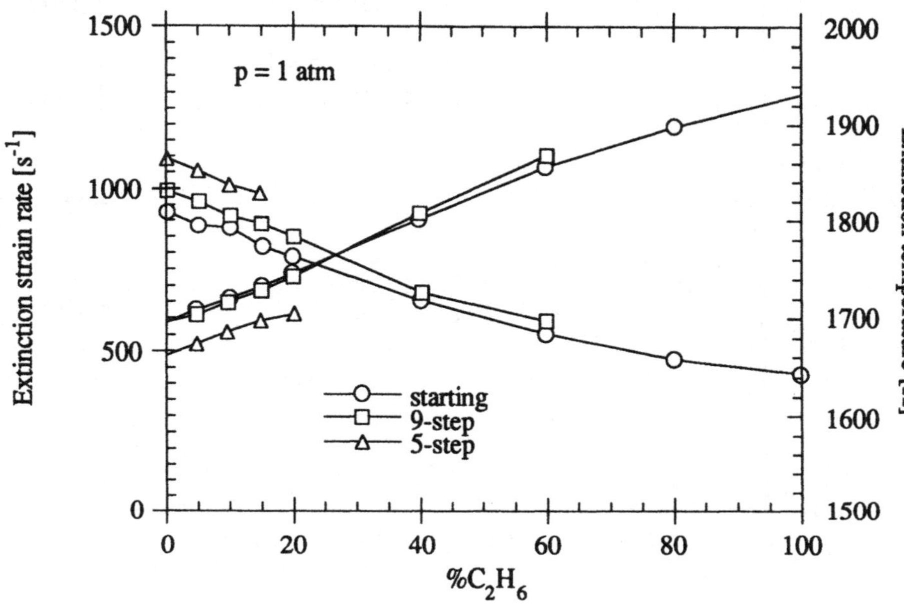

Fig. 17.10. Extinction strain rates and temperatures as a function of ethane mole fraction in the fuel mixture for the starting and reduced mechanisms.

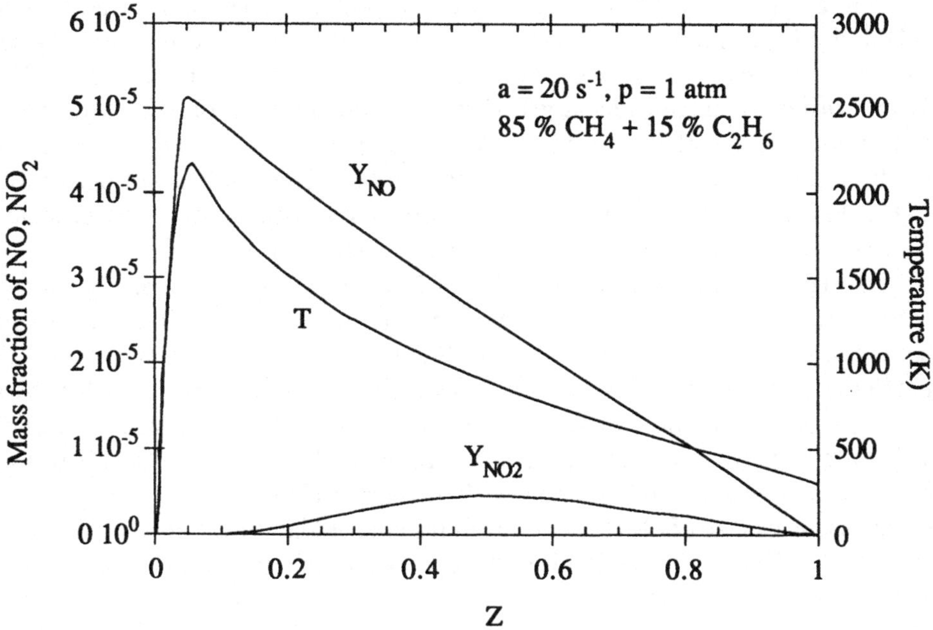

Fig. 17.11. Profiles of NO and NO_2 as a function of mixture fraction at $a = 20\ s^{-1}$.

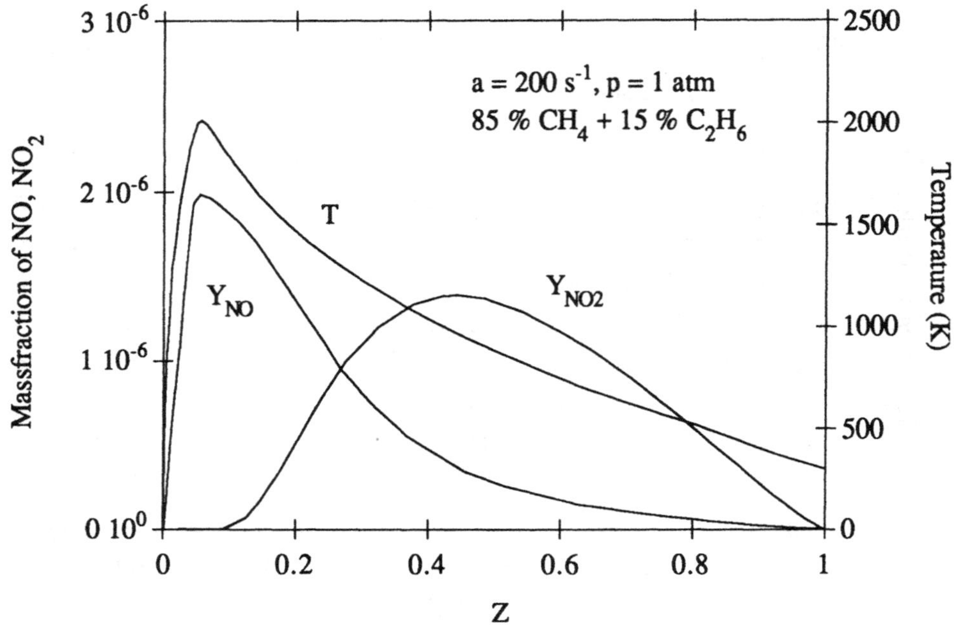

Fig. 17.12. Profiles of NO and NO_2 as a function of mixture fraction at $a = 200\ s^{-1}$.

For a mixture of 85% CH_4, 15% C_2H_6 profiles of the mass fractions of NO and NO_2 are plotted in Figs 17.11 and 17.12 for $p = 1$ atm at $a = 20$/sec and $a = 200$/sec, respectively. For the low strain rate the maximum of NO is one order of magnitude larger than the two maxima of NO_2 which are approximately equal and occur at lower temperatures on the fuel and the oxidizer side of the flame. (The maximum on the oxidizer side is not detectable in this figure due to the limitations in graphical resolution). At the large strain rate the maxima of NO and NO_2 are nearly equal in magnitude.

In Fig. 17.13 the production rates of NO and NO_2 are shown. For small rates of $a = 20\,s^{-1}$ ($p = 1$ atm), the production rates of NO in the flame zone exceed the destruction rates of NO in the low temperature zone. In particular, the destruction rates of NO on the fuel side is low, which makes the Y_{NO} profile on the fuel side in Fig. 17.11 linear. For high strain rates of $a = 200\,s^{-1}$ ($p = 1$ atm), the destruction rates of NO and production rates of NO_2 on the fuel side are the larger with increasing the strain rates (Fig. 17.13) because the leakage of O_2 through the flame zone is significant and the reaction rate of $NO+HO_2 \to NO_2+OH$ is high. The Y_{NO} profile in Fig. 17.12 on the fuel side is nonlinear because of high destruction rate of NO. The NO_2 production peak on the fuel side is higher than that on the oxidizer side and its value is larger as that for NO.

In Figs. 17.14 and 17.15 the integrated NO_x production rates are shown where

$$I_{NO_x} = I_{NO} + (W_{NO_2}/W_{NO}) I_{NO_2},$$

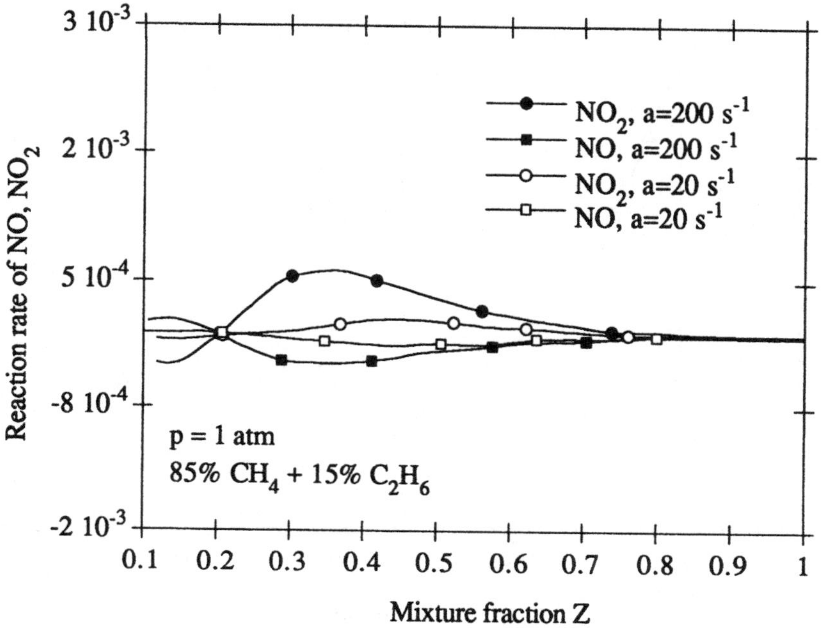

Fig. 17.13. Production rates of NO and NO_2 as a function of mixture fraction at $a = 20\,s^{-1}$ and $a = 200\,s^{-1}$.

with

$$I_{NO_i} = \int_0^1 \frac{\omega_{NO_i}(Z)}{\rho(Z)} dZ \tag{17.16}$$

This formulation is suggested by equation (18) in [17.2] and represents a measure for the NO_x production in turbulent diffusion flames. In general I_{NO_x} decreases with strain rate. The fact that I_{NO_x} increases with increasing strain rates for $20\,s^{-1} < a < 100\,s^{-1}$ for the starting mechanism results from the contribution of I_{NO_2}, which increases due to the increase of the NO_2 production rate on the fuel side.

The integrated NO_x production rates calculated from the 9-step reduced mechanism are close to those calculated with the starting mechanism. Considerable deviations of typically a factor of two occur for the five step mechanism. This larger sensitivity on approximations, in particular concerning the O concentration, was to be expected. Nevertheless, the integrated NO_x production rates are still of the same order of magnitude as those of the starting mechanism.

In Fig. 17.16 the maximum values for NO and NO_2 are shown as a function of strain rate. The NO peak values between reduced 9-step and the starting mechanism fall on the same curve for $a > 5/s$, while NO_2 peak values at small strain rates deviate somewhat. For the results of the reduced mechanism, the NO_2 peak on the fuel side is always higher than that on the oxydizer side irrespective of strain rates. Therefore, NO_x production is well predicted by

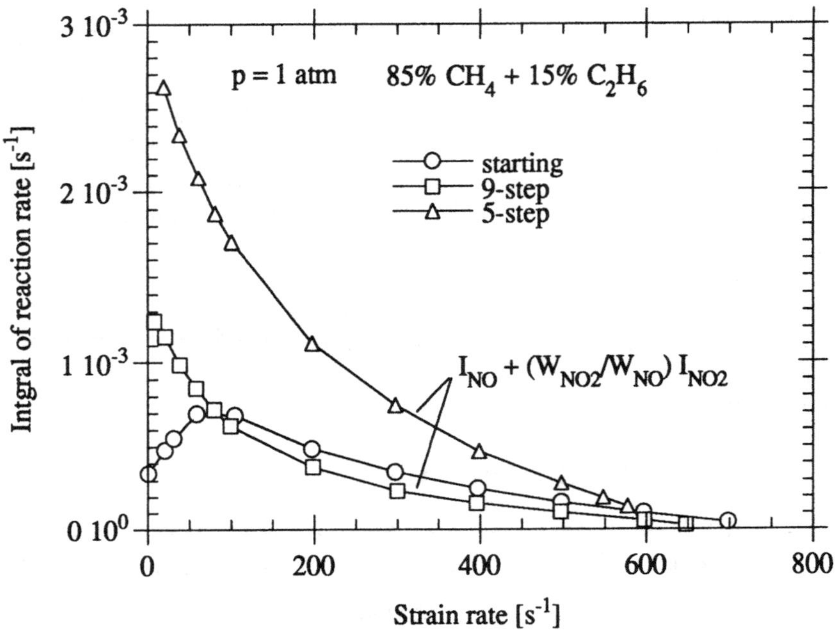

Fig. 17.14. Comparison of integrated NO_x production rates as a function of strain rate at p = 1 atm for the starting and reduced mechanisms.

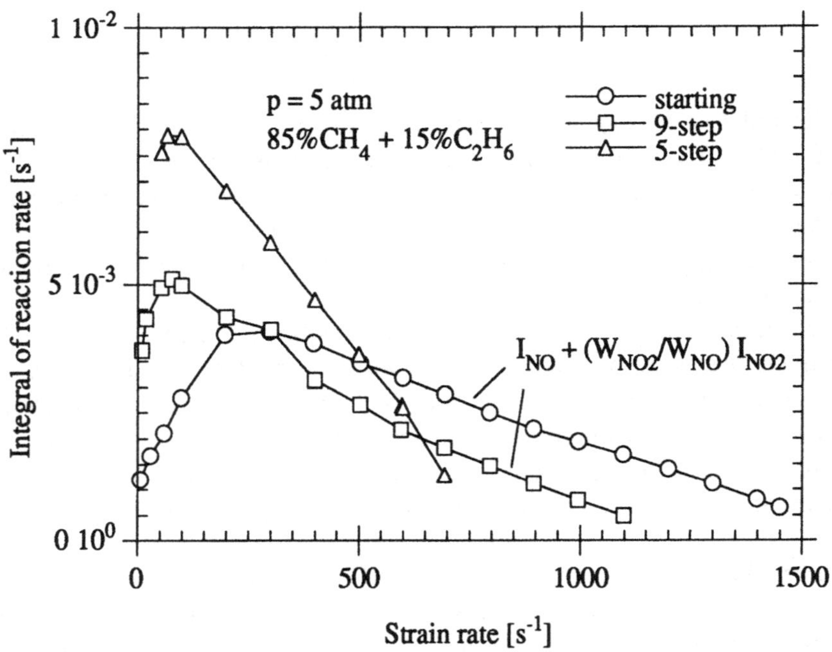

Fig. 17.15. Comparison of integrated NO_x production rates as a function of strain rate at p = 5 atm for the starting and reduced mechanisms.

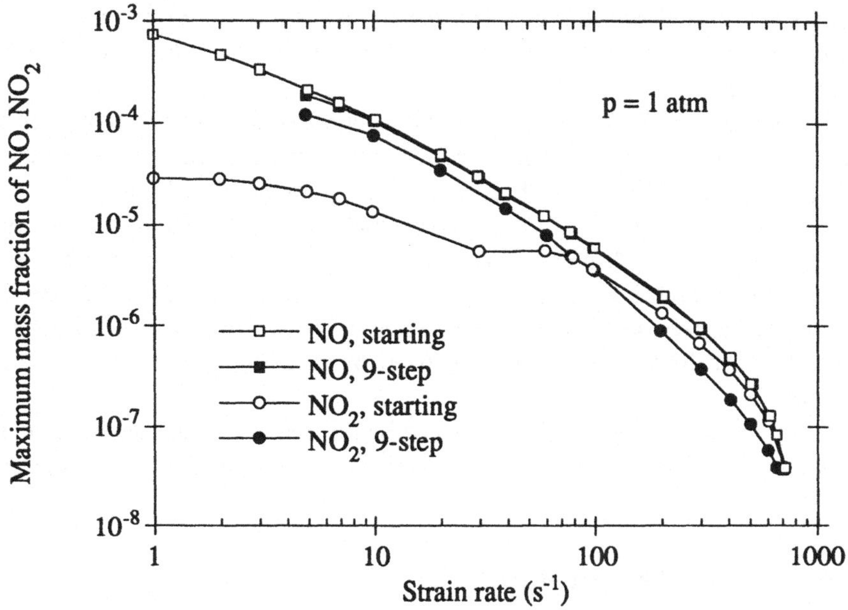

Fig. 17.16. Comparison of maximum mass fractions of NO and NO_2 as a function of strain rate at $p = 1$ atm for the starting and reduced mechanisms.

the 9-step mechanism for high strain rates but poor for small strain rates. Particularly, in the case of small strain rates, the production rate on the fuel side causes the NO_2 peak to be very high, which is very different from the case for the starting mechanism. This error is due to steady state assumption of HO_2. This also causes the difference between the 9-step reduced and the starting mechanism in Fig. 17.14 at small strain rates.

17.8 Conclusions

The diffusion flame structures of the mixture of methane and ethane were investigated using numerical calculations with the starting and reduced mechanisms to simulate the natural gas combustion. Based on the starting mechansim, the 9-step and the 5-step reduced mechanisms have been derived for the mixtures of CH_4 and C_2H_8. In addition, the nitrogen chemistry related to the thermal NO_x has been systematically reduced to two global steps.

Adding ethane to a methane diffusion flame makes the flame more resistent to flame stretch because the reaction becomes faster. The extinction strain rates and temperatures vary approximately linearly with the percentage of C_2H_6. The extinction strain rate increases with increasing pressure up to the pressure of 10 bar and then levels off.

The 9-step mechanism agrees well with the starting mechanism up to 66% of C_2H_6 in the fuel, while the 5-step mechanism up to 20% of C_2H_6 due to the steady state assumptions for C_2H_2 and C_2H_4 and to the additional truncation for O and OH.

The integrated NO_x production rates calculated from the 9-step reduced mechanism agree well with those from the starting mechanism whereas those from the 5-step reduced mechanism are less satisfactory. This is due to the steady state assumptions of O and OH which have influence on the formation of NO_x. The peak of NO concentration among these reduced mechanism and the starting mechanism agrees well for the full range of the strain rate, while that of NO_2 agrees well except for relatively low strain rate. The discrepancy for the low strain rate could be improved by relaxing the steady assumption of HO_2.

Comparisons between these reduced and the starting mechanism show that the reduced mechanism can successfully be derived for the mixture fuel.

17.9 Acknowledgements

S. H. Chung thanks the Korea Science and Engineering Foundation (KOSEF) and the Deutsche Forschungsgemeinschaft (DFG) for supporting this project.

References

[17.1] Giovangigli, V., Smooke, M. D., "Extinction of Strained Premixed Laminar Flames with Complex Chemistry", Comb. Sci. and Tech., **53**, p. 23, 1987.

[17.2] Peters, N., Donnerhack, S., Structure and Similarity of Nitric Oxide Production in Turbulent Diffusion Flames, Eighteenth Symposium (Int.) on Combustion, The Combustion Institute, pp. 33–42, Pittsburgh 1981.

[17.3] Drake, M. C., Blint, R. J., "Structure of Laminar Opposed-Flow Diffusion Flames with $CO/H_2/N_2$ Fuel", Comb. Sci. and Tech., **61**, p. 187, 1988.

Appendices

Appendix A

Flamelet Libraries – Description of the Data Format and Catalogue of Existing Libraries

Josef Göttgens and Fabian Mauss
Institut für Technische Mechanik, RWTH Aachen, W-5100 Aachen, Germany

A.1 Introduction

Numerical computations of flames are notorious for the amount of data they produce and having means to deal with them in a managable way seems desirable. Part of this is a data storage scheme for flame and flamelet computations that takes away much of the burden one usually faces when working with all those numbers. All flamelet libraries described in this book are archived in the format we are going to describe and are available on request from the authors.

Besides documenting all results from a specific numerical computation, it is now easy to query the whole "flamelet data base" to find answers to questions not anticipated at the time the library was computed. One possible utilization of flamelet libraries is described in [1]. Guided by previous asymptotic analyses, burning velocities and flame thicknesses are approximated there. This kind of work could be useful in modeling turbulent chemically reacting flows. The following describes the contents of the rest of this chapter.

Section A.3 describes the data format and gives examples of how to access the data. Section A.4 deals with groups of data files. For example, it does not make sense to store a lengthy reaction scheme in each data file. What kind of information can be found in other file types is described here.

In Sect. A.5 we define what information should be supplied to characterize a flamelet data set sufficiently well. Here a set of identifiers is specified, which describe a particular type of flame. Finally, Sect. A.6 gives a list of all Flamelet Libraries that are currently available, and describes how to obtain them.

A.2 Guidelines for Flamelet Libraries

The proposed data format was chosen with the following points in mind:

1. *The data format should be as machine independent as possible.*
   Although binary files can be read into memory much faster than text files, their formats require much work to make their content accessible on a wide range of computers. Therefore we chose a format that is based on a text representation.
2. *The data format should be human-readable.*
   This allows to analyze the data without having special programs that work on the data sets. Any text editor on any computer should suffice to look at the data. This goes a bit beyond a text representation, e.g., we prefer to read `gridPoints = 101`, instead of remembering that the number of grid points is the 3rd number in the 7th line.
3. *The format should not depend on features of a certain programming language.*
   Today, probably most numerical simulations of chemical processes are still written in FORTRAN, a language, which is rich in number formatting support, but support of text-files is line oriented and character processing is weak. On the other hand, there is the C programming language, which is popular on UNIX systems. Its file handling builds on the notion of a stream, where a new line just means that a special newline character appears in the input stream. Furthermore, Pascal does not have the library support for formatting numbers as C or Fortran have. To summarize, text file processing can be very different between different languages. By building too much on special features of a language, it happens easily that data files are very difficult to interpret in a different language, which we'd like to avoid.
4. *The data format should be flexible.*
   A good data format describes how data are stored, but not what is stored. This separation simplifies later changes to a data set. Consider a data file, where at the beginning there is a short description of what is in the file. There might be lines like

   ```
   Title = "Unstrained Premixed Flame"
   Pressure = 12 [atm]
   ...
   ```

   but a hint about the fuel has been forgotten (or was not necessary to specify at the time the program ran). Just by simply adding later another line of the pattern "symbol, assignment operator, value", here perhaps

   ```
   Fuel = "CH4"
   ```

   would make this information accessible to those who want it, but would not require source code changes in an older program that does not require this information. Having a format based on a simple grammar looks very attractive to us.

5. *The data should not be accessible to wizards only.*
   There is a contradiction between a largely self-explaining data set, based on a "context free grammar", and easy programming. We will try to resolve this by providing libraries that make the contents of a file easily accessible and by somewhat relaxing the rules of a "pure" format. Since the major part of a flamelet library are concentration and temperature profiles, they are stored in a free text format, i.e., an entire profile (vector) can be read into memory with a single FORTRAN statement like

   ```
   dimension yh2o(96)
   ...
   read(12,*) yh2o
   ```

   This requires some manual work to obtain the array dimensions, the order of the profiles, and the number of lines to skip (which carry information that cannot be processed easily with this simple technique), but one doesn't need hours to "get the data" into memory. Of course, skilled programmers would want to write routines that parse the entire file to eliminate any manual work.

A.3 Description of the Data Format

A.3.1 Lexical Details

The content of the data files should be handled in a case insensitive manner. This means that the conversion from uppercase to lowercase letters or vice versa should be handled by the routines that parse the data files. Except for separating tokens, white space characters are not significant.

Examples:

```
pressure = 12 [atm]
Pressure   =   12 [atm]
```

The assignment statements above are equivalent because symbols are case insensitive and additional white space characters are allowed, but

```
pressure=12
```

is wrong, since the tokens `pressure`, =, and `12` should be separated by at least one white space character. The reason to require that tokens be separated by at least one space character is that it facilitates parsing the header section (see below) of a file.

Identifiers and numbers are written exactly as in programming languages and strings must be quoted (e.g., `Title = "Premixed Flame"`). Dashes "-" are allowed in identifier names.

Physical quantities should be tagged with their units. The units enclosed in brackets may either follow the number or the identifier.

Examples:

```
pressure = 12 [atm]
pressure [atm] = 12
```

A.3.2 The Structure of a File

A data file consists of 3 parts: the **header**, the **body** and a **trailer**. The last part, the trailer, is optional. The idea is, that the header contains scalar data that describe the flame in a format, which is easy to read for a human, but still managable by a computer program. The body then contains all the profiles (concentrations, temperatures, ...). The main reason to separate this part from the header is to allow a data format that is easier to handle from a programmer's point of view. The purpose of the trailer is to provide a place to put additional information, which doesn't fit into the header. For example one might put personal notes into the trailer, like a reference to a publication, or someone, who runs a code to generate flamelet data, might want put the name of a file with starting profiles into it. The format of the trailer is free—anything in any way may be written here. The trailer is supposed to be a place for things that don't fit anywhere else.

The individual parts of a data file are prefaced with the keywords `Header`, `Body`, and `Trailer`, such that the structure of a file is:

```
Header
    <info about the data sets>
Body
    <data sets>
Trailer
    <some notes>
```

A.3.3 The Header

The header contains information that characterizes the flame. The order in which the assignment statements occur is not important.

Example:

```
Title = "premixed freely propagating unstrained flame"
Date = "September 18, 1990"
Fuel = "C2H2"
Pressure =  2.000 [bar]
```

Sometimes it is desirable to group variables. A good example is the specification of the unburnt properties of a premixed flame. Similar to records in Pascal or structs in C, variables may be grouped by specifying a group name followed by the keyword `Begin`. Then an arbitrary number of assignment statements can be given and the group is terminated by the keyword `End`.

Example:

```
Unburnt
  Begin
    Temperature        = 298.0000 [K]
    Molefraction-C2H2 =     .0650
    Molefraction-O2    =     .1963
    Molefraction-N2    =     .7386
    ...
  End
```

To ease reading of a group, the group name may be followed optionally by an identifier, giving something like

```
Unburnt mixture
  Begin
      Temperature        = 298.0000 [K]
      Molefraction-C2H2 =     .0650
      Molefraction-O2    =     .1963
      Molefraction-N2    =     .7386
    ...
  End
```

A.3.4 The Body

The body contains the actual data. Each profile is stored as a simple list of data and the profiles are stored on after the other. The identifier part should be on a line by itself (to allow FORTRAN list directed reads).

Example:

```
Temperature [K] =
  .298000E+03 .298000E+03 .298003E+03 .298006E+03 .298012E+03
  .298021E+03 .298039E+03 .298069E+03 .298125E+03 .298218E+03
  ...
```

The assignment operator "=" may be dropped.

One thing to watch out for is that the number of characters per line does not exceed 80, since on some computers (usually mainframes) the default length of lines is 80 characters. It is usually possible to allow larger records, but this may require additional work.

A.3.5 The Trailer

Put here whatever you want. This part should never contain vital information. For instance, informative stuff, that one wouldn't want to put into the header, could go here.

A.3.6 A Sample Flamelet Library File

This is a sample flamelet library file in an abbreviated form (omitted data are marked by ellipsis):

```
Header
    Title = "unstretched freely propagating premixed flame"
    author = "F. Mauss, ITM Aachen, Germany"
    date = "September 18 1990"
    fuel = "C2H2"
    fuel-air-equivalence-ratio = .828
    pressure = 2.000 [bar]
    unburnt mixture
      begin
       temperature        = 298.000 [K]
       molefraction-C2H2 =     .0650
       molefraction-O2   =     .1963
       molefraction-N2   =     .7386
       massfraction-C2H2 =     .0650
       massfraction-O2   =     .1963
       massfraction-N2   =     .7386
      end
     burningVelocity = 93.290 [cm/sec]
     numOfSpecies    = 22
     gridPoints      = 99
Body
x [m]
 -.131485E-03 -.125171E-03 -.118856E-03 -.112541E-03 -.106226E-03
 -.999114E-04 -.935966E-04 -.872819E-04 -.809671E-04 -.746523E-04
   ...
Temperature [K]
  .298000E+03  .298000E+03  .298003E+03  .298006E+03  .298012E+03
  .298021E+03  .298039E+03  .298069E+03  .298125E+03  .298218E+03
   ...
massfraction-H
  .100000E-41  .703249E-17  .170441E-16  .311860E-16  .510716E-16
  .795201E-16  .121779E-15  .187185E-15  .291888E-15  .464684E-15
   ...
massfraction-O
  .558099E-40  .355470E-14  .873300E-14  .162813E-13  .272799E-13
  .432853E-13  .665242E-13  .100130E-12  .148420E-12  .217156E-12
Trailer
Chemical-kinetics mechanism: see file "abc"
Start profiles from /.../.../xxx
```

A.4 Information about Data Series

With the term data series we mean a collection of related data files. Typically, within a data set one or more paramters are changed, like pressure or unburnt mixture composition. Besides the information given in each data file, some information is relevant for the whole set of data. In particular the reaction scheme and a list of the species are given in separate files.

It is probably not necessary to specify a data format for this kind of information, since it is unlikely that a many people need to parse these files.

A.5 Description of Various Flame Types

In this section we describe identifiers used to characterize certain flames. All identifiers are chosen to be descriptive rather than short. The identifiers given here are mandatory, but other identifiers may be added, if appropriate. If the item's data type is "string", we say that explicitly.

It is likely that this section changes in the future as more flamelet libraries become available.

A.5.1 Unstretched Freely Propagating Premixed Flames

We consider the following configuration: a 1-dimensional unstrained premixed fuel stream enters from the left, is burnt, and the burnt gases leave to the right. The upstream velocity (unburnt gas) is equal to the laminar flame speed of the mixture.

This **unstretched freely propagating premixed flame** is described by the following identifiers:

`Title`
The title of the data series, here the string is `"unstretched freely propagating premixed flame"` (string).

`Author`
The person's name who generated the data and his/her affiliation (string).

`Date`
The date when the data were generated (string).

`Fuel`
The name of the fuel (string).

`Fuel-Air-Equivalence-Ratio`
The fuel-air equivalence ratio, often denoted with the symbol Φ.

`Pressure`
Pressure in atm.

`Unburnt`

A group specifying the unburnt state of the mixture. The individual items are

Temperature

MoleFraction-*Fuel* (e.g., MoleFraction-C2H2)

MoleFraction-O2
MoleFraction-N2
MassFraction-*Fuel*
MassFraction-O2
MassFraction-N2

Some of the entries are redundant, but a little redundancy gives security and is sometimes convenient.

BurningVelocity
Usually given in cm/s.

NumOfSpecies
The number of species. The names of the species are given in an extra file (see Sect. A.4) or can be obtained by parsing the rest of the file.

GridPoints
The number of grid points used in the numerical simulation.

Example:

```
header
   Title = "unstretched freely propagating premixed flame"
   author = "F. Mauss, ITM Aachen, Germany"
   date = "February 1991"
   fuel = "CH4"
   pressure =  1.000 [bar]
   fuel-air-equivalence-ratio =  0.900
   unburnt
   begin
        temperature = 298.000 [k]
        molefraction-O2       = 0.1919
        molefraction-CH4      = 0.0863
        molefraction-N2       = 0.7218
        massfraction-O2       = 0.2213
        massfraction-CH4      = 0.0499
        massfraction-N2       = 0.7288
   end
   burningVelocity =  33.043 [cm/sec]
   numOfSpecies =  24
   gridPoints = 100
```

```
  body
x [m]
   -0.110748E-02   -0.105315E-02   -0.998823E-03   -0.944492E-03
   ...
    0.147387E-01    0.152537E-01    0.157688E-01    0.162838E-01
Temperature [K]
    0.298000E+03    0.298000E+03    0.298000E+03    0.298000E+03
    ...
    0.214435E+04    0.214522E+04    0.214603E+04    0.214683E+04
massfraction-H
    0.128008E-21    0.225468E-23    0.101633E-22    0.436973E-22
    ...
    0.382100E-05    0.364120E-05    0.348195E-05    0.335311E-05
massfraction-HO2
    0.000000E+00    0.518394E-12    0.181532E-11    0.506000E-11
    ...
    0.290735E-06    0.278449E-06    0.267455E-06    0.258348E-06
...
massfraction-CH4
    0.499245E-01    0.499244E-01    0.499243E-01    0.499241E-01
    ...
    0.812017E-25    0.104765E-24    0.292520E-24    0.287948E-24
massfraction-N2
    0.728784E+00    0.728783E+00    0.728781E+00    0.728780E+00
    ...
    0.728988E+00    0.728987E+00    0.728985E+00    0.728981E+00
velocity (m/s)
    0.330426E+00    0.330431E+00    0.330438E+00    0.330447E+00
    ...
    0.238338E+01    0.238421E+01    0.238498E+01    0.238576E+01
```

A.5.2 Counterflow Diffusion Flames

We consider a configuration, where a flat (i.e. 1-D) non premixed fuel stream enters from the left ($-\infty$) and a flat oxidizer stream from the right ($+\infty$). The stagnation point is chosen to be the origin of the spatial coordinate. The fuel and the oxidizer diffuse towards each other and a steady state laminar flame establishes at a position where the mixture is stoichiometric. Along the stagnation point streamline the flame can be described mathematically by a one-dimensional two point boundary value problem. We consider planar and axisymmetric counter flow diffusion flames.

A **counterflow diffusion flame** is described by the following identifiers:

`Title`

The title of the data series, here the string is `"planar counterflow diffusion flame"` or `"axisymmetric counterflow diffusion flame"` (string).

Author
The person's name who generated the data and his/her affiliation (string).

Date
The date when the data were generated (string).

Fuel
The name of the fuel (string).

Strain Rate
The velocity gradient on the oxidizer side in [1/s].

Pressure
Pressure in atm.

Oxidizer Side
A group specifying the oxidizer side of the mixture. The individual items are

Temperature
MoleFraction-O2
MoleFraction-N2
MassFraction-O2
MassFraction-N2

Some of the entries are redundant, but a little redundancy gives security and is sometimes convenient.

FuelSide
A group specifying the fuel side of the mixture. The individual items are

Temperature
MoleFraction-*Fuel* (e.g., MoleFraction-C2H2)
MoleFraction-N2
MassFraction-*Fuel* (e.g., MassFraction-C2H2)
MassFraction-N2

NumOf Species
The number of species. The names of the species are given in an extra file (see Sect. A.4) or can be obtained by parsing the rest of the file.

GridPoints
The number of grid points used in the numerical simulation.

Example:

```
HEADER
  Title = "planar counterflow diffusion flame"
  Author = "C. Mueller, ITM Aachen"
  Date = "3/21/1991"
  Fuel = "H2"
  StrainRate = 100 [1/s]
  Pressure = 1 [bar]
  OxidizerSide
    Begin
      Temperature = 298 [K]
      MoleFraction-O2 = 0.21
      MoleFraction-N2 = 0.79
      MassFraction-O2 = 0.232617
      MassFraction-N2 = 0.767383
    End
  FuelSide
    Begin
      Temperature = 298 [K]
      MoleFraction-H2 = 1
      MoleFraction-N2 = 0
      MassFraction-H2 = 1
      MassFraction-N2 = 0
    End
  NumOfSpecies = 7
  GridPoints = 119

BODY
y [m]
  -0.684900E-02  -0.680882E-02  -0.676844E-02  -0.672778E-02
  ...
   0.264616E-02   0.270368E-02   0.276112E-02   0.281851E-02
Z
   0.000000E+00   0.215574E-05   0.508924E-05   0.906375E-05
   ...
   0.998946E+00   0.999394E+00   0.999738E+00   0.100000E+01
Massfraction-H
   0.237338E-12   0.253295E-12   0.397682E-12   0.636252E-12
   ...
   0.133702E-05   0.798587E-06   0.358407E-06   0.100000E-59
Massfraction-O
   0.277345E-06   0.347446E-06   0.435879E-06   0.547398E-06
   ...
   0.263901E-14   0.141931E-14   0.577748E-15   0.793651E-59
...
```

```
TEMPERATURE [K]
   0.298000E+03   0.299233E+03   0.300910E+03   0.303182E+03
   ...
   0.299080E+03   0.298620E+03   0.298268E+03   0.298000E+03
V [M/S]
   0.206380E+00   0.203227E+00   0.200339E+00   0.197798E+00
   ...
  -0.826956E+00  -0.847456E+00  -0.868191E+00  -0.889117E+00
```

A.6 Available Flamelet Libraries

A.6.1 Distribution of Flamelet Libraries

Flamelet Libraries are distributed by

Prof. Dr.-Ing. N. Peters
Institut für Technische Mechanik
RWTH Aachen
Templergraben 64
W-5100 Aachen, Germany

The distribution media are 3.5" floppy discs in Apple Macintosh format. Due to the considerable size of the libraries we compress all files with a public domain program, which we also provide. On request we can also provide the data on SyQuest cartridges.

A.6.2 Premixed Flamelet Libraries

A.6.2.1 Premixed Methane Library

This Library consists of 1038 files of numerical computations of premixed freely propagating unstrained methane flames. The size of the library is about 46 MB. The pressure ranges from 1 atm to 40 atm, the unburnt temperature between 298 K and 800 K. The fuel-air equivalence ratio, Φ, varies between a lean limit, defined by the criterion burning velocity < 1 cm/s, and a rich limit of $\Phi = 2.5$.

The computations were performed by F. Mauss (ITM, RWTH Aachen) in February 1991. His program is based on a program provided by B. Rogg (Cambridge).

A.6.2.2 Premixed Propane Library

This Library consists of 828 files of numerical computations of premixed freely propagating unstrained propane flames. The size of the library is about 46 MB. The pressure ranges from 1 atm to 40 atm, the unburnt temperature between

298 K and 800 K. The fuel-air equivalence ratio, Φ, varies between a lean limit, defined by the criterion burning velocity < 1 cm/s, and a rich limit of $\Phi = 2.5$.

The computations were performed by F. Mauss (ITM, RWTH Aachen) in February 1991. His program is based on a program provided by B. Rogg (Cambridge).

A.6.2.3 Premixed Acytelene Library

This Library consists of 1010 files of numerical computations of premixed freely propagating unstrained acytelene flames. The size of the library is about 57 MB. The pressure ranges from 1 atm to 40 atm, the unburnt temperature between 298 K and 600 K. The fuel-air equivalence ratio, Φ, varies between a lean limit, defined by the criterion burning velocity < 1 cm/s, and a value of 3 on the rich side.

The computations were performed by F. Mauss (ITM, RWTH Aachen) in February 1991. His program is based on a program provided by B. Rogg (Cambridge).

A.6.2.4 Premixed Hydrogen Library

This Library consists of 820 files of numerical computations of premixed freely propagating unstrained hydrogen flames. The size of the library is about 18 MB. The pressure ranges from 1 atm to 40 atm, the unburnt temperature between 298 K and 500 K. The fuel-air equivalence ratio, Φ, varies between a lean limit, defined by the criterion burning velocity < 1 cm/s, and a rich limit, defined by either $\Phi = 11$ or a burning velocity < 10 cm/s.

The computations were performed by F. Mauss (ITM, RWTH Aachen) in February 1991. His program is based on a program provided by B. Rogg (Cambridge).

A.6.2.5 Premixed Hydrogen 2-Step Library

This library consists of 829 files of numerical computations of premixed freely propagating unstrained hydrogen flames. The size of the library is about 17 MB. The pressure ranges from 1 atm to 40 atm, the unburnt temperature between 298 K and 500 K. The fuel-air equivalence ratio, Φ, varies between a lean limit, defined by the criterion burning velocity < 1 cm/s, and a rich limit, defined by either $\Phi = 11$ or a burning velocity < 10 cm/s.

The computations were performed by F. Mauss (ITM, RWTH Aachen) in February 1991. His program is based on a program provided by B. Rogg (Cambridge).

A.6.2.6 Premixed Ethylene Libraries

Ethylene libraries were supplied by W. Wang (Cambridge) and F. Mauss (Aachen).

Mr. Wang's library consists of 20 files of numerical computations of stoichiometric premixed freely propagating unstrained ethylene flames. The pressure ranges from 0.2 atm to 10 atm. Half of the data were computed using a "full mechanism", the other half using a reduced mechanism. The size of the library is about 600 KB. The computations were performed using the Cambridge Laminar Flame Code RUN-1DL developed by B. Rogg.

Mr. Mauss' library consists of 1009 files of numerical computations of lean to rich premixed freely propagating unstrained ethylene flames. The pressure ranges from 1 atm to 40 atm and the unburnt temperature from 298 K to 600 K. The size of the library is about 56 MB.

A.6.2.7 Premixed Ethane Libraries

Ethane libraries were supplied by W. Wang (Cambridge) and F. Mauss (Aachen).

Mr. Wang's Library consists of 20 files of numerical computations of stoichiometric premixed freely propagating unstrained ethane flames. The pressure ranges from 0.2 atm to 10 atm. Half of the data were computed using a "full mechanism", the other half using a reduced mechanism. The size of the library is about 600 KB. The computations were performed using the Cambridge Laminar Flame Code RUN-1DL developed by B. Rogg.

Mr. Mauss' library consists of 881 files of numerical computations of lean to rich premixed freely propagating unstrained ethane flames. The pressure ranges from 1 atm to 40 atm and the unburnt temperature from 298 K to 800 K. The size of the library is about 52 MB.

A.6.3 Diffusion Flame Libraries

A.6.3.1 CO-H_2-N_2/Air Diffusion Flame Library

This Library consists of 81 files of numerical computations of planar CO-H_2-N_2/air diffusion flames. The size of the library is about 1.8 MB. The temperature on the rich and lean side is always 300 K and the pressure ranges from 1 atm to 10 atm. At each pressure, the library data were computed for various strain rates.

The computations were performed by Y. Liu (Cambridge) in July 1991 using the Cambridge Laminar Flame Code RUN-1DL developed by B. Rogg.

A.6.3.2 Axisymmetric Hydrogen Counterflow Diffusion Flame

This library consists of 456 files of numerical computations of axisymmetric hydrogen counterflow diffusion flames with varying strain rates and pressures. The temperatures of the fuel and oxidizer side were fixed at 298 K. The size of the library is about 7.2 MB and was supplied by G. Balakrishnan from UC San Diego.

References

[A.1] Göttgens, J., Mauss, F., Peters, N.: "Analytic Approximations of Burning Velocities And Flame Thicknesses of Lean Hydrogen, Methane, Ethylene, Ethane, Acetylene, And Propane Flames", Twentyfourth Symposium (International) on Combustion, The Combustion Institute, to appear.

Appendix B

RedMech – An Automatic Reduction Program

J. Göttgens and P. Terhoeven
Institut für Technische Mechanik, RWTH Aachen, W-5100 Aachen
Germany

Reducing mechanisms by hand is an error prone and tedious process. Fortunately, it is not very difficult to write a little computer program that does the reduction automatically. Here we will outline the necessary procedures and describe a program that implements the algorithms. The source code for the program is in the public domain and can be obtained from the address given at the end of this appendix. The length of the program is about 1500 lines and much too long to be printed here.

While the numerical algorithms can be expressed in a few lines, the same is not true for the code that handles user input. Since the program is targeted at avoiding silly mistakes, we didn't want to compromise this goal by requiring a clumsy input format that is adjusted to the needs of the computer and not to human beings. In essence, you would specify your mechanism in standard notation and a list of steady-state species with reaction rates to eliminate and obtain a resulting reduced mechanism. Therefore the bulk of the code deals with turning text information into matrices and vectors for easy numerical manipulation.

Recall that chemical reactions are described by

$$\nu'_{\mathrm{A}}\,\mathrm{A} + \nu'_{\mathrm{B}}\,\mathrm{B} \rightleftharpoons \nu''_{\mathrm{C}}\,\mathrm{C} + \nu''_{\mathrm{D}}\,\mathrm{D} \tag{1}$$

and the production/consumption of each species A–D with $\nu_i = \nu''_i - \nu'_i$ being the stoichiometric coefficient of species i is

$$\frac{\mathrm{d}C_i}{\mathrm{d}t} = \nu_i w. \tag{2}$$

Similarly, for a set of r reactions with n species, we can write a conservation equation (using the more general operator $\mathcal{L}(C_i)$ instead of $\mathrm{d}C_i/\mathrm{d}t$)

$$\mathcal{L}(C_i) = \sum_{k=1}^{r} \nu_{ik}\, w_k \quad \text{for} \quad i = 1, \ldots, n \tag{3}$$

for every species i out of n. Equation (3) is a system of n equations that is linear in the reaction rates w_k. ν_{ik} is known as the coefficient matrix of the mechanism. Its i-th row corresponds to the i-th species' conservation equation, the k-th column holds the stoichiometric coefficients of the k-th reaction.

Introducing a steady-state assumption for a species, the operator $\mathcal{L}$ is set to zero. Therefore one can eliminate one reaction rate from the system (3) and evaluate it from the remaining reaction rates instead. This is equivalent to zeroing out the corresponding column in the coefficient matrix except for the row corresponding to the related steady-state species.

This may be implemented in the programming language C as follows: for a chosen reaction `l` related to the steady-state-species `m`, eliminate all entries in column `l` except for the one in row `m` in the coefficient matrix `a[i][k]`,

```
for ( i = 0; i < rows; ++i ) {
    if ( i != m ) {
        factor = -a[i][l] / a[m][l];
        for ( k = 0; k < columns; ++k )
            a[i][k] += factor * a[m][k];
    }
}.
```

The number of linearly independent global reactions, n_g, for a chemical reaction mechanism consisting of r reactions can be computed from

$$n_g = n - n_s - n_e ,$$

where n denotes the number of chemical species, n_s the number of species assumed to be steady-state, and n_e the number of elements present. After eliminating one reaction rate for each steady-state species, we end up with $r - n_s$ remaining reactions, out of which n_g linearly independent reations have to be chosen as global reactions.

As an illustrative example, we present the reduction of a mechanism for H_2. The original mechanism consists of the first eight reactions of Table 1 in Chap. 1:

1 : $H + O_2 \rightleftharpoons OH + O$

2 : $O + H_2 \rightleftharpoons OH + H$

3 : $H_2 + OH \rightleftharpoons H_2O + H$

4 : $OH + OH \rightleftharpoons O + H_2O$

5 : $H + O_2 + M \rightleftharpoons HO_2 + M$

6 : $H + HO_2 \rightleftharpoons OH + OH$

7 : $H + HO_2 \rightleftharpoons H_2 + O_2$

8 : $OH + HO_2 \rightleftharpoons H_2O + O_2$.

As far as the reduction process is concerned, only $w_i = w_{fi} - w_{bi}$ enters the algorithm. Currently, forward and backward reactions are treated as independent reactions and would not handled correctly by the program, therefore we do not add forward **and** backward reaction separately to the mechanism. We assume that O, OH, and HO_2 are in steady state and choose the reactions 2, 3, and 7, respectively, for elimination.

The program takes two input files: the first file describes the original mechanism to be reduced. Since the assembly of the coefficient matrix is error prone (imagine 85 reactions and count all zero fill-ins), the mechanism is entered as plain text in a human readable format, very similar to the standard notation used for chemical reactions. The input file containing the mechanism for our example is the following (the hash mark "#" signals a line comment):

```
# Hydrogen mechanism
1: H + O2 -> OH + O
2: O + H2 => OH + H
3: H2 + OH -> H2O + H
4: OH + OH -> O + H2O
5: H+O2+M->HO2+M
6: 1H + 1 HO2 = 1OH + 1 OH
7: H + HO2 == H2 + O2
8: OH + HO2 -> H2O + O2
```

The second file lists all species, for which steady state is assumed, and the rates of the corresponding reaction to eliminate. The input file used for this example is

```
# steady-state species
OH   w3
O    w2
HO2  w7
```

Given these input files, the program generates the following output: First, the species that have been found in the mechanism are listed, together with the number of elements.

```
7 species found in 8 reactions
H
O2
OH
O
H2
H2O
HO2
2 different elements found.
```

The program prints the mechanism read from the input file and computes the number of linearly independent equations.

```
Original mechanism:
1: H + O2 -> OH + O
2: O + H2 -> OH + H
3: H2 + OH -> H2O + H
4: OH + OH -> O + H2O
5: H + O2 -> HO2
6: H + HO2 -> OH + OH
7: H + HO2 -> H2 + O2
8: OH + HO2 -> H2O + O2
There are 2 linearly independent global reactions.
```

Next, the conservation equations for all species, separated in non-steady-state and steady-state species, is printed.

```
Conservation equations (original mechanism):
L([H]) = - w1 + w2 + w3 - w5 - w6 - w7
L([O2]) = - w1 - w5 + w7 + w8
L([H2]) = - w2 - w3 + w7
L([H2O]) = w3 + w4 + w8

L([OH]) = 0 = w1 + w2 - w3 - 2 w4 + 2 w6 - w8
L([O]) = 0 = w1 - w2 + w4
L([HO2]) = 0 = w5 - w6 - w7 - w8
```

After processing the input files, the program now reduces the mechanism and prints the following remaining reactions and conservation equations.

```
Reduced mechanism:
1: O2 + 3 H2 -> 2 H + 2 H2O
5: 2 H -> H2
6: O2 + 3 H2 -> 2 H + 2 H2O
Conservation equations (reduced mechanism):
L([H]) = 2 w1 - 2 w5 + 2 w6
L([O2]) = - w1 - w6
L([H2]) = - 3 w1 + w5 - 3 w6
L([H2O]) = 2 w1 + 2 w6

L([OH]) = 0 = 2 w1 - w3 - w4 + 2 w6 - w8
L([O]) = 0 = w1 - w2 + w4
L([HO2]) = 0 = w5 - w6 - w7 - w8

Choose 2 global reactions:
```

Now the user is prompted to choose the required number of linearly independent global reactions out of the remaining reactions, in this example these are the reactions 1, 5, and 6. The user's choice is checked for linear independence using an SVD algorithm taken from [A.1] and, if necessary, the user is asked for a correction of the reaction choice.

We choose reactions 1 and 5 as linearly independent global reactions, and the program finally produces the following output:

```
I: O2 + 3 H2 -> 2 H + 2 H2O
II: 2 H -> H2
Conservation equations:
L([H]) = + 2 wI - 2 wII
L([O2]) = - 1 wI
L([H2]) = - 3 wI + 1 wII
L([H2O]) = + 2 wI
Global rates:
wI = w1 + w6
wII = w5
```

As result we obtain the global reactions

$$\text{I}: \quad O_2 + 3H_2 = 2H + 2H_2O$$

$$\text{II}: \quad 2H = H_2\,,$$

the well known 2-step mechanism for H_2 with the global rates

$$w_{\rm I} = w_1 + w_6 \quad \text{and} \quad w_{\rm II} = w_5\,.$$

The program RedMech was written by Josef Göttgens, Peter Terhoeven, and Robert Korff. The source code is in the public domain and available from:

Josef Göttgens, Institut für Technische Mechanik, RWTH Aachen,
W-5100 Aachen, Germany (email: ci010go@dacth11).

References

[B.1] Press, W.H., Flannery, B.P., Teukolsky, S.A., Vetterling, W.T., "Numerical Recipes in C", Cambridge University Press, 1988

Appendix C

RUN-1DL – The Cambridge Universal Laminar Flamelet Computer Code

B. Rogg
University of Cambridge, Department of Engineering
Trumpington Street, Cambridge CB2 1PZ, England

In this Appendix the Cambridge Universal Laminar Flamelet Computer Code RUN-1DL is briefly described. The basic version of this code, which is sufficient to perform calculations of the kind presented in the present book, is distributed for free by

Dr. B. Rogg
University of Cambridge
Department of Engineering
Trumpington Street
Cambridge CB2 1PZ
England.

The distribution media are 3.5" HD floppy disks in Apple Macintosh format. The code is written in standard FORTRAN 77.

To model and numerically predict laminar reacting-flow phenomena in the variety of possible one-dimensional and quasi one-dimensional flow configurations, a sophisticated but user-friendly, versatile and powerful computer program is required. Program RUN-1DL is such a code. It has been developed for the numerical simulation of steady or unsteady laminar, one-dimensional and quasi one-dimensional, chemically reacting flows, viz.,

- unstrained, premixed, freely propagating flames,
- unstrained, premixed, burner-stabilized flames,
- strained premixed flames,
- strained diffusion flames,
- strained partially premixed diffusion flames,
- tubular flames,

and a variety of other chemically reacting flow problems including two-phase flames involving single droplets or sprays. At present two-phase problems can not be handled by the publicly released version of the code. Various chemistry models are implemented in RUN-1DL, viz.,

- detailed mechanisms of elementary reactions,
- systematically reduced kinetic mechanisms,
- one-step global finite-rate reactions, and the
- flame-sheet model for diffusion flames.

Trivially, RUN-1DL is able to simulate inert flow of multi-component mixtures, but truly time-dependent ignition and extinction phenomena as well as other non-trivial combustion phenomena can also be simulated.

In general the code employs detailed multi-component models of chemistry, thermodynamics and molecular transport, but simpler models are also implemented. Other models for chemistry and molecular-transport are easily implemented through user interfaces provided by RUN-1DL. In particular, RUN-1DL can be used together with the CHEMKIN-II package developed by Sandia National Laboratories.

Index

Lecture Notes in Physics

For information about Vols. 1–374
please contact your bookseller or Springer-Verlag

Vol. 375: C. Bartocci, U. Bruzzo, R. Cianci (Eds.), Differential Geometric Methods in Theoretical Physics. Proceedings, 1990. XIX, 401 pages. 1991.

Vol. 376: D. Berényi, G. Hock (Eds.), High-Energy Ion-Atom Collisions. Proceedings, 1990. IX, 364 pages. 1991.

Vol. 377: W. J. Duschl, S.J. Wagner, M. Camenzind (Eds.), Variability of Active Galaxies. Proceedings, 1990. XII, 312 pages. 1991.

Vol. 378: C. Bendjaballah, O. Hirota, S. Reynaud (Eds.), Quantum Aspects of Optical Communications. Proceedings 1990. VII, 389 pages. 1991.

Vol. 379: J. D. Hennig, W. Lücke, J. Tolar (Eds.), Differential Geometry, Group Representations, and Quantization. XI, 280 pages. 1991.

Vol. 380: I. Tuominen, D. Moss, G. Rüdiger (Eds.), The Sun and Cool Stars: activity, magnetism, dynamos. Proceedings, 1990. X, 530 pages. 1991.

Vol. 381: J. Casas-Vazquez, D. Jou (Eds.), Rheological Modelling: Thermodynamical and Statistical Approaches. Proceedings, 1990. VII, 378 pages. 1991.

Vol. 382: V.V. Dodonov, V. I. Man'ko (Eds.), Group Theoretical Methods in Physics. Proceedings, 1990. XVII, 601 pages. 1991.

Vol. 384: M. D. Smooke (Ed.), Reduced Kinetic Mechanisms and Asymptotic Approximations for Methane-Air Flames. V, 245 pages. 1991.

Vol. 385: A. Treves, G. C. Perola, L. Stella (Eds.), Iron Line Diagnostics in X-Ray Sources. Proceedings, Como, Italy 1990. IX, 312 pages. 1991.

Vol. 386: G. Pétré, A. Sanfeld (Eds.), Capillarity Today. Proceedings, Belgium 1990. XI, 384 pages. 1991.

Vol. 387: Y. Uchida, R. C. Canfield, T. Watanabe, E. Hiei (Eds.), Flare Physics in Solar Activity Maximum 22. Proceedings, 1990. X, 360 pages. 1991.

Vol. 388: D. Gough, J. Toomre (Eds.), Challenges to Theories of the Structure of Moderate-Mass Stars. Proceedings, 1990. VII, 414 pages. 1991.

Vol. 389: J. C. Miller, R. F. Haglund (Eds.), Laser Ablation-Mechanisms and Applications. Proceedings. IX, 362 pages, 1991.

Vol. 390: J. Heidmann, M. J. Klein (Eds.), Bioastronomy - The Search for Extraterrestrial Life. Proceedings, 1990. XVII, 413 pages. 1991.

Vol. 391: A. Zdziarski, M. Sikora (Eds.), Ralativistic Hadrons in Cosmic Compact Objects. Proceedings, 1990. XII, 182 pages. 1991.

Vol. 392: J.-D. Fournier, P.-L. Sulem (Eds.), Large-Scale Structures in Nonlinear Physics. Proceedings. VIII, 353 pages. 1991.

Vol. 393: M. Remoissenet, M.Peyrard (Eds.), Nonlinear Coherent Structures in Physics and Biology. Proceedings. XII, 398 pages. 1991.

Vol. 394: M. R. J. Hoch, R. H. Lemmer (Eds.), Low Temperature Physics. Proceedings. X,374 pages. 1991.

Vol. 395: H. E. Trease, M. J. Fritts, W. P. Crowley (Eds.), Advances in the Free-Lagrange Method. Proceedings, 1990. XI, 327 pages. 1991.

Vol. 396: H. Mitter, H. Gausterer (Eds.), Recent Aspects of Quantum Fields. Proceedings. XIII, 332 pages. 1991.

Vol. 398: T. M. M. Verheggen (Ed.), Numerical Methods for the Simulation of Multi-Phase and Complex Flow. Proceedings, 1990. VI, 153 pages. 1992.

Vol. 399: Z. Švestka, B. V. Jackson, M. E. Machedo (Eds.), Eruptive Solar Flares. Proceedings, 1991. XIV, 409 pages. 1992.

Vol. 400: M. Dienes, M. Month, S. Turner (Eds.), Frontiers of Particle Beams: Intensity Limitations. Proceedings, 1990. IX, 610 pages. 1992.

Vol. 401: U. Heber, C. S. Jeffery (Eds.), The Atmospheres of Early-Type Stars. Proceedings, 1991. XIX, 450 pages. 1992.

Vol. 402: L. Boi, D. Flament, J.-M. Salanskis (Eds.), 1830-1930: A Century of Geometry. VIII, 304 pages. 1992.

Vol. 403: E. Balslev (Ed.), Schrödinger Operators. Proceedings, 1991. VIII, 264 pages. 1992.

Vol. 404: R. Schmidt, H. O. Lutz, R. Dreizler (Eds.), Nuclear Physics Concepts in the Study of Atomic Cluster Physics. Proceedings, 1991. XVIII, 363 pages. 1992.

Vol. 405: W. Hollik, R. Rückl, J. Wess (Eds.), Phenomenological Aspects of Supersymmetry. VII, 329 pages. 1992.

Vol. 406: R. Kayser, T. Schramm, L. Nieser (Eds.), Gravitational Lenses. Proceedings, 1991. XXII, 399 pages. 1992.

Vol. 407: P. L. Smith, W. L. Wiese (Eds.), Atomic and Molecular Data for Space Astronomy. VII, 158 pages. 1992.

Vol. 408: V. J. Martínez, M. Portilla, D. Sàez (Eds.), New Insights into the Universe. Proceedings, 1991. XI, 298 pages. 1992.

Vol. 409: H. Gausterer, C. B. Lang (Eds.), Computational Methods in Field Theory. Proceedings, 1992. XII, 274 pages. 1992.

Vol. 410: J. Ehlers, G. Schäfer (Eds.), Relativistic Gravity Research. Proceedings, VIII, 409 pages. 1992.

Vol. 411: W. Dieter Heiss (Ed.), Chaos and Quantum Chaos. Proceedings, XIV, 330 pages. 1992.

New Series m: Monographs

Vol. m 1: H. Hora, Plasmas at High Temperature and Density. VIII, 442 pages. 1991.

Vol. m 2: P. Busch, P. J. Lahti, P. Mittelstaedt, The Quantum Theory of Measurement. XIII, 165 pages. 1991.

Vol. m 3: A. Heck, J. M. Perdang (Eds.), Applying Fractals in Astronomy. IX, 210 pages. 1991.

Vol. m 4: R. K. Zeytounian, Mécanique des fluides fondamentale. XV, 615 pages, 1991.

Vol. m 5: R. K. Zeytounian, Meteorological Fluid Dynamics. XI, 346 pages. 1991.

Vol. m 6: N. M. J. Woodhouse, Special Relativity. VIII, 86 pages. 1992.

Vol. m 7: G. Morandi, The Role of Topology in Classical and Quantum Physics. XIII, 239 pages. 1992.

Vol. m 8: D. Funaro, Polynomial Approximation of Differential Equations. X, 305 pages. 1992.

Vol. m 9: M. Namiki, Stochastic Quantization. X, 217 pages. 1992.

Vol. m 10: J. Hoppe, Lectures on Integrable Systems. VII, 111 pages. 1992.

Vol. m 11: A. D. Yaghjian, Relativistic Dynamics of a Charged Sphere. XII, 115 pages. 1992.

Vol. m 12: G. Esposito, Quantum Gravity, Quantum Cosmology and Lorentzian Geometries. XVI, 326 pages. 1992.

Vol. m 13: M. Klein, A. Knauf, Classical Planar Scattering by Coulombic Potentials. V, 142 pages. 1992.

Vol. m 14: A. Lerda, Anyons. XI, 138 pages. 1992.

Vol. m 15: N. Peters, B. Rogg (Eds.), Reduced Kinetic Mechanisms for Applications in Combustion Systems. X, 360 pages. 1993.

GPSR Compliance

The European Union's (EU) General Product Safety Regulation (GPSR) is a set of rules that requires consumer products to be safe and our obligations to ensure this.

If you have any concerns about our products, you can contact us on ProductSafety@springernature.com

In case Publisher is established outside the EU, the EU authorized representative is:

Springer Nature Customer Service Center GmbH
Europaplatz 3
69115 Heidelberg, Germany

Batch number: 10640288

Printed by Printforce, the Netherlands